T0344884

Power Integrity for Electrical and Computer Engineers

Power Integrity for Electrical and Computer Engineers

J. Ted Dibene II
Watsonville
CA, US

David Hockanson
Boulder Creek
CA, US

This edition first published 2020
© 2020 John Wiley & Sons, Inc.

The right of J. Ted Dibene II and David Hockanson to be identified as the authors of this work has been asserted in accordance with law.

Registered Office
John Wiley & Sons, Inc., 111 River Street, Hoboken, NJ 07030, USA

Editorial Office
111 River Street, Hoboken, NJ 07030, USA

For details of our global editorial offices, customer services, and more information about Wiley products visit us at www.wiley.com.

Wiley also publishes its books in a variety of electronic formats and by print-on-demand. Some content that appears in standard print versions of this book may not be available in other formats.

Library of Congress Cataloging-in-Publication Data

Names: DiBene, J. Ted., II (Joseph Ted), author. | Hockanson, David, author.
Title: Power integrity for electrical and computer engineers / J. Ted Dibene, II (Watsonville, CA, US), David Hockanson (Boulder Creek, CA, US).
Description: Hoboken, NJ : Wiley, 2020. | Includes bibliographical references and index. |
Identifiers: LCCN 2019014832 (print) | LCCN 2019020732 (ebook) | ISBN 9781119263265 (Adobe PDF) | ISBN 9781119263296 (ePub) | ISBN 9781119263241 (hardcover)
Subjects: LCSH: Electric power system stability. | Electric power systems–Quality control.
Classification: LCC TK1010 (ebook) | LCC TK1010 .D5355 2019 (print) | DDC 621.319–dc23
LC record available at https://lccn.loc.gov/2019014832

Cover design by Wiley
Cover image: © Suchart Doyemah/EyeEm/Getty Images

Set in 10/12pt WarnockPro by SPi Global, Chennai, India

Printed in the United States of America

V10013252_082119

Contents

Part I

Power Integrity Fundamentals

1

Introduction

1.1   Introduction to Power Integrity for Computer Engineers

Power integrity as a discipline has evolved over the past 25 years. Once thought of as just a branch of "Signal Integrity," the area now encompasses many facets of the platform design for computers and other electronic devices and stands on its own as a distinct and necessary discipline. Even as little as 10 years ago, most technologists viewed the study of the power distribution network (PDN) and accompanying areas as simply a verification of the voltage budget of the delivered power to the device. Today, many companies dedicate entire groups with multiple engineers focusing on various aspects of power integrity as it relates to their product. Power integrity is quickly beginning to stand on its own, while conferences are now dedicating significant time and effort to advancing the state of the art.

Possibly the most important change in power integrity is the transition of the issues from being focused mainly on the motherboard to manifesting themselves within the package and silicon. As the complexity of the electrical behavior of the load has increased, so have the intricacies in solving the power delivery between the power source and the load. It is not by accident that this evolution has occurred, but by necessity. The urge to produce electronics with more capability than previous generations has led to these advancements. More performance within a device will invariably lead to higher bandwidth requirements. The consequence of this is that as the frequencies in the devices have increased, so has the rate of change in the current into and out of those devices. Moreover, as silicon processes have continued to shrink, so have the voltage margins required by the transistors to switch properly [1]. This evolutionary change will be discussed in some detail in Section 1.2.

Fundamentally, the study of power integrity has grown out of the need to understand more carefully three key elements: the power source, distribution network (or interconnect), and the load. The complexity of these components,

Power Integrity for Electrical and Computer Engineers, First Edition. J. Ted Dibene II and David Hockanson.
© 2020 John Wiley & Sons, Inc. Published 2020 by John Wiley & Sons, Inc.

however, has been governed by the requirements of the silicon as mentioned previously. The never-ending drive to lower the power of the devices (both for battery- and non-battery-powered applications), along with performance, has forced the voltages in the silicon to decrease linearly over time, while, at the same time, the currents and rate of change of currents going into and out of these components have increased geometrically. However, this is only one of the factors leading to the expansion of the power-integrity field. The other is that the physics of advanced silicon devices has imposed more stringent requirements on the power delivered to them. To ensure the integrity of the data propagating through the logic, and therefore, the transistors switching in the device, the voltage and currents must be maintained within very stringent limits at all times. To accommodate this, a highly tuned passive network of capacitors, along with the resistance and inductance of the interconnections, is composed of between the load and the power source. Power integrity engineers call this the PDN. Nevertheless, often, in high-performance devices such as CPUs, this is not enough to ensure proper regulation, and the power supply itself must share in the burden in delivering a suitable voltage and current. Moreover, the effects of developing advanced PDNs and power converters encroach on the entire system design which requires power integrity engineers to often be well versed in many aspects of the platform outside the scope of their discipline.

It is true that engineers are in fact concerned with all aspects of power integrity; from large server systems with alternating current (AC) wall inputs to smaller devices, such as *wearables*, with small batteries delivering power to one or more pieces of silicon, power integrity is becoming a necessary part of the development process. However, the issues, whether in small or large platforms, are still the same. In many ways, the study of power integrity has allowed the advancement of these electronic devices where just a decade ago, their creation would not have been possible. As the requirements of power silicon progress, it is apparent that as these remarkable devices advance, the study of power integrity will surely evolve along with it.

Because of the changes that have occurred over the past years in the discipline of power integrity, it was evident to the authors that a more comprehensive treatment of the subject was required. Our intention was therefore to present material that would extend beyond many of the more common texts [2] and other books to help those in the field take further steps in their education on this subject. Since there are few texts that focus on the detailed aspects of silicon in the area of power integrity, it was decided that we emphasize this in a practical manner to broaden the field of study. Thus, the purpose was to create a book intended for engineers who are interested in increasing their knowledge in this field beyond an introductory treatment. Consequently, this book was written with a focus towards developing an intermediate text on the subject. As stated in the foreword, the student should already have an advanced undergraduate or beginning graduate level education in engineering. This will be required for the subject matter in many of the chapters since it is believed that the student has a relatively strong mathematics and physics foundation.

At the end of this chapter, we review in more detail the scope of the text to help guide the reader through the different aspects of the study. A quick overview of the contents though is warranted here. Part 1 of the text will form the basis for the other sections of the book. A treatment of power conversion for computer systems as well as the system aspects of a whole platform will give the reader a good description of the electronics and issues that influence the development of the PDN and its aspects. This will be evident in the discussion around signal and numerical analysis for power integrity in Part 2 where an emphasis on toolsets is presented. Much of the foundational math and analysis should already be familiar to the student from their prior engineering education here. These tools will then be used in a number of the examples and analyses later on in Part 3 when frequency- and time-domain analytics are discussed. In addition, the student should be well versed in utilizing math programs and Simulation Program with Integrated Circuit Emphasis (SPICE). Though a review of the basics of SPICE is presented in the appendix, this is only to serve as a refresher since it is believed that the student will have a good foundation in both. The authors here have chosen Mathcad, MATLAB, and LTSPICE as their tools; other programs will suffice for the student should they not be obtainable or familiar with them. However, many of the examples and code will be made available in these tools should the student wish to use them.

In the following chapters of this text, we introduce "Power Integrity" from a brief historical perspective and then discuss some fundamental practices that are used throughout the book.

1.2   Some Advancements in Power Integrity

As discussed earlier, power integrity is, and has been, mainly focused around the three main components shown in Figure 1.1. When the voltages delivered to the loads were in the 1.8 V and greater regime, many power delivery problems were solved with basic layout and decoupling practices [3].

This was due to a number of factors but mostly because the load currents and voltage margins were manageable with the given capacitor technology at

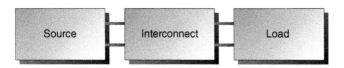

Figure 1.1 Basic components of power integrity.

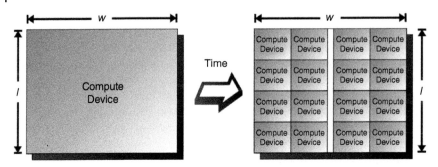

Figure 1.2 Change in computing devices over years.

the time and the design limitations of the circuit board. Moreover, many of the devices in a platform only required a single power supply rail and the real estate for the decoupling capacitors, along with the construction of low-impedance planes, was less of a challenge than it has been in recent years. However, as voltages have decreased on high-performance silicon, so has the complexity of the electrical network required to deliver power to them increased. Figure 1.2 shows a simple illustration of how things have changed to devices over the years. Many components, such as microprocessors, have stayed relatively constant in size, but their computational power has increased significantly during that time. This has been driven by the amazing advancements we have seen from silicon process manufacturing. However, the increase in computing performance has come about not only through increasing the speed of the device, but also, in part, through logic density. Today, many processors have billions of transistors, which contribute mightily to their processing capability. Figure 1.3 shows the relative increase in transistor density that has occurred relative to the process node size [7].

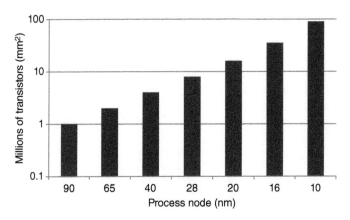

Figure 1.3 Transistor density vs. process node size.

Additionally, advancements in computing power, in recent years, have been partially attributable to architectural changes. The advancement of the multi-core processor also attributed to this evolution. A number of years ago, it was discovered that paralleling multiple processing units (called cores) and grouping them into a single silicon device offered more computing capability than if one were to simply build a single processing unit with that same area of silicon [6]. Additionally, it has been common practice in the computer industry for some time to take the previous generation of a computing core and add more of them to the space that was previously available in an earlier generation [4]. This, as already mentioned, has been achieved through silicon process scaling which has typically reduced the area for a given piece of silicon (for a similar density of logic) by approximately 1.4× every generation [5]. The final assembled device becomes a conglomeration of high-performance computational devices that were based on this previous *core* architecture. This segmentation of components within a singular device has resulted in more voltage rail proliferation as well. However, this has also reduced the physical area available to deliver the power into the device. Shrinking the available space, while also decreasing the voltage to the device, and increasing the rate of change of current into the device have placed a very large burden on the power delivery to each processing block within the silicon. This evolution has required power integrity engineers to become more critical in their thinking when it has come to designing a proper power delivery network for such subsystems. Thus, power integrity engineers are now examining closely the aspects of on-silicon power integrity, which has brought about a completely new field of study.

Figure 1.4 shows an illustration of how the change to continuously shrinking silicon devices has exacerbated the power delivery problem. If each unit shown in Figure 1.4 requires an individual voltage rail to power it, then it is clear that the proliferation of these additional power sources makes up for increased complexity. The first question one might ask is "Why not only use a single power source to all of the computing devices instead of trying to power each processing unit individually?" The answer to this question has to do with the competing

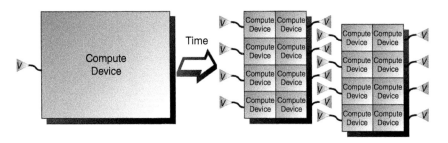

Figure 1.4 Illustration of voltage rail proliferation.

trends to reduce or maintain the overall system power while increasing performance. To ensure that the power does not increase for the overall device from one generation to the next (or ensure the power decreases), it is often required to not only control the power to each computational block independently but also provide it with its own power source. For most software workloads, the time in which the work is sent to the processing unit is very indeterminate. That is, the processing units do not know exactly when software will require them to perform actual processing. To combat this problem, architects and engineers figured out that growing the number of processing units on a die and joining them via a common memory and data bus allowed them to select which processing unit or core does the actual processing and to find the one that was idle and available to perform the work [6]. This is partly the rationale behind the multiprocessor architecture. As long as the memory bus feeding the individual cores had sufficient bandwidth to fill the cores adequately, this architecture resulted in many advantages over a larger core utilizing the same space. However, if each processing unit is not active and sits waiting to perform work, due the silicon current leakage [8], particularly in sub-nanometer processes, the additional power burned when the cores were *not* active could outweigh the benefit of the extra computing capability. To combat this problem, engineers determined that by either gating off these cores through large-power metal-oxide semiconductor field-effect transistors (MOSFETs), called power gates [9], or through providing individual power sources to each core, they could control the voltage to each unit within the same silicon and reduce the effective leakage to very low values, thus improving the performance/watt for the overall system. Since most processors do actual work at only a small portion of the time when they are active, this reduces the effective power being consumed by the cores and processor over a given workload period to a much lower value than would otherwise be possible.

An example of how this architecture has come about is shown in Figure 1.5. Note that the increases in transistor count (log scale on the left axis) have grown steadily over the years shown. However, the average power consumption has changed over this time and has not scaled with transistor density. This is very evident in the graph in Figure 1.6. Note that from 2005 onward, a change occurred in the architecture of the microprocessor and the number of cores increased. This increase in cores demanded a change in the power consumed and one can see that the power in these devices tended to fluctuate from this time onward. The reason for this is evident once one examines the work that was put into managing power in these cores during this time. It is clear over this more recent generation of processors that power increases within the device forced engineers to become more creative in managing the power delivered to them [12]. It is also evident by examining the same figure that the processing unit has proliferated within the silicon device at a similar rate. However, note that the power consumption either has been capped over those

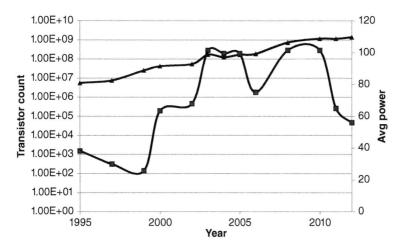

Figure 1.5 Trends for microprocessor transistor count (▲) and average power (■).

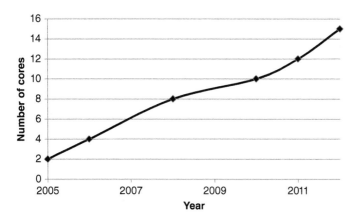

Figure 1.6 Trends in number of cores in microprocessors over time.

years or has even gone down. This is due in part to the changes in architecture over this time in moving from a single processing unit to a multiprocessing unit. Not only has this change occurred within the microprocessor, but it has now made its way into other units such as graphics processors and SoCs. The ability to extend the capability of processing on silicon has helped fuel the increased processing capability of not just larger systems such as servers but also smaller devices such as laptops and handhelds.

This change though has come with a price as shown in Figure 1.4. To manage power in individual components on silicon requires individual power management. This necessitates additional circuitry. As the silicon device shrinks, so must the components that circumscribe the device to power it. If not, then the system begins to grow outward as illustrated. Thus, the power source and distribution must begin to shrink with the device as well. Unfortunately, this facet of the design competes with the performance of the device as well. Even if the component shrinks, there will still be a requirement to boost its capability through frequency changes. This increases the current into the component, resulting in larger devices converting power for this component.

Over the next decade or more, the power integrity engineer will be faced with a number of growing challenges, many of which will be due to rapid increases in voltage proliferation driven by the advancements in process scaling and the need to control power consumption in a device. This will require them to be well versed in the fundamentals of power integrity and all its aspects. Section 1.3 discusses an important aspect to this education: first principles analysis (FPA).

1.3 First Principles Analysis

As discussed in Ref. [2], the aspects of FPA begin with the fundamentals and understanding of the boundary conditions of a problem. Too often engineers are tasked with a problem in which they require a reasonable answer within a short amount of time which requires them to not only be thorough but also give a reasonable answer to a problem. Unfortunately, most engineers forego the understanding and lean too heavily on the software tools to give them the answers they desire. This is where an iterative process of program data analysis forces the engineer into loop after loop of trying to come up with the correct answer based on the output of the tools. The software eventually gives them something that they frequently realize is erroneous or nonsensical, and consequently useless.

In such cases, this is where FPA becomes an integral part to one's analytical education. FPA is simply a variation of the scientific method. It is intended to help the engineer determine the best direction to take when tackling a problem. It also starts with an analysis that is solely based on a strong fundamental understanding of the problem. This requires that the engineer be well versed in the math and physics. In the fast-paced world of electronics product development, technologists tend to fear not having an answer more than getting the answer wrong. This is unfortunate, given that a lack of understanding typically results in a delay or an unnecessary additional number of resources added to the problem. In virtually all of these cases, it is better to gain the understanding first, rather than rely on a computer program to tell one the answer in the beginning. This is not to say that numerical software tools do not have their place. Chapter 7 is dedicated to a discussion of numerical methods for power integrity problems and reviews their role with respect to analysis in some detail.

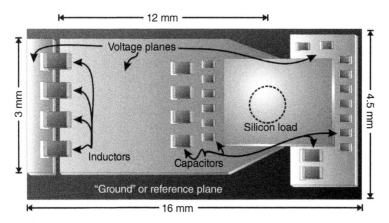

Figure 1.7 Simple board layout with power planes.

The point is that before embarking upon a detailed numerical study, it is always better to acquire a baseline understanding of the problem first.

The method of FPA is discussed in some detail in chapter 1 of Ref. [2]. However, FPA can be shown just as well with an example. Suppose that an engineer is trying to come up with a value for the plane inductance from a power source (output inductor) to a group of capacitors underneath a device close to where the load is. Furthermore, the engineer is interested mainly in the low-frequency effects (third-order droop) rather than the high-frequency droop effects. The plane structure (plan view) is illustrated in Figure 1.7. The engineer knows that if the inductance is too large, then more capacitance may be required to dampen the low-frequency voltage droop in this region. However, if it is small enough, then no further analysis would be required, and the layout person may go ahead with the board design without issue. The key is that a quick value that is accurate enough to allow the design team to continue on with their work must be found.

As a starting point, the engineer knows from experience that the inductance must not exceed 800 pH. If it does, then additional capacitance or a lower inductance path may be required. The question is, how accurate does the estimate need to be in order to ensure that the result will satisfy the design requirements? The question may be partially answered in the chart in Figure 1.8. As a general rule, the accuracy of the computation depends upon the type of analysis being done. For example, if the accuracy only needs to be within 50% (in other words, an estimate may be off by 2× on the high side in this case where it will not matter), then a simple hand calculation will usually suffice. However, if the error can be no more than 5%, then only a very good experimental test would be adequate to ensure the accuracy. In most cases, the FPA assumes that an engineer will go through the proper steps prior to dedicating the resources

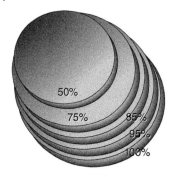

Figure 1.8 Accuracy diagram for first principles analysis.

50%
75%
85%
95%
100%

Table 1.1 Dimensions for example FPA in Figure 1.7.

Dimension	Value
Length (mm)	12
Width (mm)	3
Thickness (μm)	25

necessary to solve a given problem. Thus, given the complexity of the issue, an engineer may approach the problem differently.

Referring back to Figure 1.7, one can see that getting the proper dimensions first is required before embarking upon a study. Table 1.1 gives us some of the key measurements to allow us to proceed.

The dimensions in the table are enough to estimate the inductance of the structure. However, before doing the computation, it is best to analyze the figure to ensure that a proper computation is done. First, notice that the dimension for the length is taken to the *center* of the circle where the load is rather than say where the first capacitors are located. If this were a high-frequency analysis (relative to the load), it might be better to compute the inductance to the center of the first row of capacitors since the AC currents in a higher frequency regime would initially feed the capacitors rather than the load. This is an important distinction since often the problem statement dictates where the computation is bounded. Second, the dimensions for the width of the planes are to the outside of the planes, not to the actual current sources, which are actually from the inductors to the left. Figure 1.9 gives a better profile for the current distribution in both planes. Note that this estimated current profile better predicts the actual current distribution (or envelope) in the planes. Third, no account is given to the perforation of the planes themselves (vias). In a real design, the engineer would check to see if there were a sufficient number of vias to disrupt the flow of current through the planes.

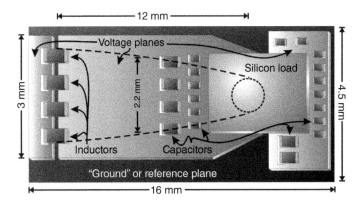

Figure 1.9 New estimate of width for Figure 1.7.

Finally, the thickness of the board must be taken into account. Since there was no tolerance given with this value, it must be assumed that this is a nominal value for the thickness. Usually, the dielectric thickness can have a 20% variation. Thus, it may be good to assume that the inductance could be 20% larger (or smaller) due to manufacturing variations and thermal cycling.

Now we can begin to compute the inductance. Chapter 4 gives a simple first principles equation for the inductance of a plane. The equation itself must also have some assumptions in it. First, and foremost, is the accuracy. For this analysis, we ask, is the estimate valid? It turns out that the equation is valid for plane lengths and widths much larger than the separation between them. In other words

$$L \cong \frac{\mu_0 ls}{w} \quad s \ll l, w$$

A quick calculation shows that this is a valid assumption. We are now ready to compute an estimate for the plane inductance.

$$L \cong \frac{\mu_0 ls}{w} = (1.2) \times \frac{(4\pi \times 10^{-7})(12 \times 10^{-3})(25 \times 10^{-6})}{(2.2 \times 10^{-3})} \cong 200 \, \text{pH}$$

It is clear that this is much less than 800 pH. The error is $\pm 2\times$ in this estimate (per Figure 1.8 as a general rule). Thus, the result is within the error band of the computation. Note also that the result is essentially two decimal places. Since the error is so large, reporting out a more accurate result would be incorrect. A simple dimensional analysis bears this out.

The next step would likely be for the power integrity engineer to perform a SPICE analysis to confirm that the droop was within reason (assuming this had not already been done). If the constraints of the problem were such that the inductance could not exceed say, 250 pH, then the engineer would have gone on to analyze the problem further since a hand calculation would not have been within the bounds of the error bar set for the problem (e.g. $400\,pH = 2 \times 200\,pH > 250\,pH$). The diagram in Figure 1.8 is a simple guide to the next steps. As a general rule, given here are the tasks that one might follow in each circle to determine which analysis to perform,

1. *50% accuracy*: Hand computation or simple spreadsheet computation.
2. *75% accuracy*: First pass numerical analysis (parameter extraction and/or in combination with SPICE or math program computation).
3. *85% accuracy*: Detailed numerical analysis over boundary or "corner" conditions along with (possibly) first pass test setup and measurement in lab.
4. *95% accuracy*: Advanced lab measurement in controlled environment. Closely calibrated conditions and baseline data from previous measurements used as checks.

The described accuracies are depicted pictorially in Figure 1.10.

The power integrity engineer should in all cases determine what will be the next best step. Though it seems repetitious for some, doing a reasonable hand computation to gain some understanding of the problem first is usually a good idea. There are cases in which the problem itself at first seems intractable. This is where it is best to break the problem down into smaller items and address them one by one rather than attempt to solve the problem all at once. This is very true of complex power distribution problems, as we shall see later on in the text. Jumping into a detailed numerical analysis will invariably have its drawbacks. This is because if the engineer does not have a good idea of what the result would be, the time it takes to get the assumptions right in the model may

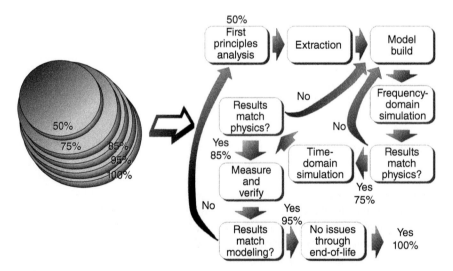

Figure 1.10 Pictorial flowchart of accuracy improvements through simulations, testing, and life cycle.

outweigh the benefit of a good numerical tool. It will be evident later on in the text where having a good software program to gain accuracy is helpful and what to look out for with regard to accuracy and other issues with these tools.

1.4 Scope of Text

It is clear that Power Integrity as a discipline today requires a broader look into a number of areas. This is mainly due to the complexity that has come about over the years in the problems that many engineers must face. It was with that perspective that the authors decided to include areas where power integrity problems crossed over to other disciplines which are now required for a better understanding of the subject matter.

Thus, this text has been broken up into three parts; the first part is focused on power integrity fundamentals. The second part emphasizes the tools required for power integrity analysis. Finally, the third part discusses analytics. From the start, our intention was to build upon the work from the previous chapter so that the power integrity engineers could increase their knowledge as they move from the subjects in one chapter to the next. However, it is also understood that many engineers will already have a strong foundation in a number of these subjects and will choose to skip certain chapters. Thus, we also tried to keep each chapter as inclusive as possible in support of those who were only looking for specific information within a chapter.

The first part which is focused on some background on power integrity as a subject starts here in Chapter 1. This is an introduction [mainly] to FPA and how to bound a problem appropriately. In Chapter 2, we introduce some concepts around power conversion as it relates to power integrity. There are many texts that do a tremendous job teaching the basics of power conversion and some are referenced at the end of the chapter to aid the reader. However, most power integrity engineers are not necessarily interested in all of the design aspects of power conversion, so this chapter mainly deals with the fundamentals around understanding computer platform regulation, focused primarily on the buck converter. The objective here was to introduce the subject from the perspective of power integrity concepts to help the engineer deal with the issues of how to model and understand the basics. We start with the connection to the PDN and work our way back into the main circuits and different aspects of the buck converter. Finally, the end of the chapter deals with analysis and modeling of the converter in some detail. An additional subject around advanced converter types (on-silicon voltage regulation) was added in part because of the movement toward silicon power integrity, which is one of the most important aspects that have come about in this field.

Chapter 3 is an introduction to platform and systems with respect to power integrity. The chapter introduces the physical elements, which make up the

PDN, load and power source that we are interested in. This is to set the stage later when these devices get analyzed in some detail. This chapter spends some time understanding both the source and the load physically and goes into understanding the load in more detail. (Chapter 2 touches on modeling the buck converter, the source.) Later in the chapter is an overview of some of the system-level issues the PI engineer normally must address, that is specifically, guardband and load line basics. This is becoming a critical aspect to understanding the effects of the PDN relative to the power source and the load. The end of the chapter is dedicated to some basic measurement techniques for the PDN as well as noise and electromagnetic interference (EMI) considerations. Some of the concepts are complex here, but the authors only touch the surface to help give the reader a better perspective of how power integrity affects different aspects of the system.

Chapter 4 introduces the fundamentals of where most of the analytical tools come from. It is an introductory chapter on electromagnetics (EMs). This is to help in analyzing fundamental circuits as well as providing a foundation for numerics. Key equations are introduced later on in the chapter to help the reader solve specific network-related problems with first principles similar to the simple problem given in this chapter. Boundary conditions are discussed for these fundamental concepts to aid the engineer in their understanding.

Part 2 of the book is focused on a number of the tools that will be needed for analyzing problems related to power integrity. Chapter 5 starts out focusing on transmission lines. Though mainly an area of study for signal integrity, transmission line theory gives the reader a good introductory foundation toward understanding the effects of the PDN. The reader is introduced to the basic theory for transverse electromagnetic (TEM) mode structures along with lumped and distributed circuits. A section on damped transmission lines is given as it relates to the PDN.

Chapter 6 is an overview of signal analysis. This is a critical chapter in the sense that it brings together a multitude of different signal techniques for analyzing problems in both the time and frequency domains. An overview of linear time-invariant (LTI) systems begins the study, which is the basis for much of the analysis. Then, a section on Laplace and Fourier transforms is discussed in preparation for future chapters. Signal analysis is a key foundational element to power integrity since not only are tools directly applicable for solving these types of problems, but they also give a strong insight into the behavior of the system which provides the engineer with the groundwork necessary to tackle some of the more complex problems they will face in the electronics world.

As mentioned earlier in this chapter, the topic of numerical method was added to this text since many engineers tend to focus on the tools rather than first principles. The importance here is not to help engineers write code for their particular problem. (The industry is replete with very good black box programs, which suffice for that.) The chapter starts out examining the main concepts around numerics for power integrity. A section then on variational methods is given which is a prelude to some of the concepts for the numerical tools, mainly the finite element method. Next, a simple introduction to conformal mapping is given which helps the reader understand the basics for numerical analysis in this space. Next, an overview of two-dimensional finite difference time-domain is given which can be easily expanded with the Yee Cube to three dimensions [15]. Some code is also provided with examples to help the students. The next section is on the "Finite Element Method" which forms the basis for many field solvers and circuit extractors [16]. The goal here is to help engineers understand the limitations of the numerical tool that they are using. More often than not, engineers will assume that the tool gives them a solution with high accuracy without understanding the fundamental math and boundaries of the tool itself. This chapter gives engineers an insight into that math and then provides an understanding of the limitations, thus helping them when they present their data and set up their problems. There is understanding of how good that data is rather than simply allowing people to take the results at face value.

Part 3 of the text is the analytical section of the book, focused primarily on how to analyze certain problems. Chapter 8 starts out with an overview of frequency-domain analysis problems and utilizes SPICE and Math programs to help the reader. A number of examples using various tools from the previous chapters are emphasized here. Since this chapter is normally the focus for most power integrity engineers, there is a detailed section on the PDN and ways to analyze it. The analysis gets into using Bode plots to analyze impedance networks along with other tools. The last section of this chapter is solely on analytical methods, particularly on how to analyze certain problems. It uses FPA to begin with and then broadens out the methods to help the student discover different aspects of analyzing PDN problems.

Chapter 9 is an introduction to time-domain analysis. Though often considered more complex than frequency domain, a number of tools are used to provide examples to the engineer on how to tackle certain problems. Again the PDN is the focus here along with utilizing both simple load and source models. For most power integrity engineers, this is a key chapter. It brings together much of the foundational knowledge from the previous chapters in an effort to quantify the efficacy of the PDN. Time-domain droop analysis is the focus and how to bound the PDN using a budget technique. Many engineers simply measure the droop without understanding the contributions from different circuits. This chapter focuses on those aspects in an effort to bound the problem. SPICE analysis is key here along with utilizing statistical methods such as Monte Carlo to determine the worst- and best-case scenarios. The final section brings together these techniques and gives examples for the reader as a guide to the problems they face.

Chapter 10 focuses on silicon power integrity. Though many of the techniques for platform-level power integrity apply, there are additional issues to be addressed at the silicon level for the PI engineer. The main issues are around local decoupling and distribution. Though the interconnect in most silicon networks has overall low inductance paths for power, the resistive portion can often dominate and has issues of its own with respect to the voltage droop budget, both globally and locally. This is often true where the silicon power distribution is split into multiple rails, where we have seen earlier in this chapter the problems this can present to the power integrity engineer. This chapter discusses these issues and gets into some of the common layer constructions in devices. Though the load modeling was discussed in Chapters 1–9, we touch upon this subject again to determine where the voltage deviations can occur under dynamic current steps and resonances. Finally, noise at the die level is discussed, and some mitigation methods are covered.

Lastly, the authors provide some important information in the appendices to make the text as self-contained as possible. A brief tutorial on SPICE is given, with specific focus on the tools that will be required for analysis in specific chapters. Basic EM derivations are provided in some detail. Selected formulae and Laplace and Fourier Transform tables are also provided for the reader.

Problems

1.1 Based on the density given in Figure 1.3, predict the density for 130 and 180 nm processes. How many transistors could one put into a device (approximately) if the die size was 200 mm^2? Assume the voltage levels to be 1.5 and 1.8 V respectively.

1.2 Examine the chart in Figure 1.5. If the process node for 90 nm was generated in 2003, estimate the peak current for the device based on a voltage level of 1.5 V. Determine the current density for a device where the core occupied 200 mm^2.

1.3 Reestimate the inductance based on Table 1.2 for the problem described in Section 1.3. If the tolerances for the dimensions were ±10% for

Table 1.2 Dimensions for example FPA in Figure 1.7 to be used in Problem 1.3.

Dimension	Value
Length (mm)	17
Width (mm)	2.5
Thickness (μm)	25

each of the dimensions, determine the inductance for each of these boundary conditions. Is it still within the tolerances of the circle chart? Why?

Bibliography

1 Semiconductor Engineering (2016). Hitting the power integrity wall at 10 nm. http://semiengineering.com/hitting-the-power-integrity-wall-at-10nm/ (accessed 22 December 2017).

2 Ted DiBene, J. II, (2013). *Fundamentals in Power Integrity for Computer Platforms and Systems*. Wiley.

3 Electronic Design (2012). What's the difference between signal integrity and power integrity? http://electronicdesign.com/boards/what-s-difference-between-signal-integrity-and-power-integrity (accessed 22 December 2017).

4 Uy, R.L. (2014). Beyond multi-core: a survey of architectural innovations on microprocessor. In: *7th IEEE International Conference HNICEM* (12–16 November 2014). Puerto Princesa, Philippines: IEEE, Inc.

5 Saraswat, K.C. (1997). Trends in integrated circuits technology. http://web.stanford.edu/class/ee311/NOTES/Trends.pdf (accessed 22 December 2017).

6 Martin, C. (2014). Multicore processors: challenges, opportunities, emerging trends, embedded world conference. https://www.hs-augsburg.de/Binaries/Binary20964/Multicore-Embeddedfinal-revised.pdf.

7 Bohr, M. (2017). *Moore's Law Leadership*. Intel.

8 Liu, T-J. (2005). MOSFET technology scaling, leakage current and other topics. http://www-inst.eecs.berkeley.edu/~ee130/sp06/chp7full.pdf.

9 Yong, L-K, Tan, F-N, and Pang, S-G (2009). Power gate design optimization and analysis with silicon correlation results. https://www.apache-da.com/system/files/DAC2009_User_Tracik_Session_5U_Intel.pdf.

10 Carroll, J.S., Peterson, T.F., and Stray, G.R. (1924). Power measurements at high voltages and low power factors. *Transactions of the American Institute of Electrical Engineers* 43: 941–949.

11 Stanford, E. (2004). Microprocessor voltage regulators and power supply trends and device requirements. In: *s.l: Proceedings of International Symposium on Power Semiconductor Devices & IC's* (24–27 May 2004). Kitakysushu: IEEE Inc.

12 Kwangok, J. (2009). A power-constrained MPU roadmap for the International Technology Roadmap for Semiconductors (ITRS). In: *IEEE SoC Design Conference (ISOCC)* (9–11 September 2009). Belfast: IEEE, Inc.

13 (2001). Technology roadmap for semiconductors. *IEEE Journals and Magazines.* https://www.semiconductors.org/wp-content/uploads/2018/08/2001-Executive-Summary.pdf.

14 Tiwari, V., Singh, D., Rajgopal, S. et al. (1998). Reducing power in high-performance microprocessors. In: *Design Automation Conference* (15–19 June 1998). San Francisco, CA: IEEE, Inc.

15 Sullivan, D.M. (2000). *Electromagnetic Simulation Using the FDTD Method.* IEEE Press.

16 Reddy, J.N. (1993). *An Introduction to the Finite Element Method*, 2e. McGraw-Hill.

2

Power Conversion for Power Integrity

The importance of power conversion to power integrity (PI) engineers has come about over only a very short period of time. The reason is that the regulator is now impacting the dynamic behavior of the power delivery between the load and the source and thus can no longer be considered as an afterthought to the system or the system designers. Because of this, the PI engineer needs to have a strong understanding of the operation of the main converters that deliver power to these critical system loads and how they can incorporate that behavior into their overall analysis. The goal of this chapter will be to develop a basic understanding of the operation of a converter system for computer platforms and thus an ability to understand that behavior as part of the overall analysis in the power integrity design.

2.1 Power Distribution Systems

As discussed in Chapter 1, one of the three fundamental building blocks in power integrity is the power converter. In the case of computer platforms, this usually means that a DC–DC switching regulator is powering one or more of the silicon loads. Though linear regulators are also used for low-power applications, the main topology is the buck regulator. How these devices are used is as important as the converters themselves. The reason is that for most systems today, the distribution of power from the source to the load dictates many aspects of the platform. The performance, power efficiency, cost, and size are all affected by the choice of architecture used to deliver power. Figures 2.1 and 2.2 show two main methods for delivering power to a set of silicon loads.

The first method is distributed; that is, the loads are *distributed* along with their own converters relative to the input power source. Basically, the topology is where one or more devices have a dedicated power source colocated with it. When the converter and load are colocated, it is often called a *point-of-load* converter design. The objective is to do the conversion as close to the load as possible. Though this technique is often more complex than others, the

Power Integrity for Electrical and Computer Engineers, First Edition. J. Ted Dibene II and David Hockanson.
© 2020 John Wiley & Sons, Inc. Published 2020 by John Wiley & Sons, Inc.

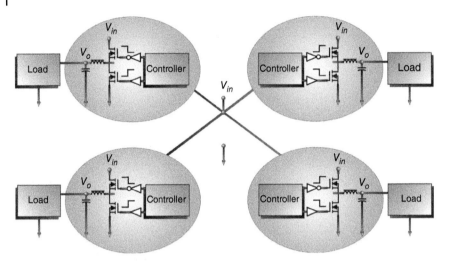

Figure 2.1 Distributed power delivery architecture.

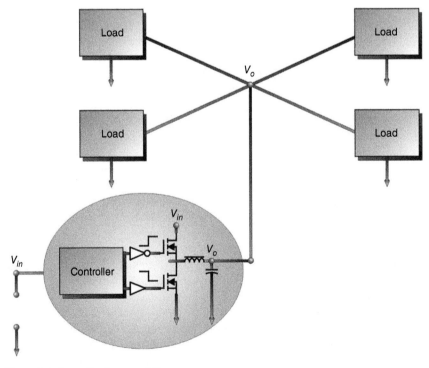

Figure 2.2 Centralized power delivery system.

advantages can be many. For instance, by locating the source and the load close to each other, the load can often control the converter, directly making response to current changes more dynamic. This can often lead to better platform power efficiency by allowing gating of the silicon off and (power gating, specifically) when it is required so that it does not continue to burn as much power when the silicon is not doing any actual work. The drawback of the distributed approach is that it often takes up more room than other methods since one now needs to build full converters for each main load in the platform. Moreover, the system power management design is more complex due to the sheer number of converters that must be brought up and controlled at any one time. Finally, since there are more components, the reliability of the platform may also suffer.

In contrast, a centralized power network distributes power to many loads from a centralized point or power source. This has the advantage of simplicity and is often used for platforms where the loads are not as dynamic and the currents are lower. Usually, if only a few converters need to be built for such platforms, the system power management design will be simpler than in a distributed topology. Such a design typically has the advantage of smaller real-estate requirements. However, having only one converter with many loads can also be an issue. If there is no activity in the other silicon, then the converter must remain on for the one that is making the system burn more power than what is really needed.

Whichever topology is used for power distribution, it is typically the load that determines what is required. Today, most designs for platforms are dictated by the silicon within them. Because silicon processes are driving power down and performance up (usually at a feverish pace), this usually means that the architects and developers of the devices have a bigger say in how their chips are powered. This is one of the reasons why more and more platforms are moving toward *distributed* power distribution for their system power solution. As the processes advance and more transistors are packed into smaller areas of silicon, the overall power usually increases to support the higher performance. However, this is counter to the direction of what the platform requires, at least for power consumption. Due to the cost of energy, most computer system designers are required to drive power *down* in their platforms. This is particularly true for battery-sourced systems where up-time[1] is a premium to the user [1]. To address this, silicon device developers usually opt to have control over the power to their device, either directly or indirectly. This allows them to modulate the voltage of their device in concert with the performance during the work it is required to deliver, rather than be dictated by a centralized power

---

1 *Up-time* is the time available to the user on a full charge of the battery for a given electronic device. Typically this is given as battery time under certain workloads or benchmarks for the platform.

network where the power needs to meet the demands of all of the loads at any given time.

Nearly all system designers for computers and electronic devices today are concerned with the power consumed in a system. This is why so many designers specify the efficiency of the converters in the platform so carefully. Efficiency is the ratio of the power into a converter to the power delivered as shown in Eq. (2.1):

$$\eta = \frac{P_{out}}{P_{in}} = \frac{P_{out}}{P_{out} + P_{LOSS}} \tag{2.1}$$

where η is the efficiency of the converter. Note that the power *loss* of the converter may also be represented as shown in the right-hand side of the equation. This is the *net* loss in conversion over a given load range that the converter consumes. The efficiency is never constant over a large load range; rather it is a curve or family of curves that typically represents the efficiency of the converter. Figure 2.3 shows the metrics that are measured in the blocks that are used to compute the efficiency.

The efficiency for a buck regulator may also be represented with the input and output voltages and currents as shown in Eq. (2.2):

$$\eta = \frac{V_O I_O}{V_I I_I} = \frac{V_O I_O}{V_O I_O + P_{LOSS}} \tag{2.2}$$

Thus, as long as the input and output currents and voltages are available (measurable), an engineer can readily determine the efficiency of a converter. The power loss is more complex than in many systems since it is made up of both silicon and passive devices. Moreover, most converters for computer systems and electronics today have fairly sophisticated power management controls to maximize the transfer of power from the input to the output in order to better serve the platform requirements. This is now a requirement since the load range for many devices has become very large relative to the changes in process technology and architecture in silicon processing units over time as discussed in Chapter 1. It is now common for a processor load to span a load range from close to 1 mA to over 10 A of current. This means that the converter must operate over 4 orders of magnitude of load current. Moreover, the converter is expected to be relatively efficient over the entire range. This is quite phenomenal given the fact that less than 10 years ago, most

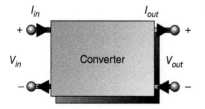

Figure 2.3 Input and outputs of converter.

converters were designed to be efficient for a limited range of currents in many computer systems (usually only two orders of magnitude). We will discuss the efficiency of the buck regulator later on in this chapter as well as address the requirements of the silicon load range later on in this text.

The efficiency of the converters in a platform is an important consideration to the power integrity engineer because not only must they provide a path from the voltage source to the load, they cannot develop a power distribution network (PDN) that causes the converter to go unstable nor increase its *inefficiency* (increase its power loss). It is common to cascade more than one converter together to determine the series efficiency of the total path. For example, if a system requires one or more converters be sourced from a single rail, then to find the total efficiency from the two requires one to compute the combination of the two efficiencies.

$$\eta_{12} = \eta_1 \eta_2 = \frac{V_{O1}I_{O1}}{V_{I1}I_{I1}} \frac{V_{O2}I_{O2}}{V_{I2}I_{I2}} = \frac{V_{O1}I_{O1}}{V_{I2}I_{I2}} \tag{2.3}$$

Note that the power into the second-stage converter is the same as the output power from the first stage. By multiplying the efficiency of one converter with another (in series), an engineer can get the resultant efficiency of the system. However, for more complex power delivery systems, it is often required to compute the overall efficiency by first determining the actual input currents of the system. This is illustrated in Example 2.1.

Example 2.1   A system designer had configured three sets of converters to power a single load. The input voltage is 5 V and the output voltage is 1 V. The combination of the two-stage converter system C1–C1′ delivers 3 A at maximum current. The middle converter delivers 1 A maximum current and the combination two-stage system C3–C3′ delivers 2 A maximum current. The mid-rail voltage to C1–C1′ (VN) is 2.5 V, while the mid-rail to C3–C3′ is 1.8 V. The efficiencies for all of the rails are in Table 2.1. What is the overall efficiency of the system at static maximum load for the entire system? Is the efficiency above 80%?

Table 2.1 Table of converter efficiencies for Figure 2.4.

Converter	Efficiency (%)
C1	93
C1′	87
C2	82
C3	92
C3′	88

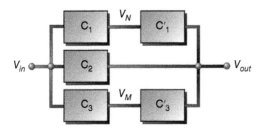

Figure 2.4 Converter configuration for Example 2.1.

Solution:

At first glance, it looks like one could simply compute the different efficiencies of each converter and then determine the overall efficiency by computing the parallel combination. However, this would be incorrect. Because each converter delivers a different current output, its effect on the static efficiency will be different. The proper method is to work backward on each converter and find the currents flowing into each converter system given the sharing of current from each of the paralleled outputs. Because C1' delivers 3 A at the output, while we know its efficiency and input voltage, we can compute its input current. Rearranging Eq. (2.2), we get the following:

$$I_{IC1'} = \frac{V_O I_O}{V_I \eta_{C1'}} = \frac{(1\ \text{V})(3\ \text{A})}{(2.5\ \text{V})(0.87)} = 1.38\ \text{A} \tag{2.4}$$

We can now compute the input current to C1 in the same way. Using a simple spreadsheet, we compute

$$I_{IC1} = \frac{(2.5\ \text{V})(1.38\ \text{A})}{(5\ \text{V})(0.93)} = 0.74\ \text{A} \tag{2.5}$$

Table 2.2 gives the input currents now for the rest of the converters. It is now simple to compute the overall efficiency of the system using Eq. (2.2).

$$\eta_{sys} = \frac{V_O I_O}{V_I I_I} = \frac{(1\ \text{V})(6\ \text{A})}{(5\ \text{V})(1.48)} = 81\% \tag{2.6}$$

Table 2.2 Table of converter efficiencies for Figure 2.4.

Converter	I_I
C1	0.74
C1'	1.38
C2	0.24
C3	0.49
C3'	1.21
Sum (C1 + C2 + C3)	1.48 A

The resulting efficiency of the system at maximum static current is above 80%, which answers the question. It is important to point out that this was only the *static* efficiency of the converter. The dynamic efficiency is much more complex and involves the operation of the converter itself under different load conditions.

2.2 The Buck Converter

In most computer platforms today there are essentially two types of DC-to-DC converters used to deliver power to the load: the linear regulator and the buck regulator. The linear regulator is basically a transistor that acts like a controlled resistor (Figure 2.5). The voltage is dropped across the controlled resistance and the output voltage is fed back to control and regulate the output. These devices are used, in general, for lower power applications, and for today's silicon devices, there are typically many of them on the platform, even on the load die. The applications are varied, but often they are used for problems where the current is small and local but the regulation is required to be very good. Voltage and current references along with phase lock loop circuits are common examples where linear regulators are often used as the output power source.

The efficiency of the linear regulator is quite simple and is proportional to the voltage drop across it.

$$P_{Loss} = P_{in} - P_{out} = (V_{in} - V_{out})I_{in} \tag{2.7}$$

Thus, if the voltage drop can be kept to a minimum, a linear regulator can be made to be very efficient. The issue is that for most applications (particularly on-die), there is limit to how much current may be sourced through a linear regulator.

Figure 2.5 Example of linear regulator.

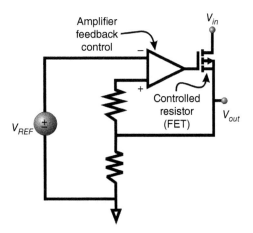

Buck regulators, on the other hand, are most commonly used for applications where the current delivered is quite high and the dynamic range is large. A buck regulator, though slightly more complex, is still a rather simple and elegant power converter. At its most basic form, it is comprised of two switches, an inductor, and a capacitor. The two switches are turned off and on alternately when transferring energy from the input to the output. As shown in Figure 2.6, the buck converter at time $t = 0$ turns on the upper switch to deliver current into the inductor.

The inductor current ramps up to a specific value, which is dependent upon the switch *on-time*. When the upper switch is turned off at time $t = t_{off}$, the lower switch is then turned on and current continues to be delivered to the load. The two modes of the buck regulator are called the *duty* and *commutating* cycles. The duty cycle of the converter is the weighted amount the upper switch is turned on, or mathematically.

$$DT = \text{upper switch on time} \tag{2.8}$$

where D is the duty cycle and T is the period of the converter switching time. This is shown in Figure 2.6. The commutating cycle is the time for which the lower switch is turned on and the upper switch is turned off. This is shown in Figure 2.7.

When the upper switch is turned off, the current through the inductor continues to flow to the load though at a decreasing rate. The lower switch essentially provides a ground return for this to happen.

The duty cycle in a buck converter has a very defined relationship to the input and output voltage. This may be easily derived by noting that the converters signals are mainly cyclo-stationary (i.e. one or more signals statistically have stationary states over a cyclical periodicity). That is, the buck regulator, at steady state, must have the inductor current come back to the same point in

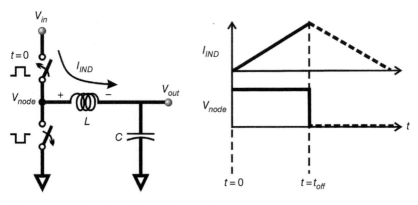

Figure 2.6 Buck regulator duty cycle operation.

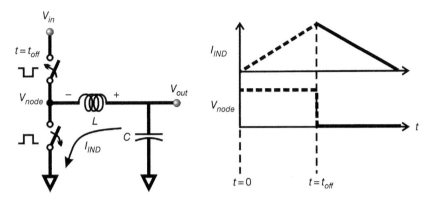

Figure 2.7 Commutating cycle of buck regulator.

the switching cycle or the system will be unstable. Thus, this indicates that the volt-seconds across the inductor must be equal during the two states:

$$(V_{in} - V_O)(DT) = (V_O)[(1 - D)T] = L\Delta I \tag{2.9}$$

The left-hand side of the equation is the relationship for the rising current through the inductor as shown in Figure 2.6. The right-hand side is the same relationship for the decaying current as shown in Figure 2.7. Because the change in current must be equal in both cases, we can set the equations equal to each other. Solving the equation, we find that the duty cycle is directly related to the ratio of the output to input voltage in a buck converter.

$$D = \frac{V_O}{V_{in}} \tag{2.10}$$

This elegant result holds true for many different topologies of buck regulators including multiphase designs. It must be pointed out that this is for *steady state* operation only. Virtually all regulators have transient requirements where the currents cause deviations in the voltages at the load and at the input. Thus, for these instances, the ratio will still be related to the input and output voltage, but depending upon the *mode* of the regulator, the point at which the upper and lower switches turn off and on may be temporarily different though the relationship in Eq. (2.10) will still hold.

The buck regulator plays an important role in the understanding of the platform power integrity for a given power delivery path in today's systems. This is because as the systems have become more complex and faster, so should the regulator in order to keep up. This will become evident in the forthcoming sections in this chapter as we build upon the understanding of the function of the buck regulator. The eventual goal is to be able to predict with some level of accuracy the behavior of the converter within a real system so that important time- and frequency-dependent information may be modeled and measured in a real system.

2.2.1 The LC Filter

The basis for the operation of the buck converter – essentially the ability to deliver a DC current – is because of the LC filter that connects to the bridge node. The LC filter does exactly as its name implies; it filters the signal from the switching point of the converter to the output load. Specifically, it is a low-pass filter that enables the DC signals to pass while reducing or attenuating the higher frequency signals.

A simple ideal LC filter is shown in Figure 2.8. Because the currents and voltages at V_{NODE} have high-frequency components with significant energy in them, it is required to filter them out so that the load only sees the DC component. The output voltage, mathematically, is simply the ratio of the impedance of the inductor to that of the capacitor. In the s-domain, this is shown in Eq. (2.11).

$$V_O(s) = \frac{Z_C(s)}{Z_C(s) + Z_L(s)} V_{node}(s) = \frac{1}{1 + s^2 LC} V_{node}(s) \tag{2.11}$$

where $s = j\omega$ in the frequency domain. Figure 2.9 shows a simple impedance plot of a typical buck LC filter for $L = 1\,\mu H$ and $C = 100\,\mu F$.

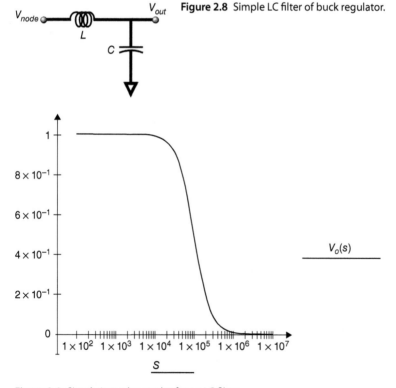

Figure 2.8 Simple LC filter of buck regulator.

Figure 2.9 Simple impedance plot for an LC filter.

The AC analysis shows that at lower frequencies, the input signal V_{NODE} will be fully transferred to the output. However, at higher frequencies, portions of the signal will be attenuated. This is the ideal case for the LC filter. In reality, there are parasitics in the path, which play a role in the filter characteristics. Figure 2.10 shows the LC filter with some of the parasitics added to the inductor and the capacitor.

Equation (2.11) is the same except that the parasitics add to the ratio of the impedances in Eq. (2.12).

$$V_{OP}(s) = \frac{1 + sCR_{pc} + s^2CL_{pc}}{1 + sC(R_{pl} + R_{pc}) + s^2C(L + L_{pc})} V_{node}(s) \tag{2.12}$$

The parasitic inductances and resistances are dependent upon both layout and construction. (As we shall see later on in the chapter, material properties – particularly for the inductor – contribute to these parasitics.) Adding in values to these elements, we may plot the new transfer function along with the ideal one. This is shown in Figure 2.11.

Figure 2.10 LC filter with parasitics added.

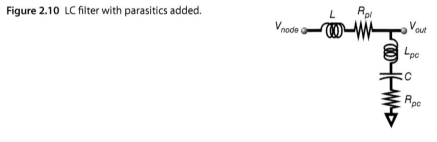

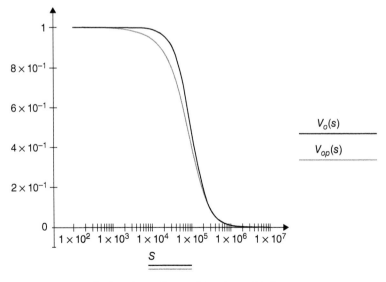

Figure 2.11 Transfer function with parasitic values added.

The additional parasitics, if significant, may often create a resonance at a frequency that is typically lower than the operating frequency of the voltage regulator (VR). In this case, as shown in Figure 2.11 for $V_{op}(s)$, the value for the parasitic inductance in the capacitor (100 pH) was low enough such that this second resonance did not appear. However, the resistances *did* create a shift in the curve and thus would attenuate the signal $V_o(s)$ as shown in the figure.

In reality, the signal that is generated from the source, V_{NODE}, has very specific spectral components to it. Figure 2.12 shows the signal at V_{NODE} for a standard square-wave function. Figure 2.13 shows the same signal using a spectral expansion. The Fourier spectrum reveals its corresponding spectral components in Figure 2.14. The first thing that is evident is that there is a strong DC component to the signal. However, there are also *high-frequency* components as well. Note that, in general, the wave is not half-wave symmetric; that is, the duty cycle is typically not 50%. This means that there are also *even* harmonics to the wave as well as *odd* harmonics. Having significant energy in the harmonic content of the waveform at these, higher frequencies may result in a more complex filter, particularly if the parasitics result in higher noise at the output. In that case, the power integrity engineer must add capacitors (and possibly reduce the inductance of the interconnections) to reduce or eliminate this noise to prevent it from coupling into the load or the system.

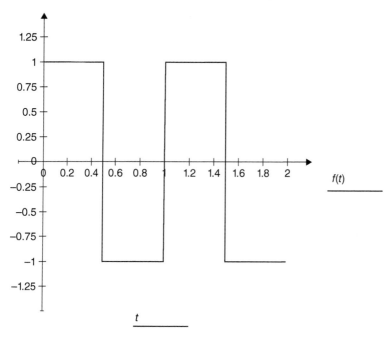

Figure 2.12 Simple plot of V_{NODE} drive voltage waveform.

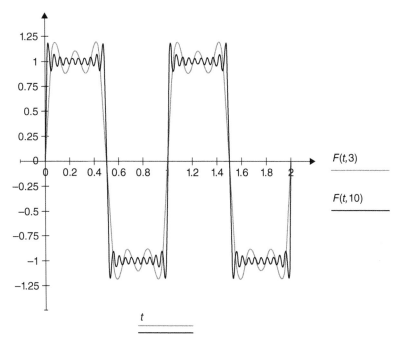

Figure 2.13 Spectral expansion for square-wave signal.

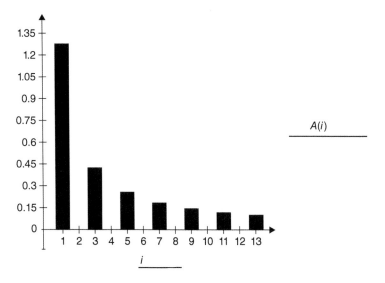

Figure 2.14 Spectral components of square-wave function.

Usually, the filter is able to reduce the noise at the PDN with respect to the high-frequency components. However, because the edge rates of the signal may be very high (as seen in the spectral plot), it is probable that noise may be generated in the system board as well. Though this can often be a conduction or emissions issue for the power converter designer and the layout engineer, the power integrity expert must be aware of these effects and understand the characteristics of these signals not only to help mitigate the noise into the PDN, but often to help reduce the system noise overall. Understanding then the facets of this signal is an important aspect to a PI engineer's job. There will be more discussion later in Chapter 6 on signal analysis with regard to understanding the waveforms such as those shown in Figure 2.12.

There is an interesting aspect to the LC filter for the buck regulator that must be mentioned here, that is, the frequency of operation. It is often misunderstood as to what defines a *good* inductor or capacitor in a buck regulator. Because the design is strictly intended to filter a trapezoidal waveform (ideally, a square wave), the best point of operation for the filter is *not* at resonance; rather, it is in a *range* that is usually higher than the resonance of the filter. Engineers often mistake the goodness of a buck regulator for the quality factor of the filter itself. For many applications, the *quality* of a filter is often dependent upon the Q of the circuit [2]. The Q, or quality factor, is a ratio of the energy transferred from the circuit to the power dissipated in the system, times the angular frequency.

$$Q(\omega) = \omega \frac{\text{Energy stored}}{\text{Power dissipated}} \qquad (2.13)$$

For the LC filter, the energy stored in both the inductor and the capacitor at any instance in time is then determined by the well-known equation

$$E = \frac{1}{2}CV^2 + \frac{1}{2}LI^2 \qquad (2.14)$$

where the first term is the energy stored in the capacitor while the second is for the inductor. The power dissipated in both the inductor and the capacitor is often more complex and involves not only the DC component of the loss but also the AC components. This will be discussed later on in the chapter as well as in other sections of the text.

It is often easier to interpret the quality factor here by redrawing the circuit in Figure 2.10 and then treating it as a circuit in series looking back at the impedance and then computing the Q. We can only do this, of course, if we assume the load at V_{out} is an open circuit for the moment and that there is no PDN network attached to it. Figure 2.15 shows the same circuit redrawn in the more conventional fashion.

Note that we can now lump the values for the resistance and inductance together. In other words, we let $L = L + L_{pc}$ and $R = R + R_{pc}$. The admittance

Figure 2.15 Circuit in Figure 2.10 redrawn for Q.

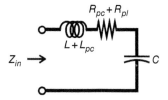

for the circuit may now be rewritten. In the frequency domain, the equation for the admittance is as follows:

$$Y(j\omega) = \frac{1}{R + j\left(\omega L - \frac{1}{\omega C}\right)} \tag{2.15}$$

where we have dropped the subscripts and notations for the extra parasitics and have simply lumped them into the "L" and "R" again. It is clear that the maximum admittance (the lowest impedance) occurs at resonance, or,

$$\omega_0 = \frac{1}{\sqrt{LC}} \tag{2.16}$$

Thus, substituting back into Eq. (2.15), we get,

$$Y(j\omega_0) = \frac{1}{R} \tag{2.17}$$

This should not be a surprising result. However, the more important thing to examine now is the Q of the circuit. The ratio of the admittance over the frequency range to that at resonance is now,

$$\frac{Y(j\omega)}{Y_0(j\omega_0)} = \frac{1}{1 + j\left(\frac{L\omega_0}{R}\right)\left(\frac{\omega}{\omega_0} - \frac{\omega_0}{\omega}\right)} \tag{2.18}$$

This is after substituting in for the resonant frequency. The Q is defined as the reactive energy of the inductor to the dissipative energy [2]

$$Y_r(j\omega) = \frac{Y(j\omega)}{Y_0(j\omega_0)} = \frac{1}{1 + j(Q)\left(\frac{\omega}{\omega_0} - \frac{\omega_0}{\omega}\right)} \tag{2.19}$$

The ratio of admittances may be easily plotted to determine where the maximum transfer of energy would be. This is shown in Figure 2.16. As expected, the maxima is at resonance where the ratio is 1. This frequency is computed from Eq. (2.16).

$$f_0 = \frac{1}{2\pi\sqrt{LC}} = \frac{1}{2\pi(10 \times 10^{-6})} \cong 16\,\text{kHz} \tag{2.20}$$

The key, however, for the circuit is to understand where this real operating point is for the buck regulator. That is, when the filter is sized for the application, does

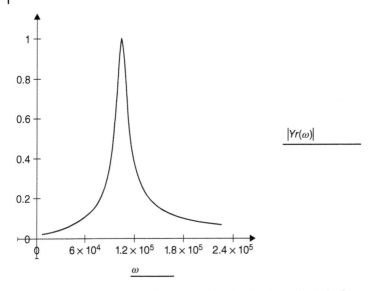

Figure 2.16 Admittance equation (2.19) plotted as function of angular frequency.

the power supply engineer truly operate the converter at this frequency? The answer is no.

In today's systems, the operating frequency of a converter with the inductance and capacitance values given would operate closer to 1 MHz [3]. If an engineer ran the circuit at its peak resonant point, the efficiency of the energy transfer would be maximized, but the loss would actually be high due to the larger ripple in the circuit (see problems at the end of this chapter).

There is actually a simple reason for this. For a given L and C value, the total AC ripple when operating at the resonant point would be so large that the losses in the circuit at the lower frequency would completely dominate and make the overall efficiency of the converter much lower than that when operating at a higher frequency. Essentially, because there are other elements in the PDN that are used for reducing noise and transferring energy from the source to the load, it is more efficient to operate the converter at a frequency that is *higher* than the resonance of the filter to help reduce these losses in the system. Moreover, the power designer will often choose a filter design that appears to have a *lower* Q than what would normally be prescribed at the peak resonant point. The reason, as stated previously, is that the objective is to end up with a design that results in the *highest* efficiency, or lower loss, rather than a circuit with the highest quality factor. In the end, it is the job of the power integrity engineer to work with the power converter designer to help mitigate these additional noise elements while simultaneously helping to maximize energy transfer to the load.

2.2.2 Silicon Power Devices in a Buck Regulator

2.2.2.1 Power MOSFETs

If the L–C filter in a buck regulator is the main storage element, the bridge is the mechanism through which energy is transferred to them. The bridge in a buck converter is simply the two switches (power MOSFETs) which help to transfer energy from a higher input voltage to the output (Figure 2.17). This is where the simplicity, however, ends. As the development of integrated systems has evolved, so has the complexity of the bridge and the driver circuits, which pass current to them. As stated in Chapter 1, much of this evolution has come about due to process shrinkage in silicon devices, and this includes power devices. Because the process has reduced the size of the loads in devices, so has the need to lower the voltage delivered to them, which has resulted in minimizing the power loss through the devices as well. Nonetheless, the converters external to the load silicon have had to keep up. This has occurred not only through lowering the output voltage, but also through the size of the power converters. Usually, the result has been higher switching frequencies and greater amounts of circuit integration.

The bridges and drivers of today are typically highly integrated in most buck converter designs. This is because the silicon manufacturing process to develop them has also shrunk resulting in the need to combine different circuits into one monolithic device. This trend has been going on for much of the last decade where now most power circuits that drive advanced silicon have virtually all of the analog and digital control integrated with the bridges and drivers in them. These devices, sometimes called power management integrated circuits (PMICs), have essentially become the one-stop shop for delivering power onto many platforms [1].

From a design perspective, power circuit engineers are mostly interested in the size, cost, efficiency, and the reliability of the bridges and drivers when constructing an integrated power converter. Though this does not conflict entirely with the power integrity engineer's objectives, it certainly can go beyond their scope. The PI engineer is more often interested in modeling such a device

Figure 2.17 Bridge and driver for buck regulator.

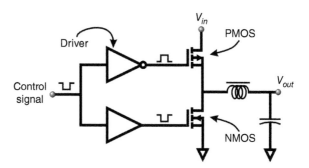

to determine how it responds within the overall PDN and system design. Thus, understanding the behavior of the bridges is the first and foremost goal since the end goal is usually some type of model to be used in time-domain simulations.

Referring to Figure 2.17, the main drive-train of the buck regulator consists of some type of driver circuit sending a signal to either the upper or lower FET of the bridge. The upper FET is usually a PMOS-based device.[2] This connection is shown explicitly in Figure 2.18.

There are multiple types of power MOSFETs that are available in the industry today. The three more common types of power MOSFETs are the vertical, or *VDMOSFET*, lateral, and gallium-nitride, or GaN based. Since VDMOSFETs are the more common of the three for computer systems, we will focus our attention on these devices. For highly integrated systems, the PMOS high-side FET (and the NMOS for that matter) is typically constructed of many devices in parallel rather than just one [4]. Thus, this means that the *set* of PMOS devices usually have some type of fan-out signal from the driver going to them along with a timing relationship to synchronize their turn-on. The PMOS device is connected to the input voltage rail on the platform at the *source* of the device.[3] The drain is connected to the node voltage where the inductor is connected. Note that in Figure 2.17, the signal into the PMOS device is *inverted* from that of the NMOS. A PMOS device is active when the *gate* of the device is pulled low; that is, the current *sinks* into the driver circuit when the switch is turned *on*. This is important since the drivers usually have to both source and sink current. The NMOS device is active when the signal is high at the gate, which turns the NMOS switch *on*. As discussed in Section 2.2 on the buck regulator

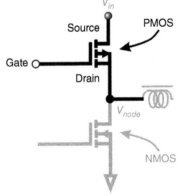

Figure 2.18 PMOS device in bridge.

---

2 Though, many buck converter designs still utilize NMOS based transistors for their upper switch even today.
3 This rail may be either a general voltage on a server platform or may even be the battery voltage for a handheld electronic device.

operation, the switches are turned off and on at opposite times when the buck is operational.

The current and voltage (idealized) across the PMOS device (source to drain) is shown in Figure 2.19. As the device is turned on, current ramps into the inductor from time $t = 0$ to $t = t_1$.

From $t = t_1$ to $t = t_2$, the upper switch is off (Figure 2.20) and the current is allowed to commutate through the lower NMOS device which is now turned on (as described in Section 2.2). This is shown in the dashed line for the current through the inductor during this interval when the PMOS device is off and in the solid line in Figure 2.20 for the same time interval. The current through the PMOS (ideally), however, is zero during this interval as indicated in the figures.

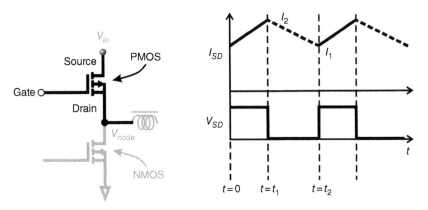

Figure 2.19 PMOS bridge operation.

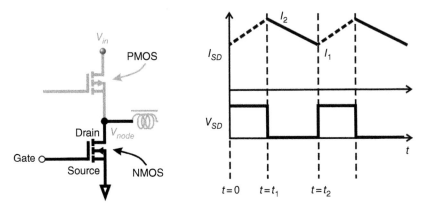

Figure 2.20 NMOS operation waveforms.

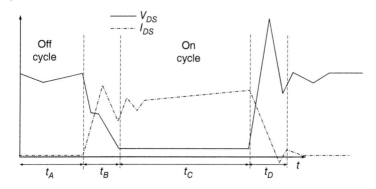

Figure 2.21 Power loss in full switching cycle for bridge.

The power dissipated is the loss at any instant in time for the circuit in Figure 2.10. However, with regard to the bridges themselves, the losses may be slightly more complicated than in a typical passive device. There are actually three components to the power loss that one needs to consider in a real power device: the on-resistance, the switching losses (AC losses), and the transition losses. These regions are illustrated in Figure 2.21. With regard to the NMOS, during the off-cycle in the time t_A window, the voltage V_{DS} is high during this time of the operation. Moreover, the voltage at V_{NODE} is high – meaning the PMOS is on. Thus, it is clear that the current is essentially zero going through the device at this time. During t_B time, however, the device transitions and the current rises. This results in the transition losses between the turn-on of the NMOS and the turn-off of the PMOS. During the t_C time window, steady state current flows through the device and the losses are due mainly to the on-resistances of the MOSFET. Finally, in the t_D time frame, the device turns off and the cycle starts again. Note that the voltages and currents are shown imbalanced during the transition windows t_B and t_D. This is because there will always be some type of imbalance in the circuits which limit the ability to perfectly control the switching relationships of the two FETs. This period T, which repeats itself at $T = 1/F_{SW}$, results in the switching losses of the MOSFET over a cycle.

The power loss in the PMOS and NMOS is typically different due to the operation of the converter as described earlier. Thus, the equations for the losses will also be different with respect to the bridge circuits. An approximate, but reasonably accurate, expression for the power loss in the PMOS device is

$$P_P \cong I_{RMS}^2 \cdot R_{DS_{ON}}$$
$$+ f \cdot \left(\frac{Q_{SW}}{I_G} \cdot V_{in} \cdot I_D + Q_{RR} \cdot V_{DS} + Q_G \cdot V_G + \frac{Q_{OSS}}{2} \cdot V_{in} \right) \quad (2.21)$$

where P_p is the upper PMOS device. For the NMOS device, a similar equation for the power loss is

$$P_N \cong I_{RMS}^2 \cdot R_{DS_{ON}} + f \cdot \left(Q_G \cdot V_G + \frac{Q_{OSS}}{2} \cdot V_{in} + 2 \cdot V_{BD} \cdot I_{out} \cdot t_{off} \right)$$

$$(2.22)$$

The Q or *charge* elements are related to the capacitances and voltages for the devices during the specific operation of each device and will be described in some detail later (not to be confused with the Q-factor described earlier).

The on-resistance in the power MOSFET is comprised of a number of elements, as we shall see below. They are typically considered *static* losses. Though the root mean square (RMS) current is dynamic, the voltages and thus conduction through the devices are usually within a range to keep the resistances relatively constant.

The on-resistance or $R_{DS_{ON}}$ is the resistance of the device when it is fully on and results in loss due to the RMS current through the FET. This is the first term in Eqs. (2.21) and (2.22) and broken out as

$$P_{FET} = I_{RMS}^2 \cdot R_{DS_{on}}$$

$$(2.23)$$

For bridge power MOSFETs, the device is driven to a point when the on-resistance is at its lowest point for the transistor when it is on. This is because the designer wants to minimize the losses when it is active. This is very different from that of, say, an amplifier circuit, where often the transistors are operated in the *linear* region for many of the devices in the amplifier. The PMOS and NMOS RMS currents are slightly different. We can derive the PMOS RMS current by examining the waveform in Figure 2.22.

Figure 2.22 PMOS current waveform (top graph).

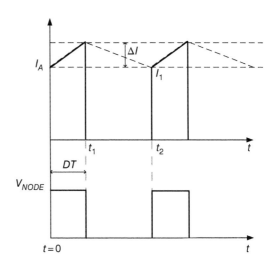

The RMS current of the power MOSFET may be computed from the following relation:

$$I_{RMS}^2 = \frac{1}{T} \int_0^T |I(t)|^2 dt \tag{2.24}$$

where T is the period. We can use linear superposition to compute the total RMS current for the PMOS portion here.

$$I_{RMS}^2 = \frac{1}{T} \int_0^T (|I_1(t)|^2 + |I_2(t)|^2) dt \tag{2.25}$$

Plugging in now for the PMOS portion of the currents, we get

$$\frac{1}{T} \int_0^{DT} (|I_A|^2 + |\Delta I|^2) dt \tag{2.26}$$

We need the equation for the line for the second portion as a function of t.

$$\frac{1}{T} \left[\int_0^{DT} I_A^2 dt + \int_0^{DT} \left(\frac{2\Delta I}{DT} t - \Delta I \right)^2 dt \right] \tag{2.27}$$

The first term is easily integrated to yield

$$\frac{1}{T} \int_0^{DT} I_A^2 dt = DI_A^2 \tag{2.28}$$

The second term we expand as the following:

$$\frac{1}{T} \int_0^{DT} \left(\frac{2\Delta I}{DT} t - \Delta I \right)^2 dt = \frac{1}{T} \int_0^{DT} \left(\frac{4\Delta I^2}{(DT)^2} t^2 - \frac{4\Delta I^2}{DT} t + \Delta I^2 \right)^2 dt \tag{2.29}$$

Integrating yields

$$\frac{1}{T} \int_0^{DT} \left(\frac{4\Delta I^2}{(DT)^2} t^2 - \frac{4\Delta I^2}{DT} t + \Delta I^2 \right)^2 dt = D \left[\frac{4\Delta I^2}{3} - 2\Delta I^2 + \Delta I^2 \right]$$

$$= D \frac{\Delta I^2}{3} \tag{2.30}$$

Adding Eqs. (2.28) and (2.30) and taking the square root yields

$$I_{RMS} = I_A \sqrt{D} \sqrt{1 + \frac{1}{3} \left(\frac{\Delta I}{I_A} \right)^2} \tag{2.31}$$

The NMOS expression is left as an exercise at the end of the chapter.

In Eq. (2.23), we see that we need both the on-resistance and the RMS current through the device to compute the power loss. Depending upon the complexity of the package and silicon construction, the on-resistance may be comprised of

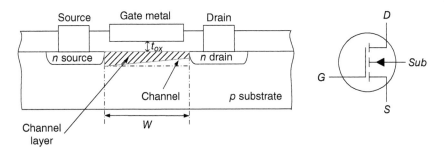

Figure 2.23 Simple cross-section of N-channel MOSFET device.

a number of different components. For the power integrity engineer, however, it is usually lumped into a single value for modeling purposes.

The on-resistance of the power MOSFET may be better understood by examining a cross-section of the construction for the device itself. This view is shown for an N-channel MOSFET in Figure 2.23 along with its model symbol. In a power converter system, the device operates almost exclusively as a switch. When a voltage is applied across the gate to source, charge accumulates in a layer between the source and drain. As the voltage goes above the threshold (V_T), current starts to flow through the channel region between the drain and the source (Figure 2.23). The transport of charge allows *hole* mobility to take place across the depletion region of the MOSFET. The mobility is *enhanced* due to increasing voltage at the gate, which is why this type of transistor is called an *enhancement* mode MOSFET. This allows electrons to flow from the drain. At this time, the charge in the region saturates and the device is fully on.

When the device is fully on, whether it is the upper or lower bridge FET, the *on-resistance* of the device is the component that dominates the losses. The resistive components are a combination of the *ohmic* losses in the MOSFET itself, and the interconnection to the platform devices and interconnects. Figure 2.24 shows a simplified diagram with the connections in place for clarity. The three main contact points are the *gate, drain*, and *source*. The gate is typically made from a *polysilicon* construction and is the control point for the switch [14]. The source metal conforms over the gate oxide to create the connection. Finally, on the third connection, the drain metal is connected to the $n+$ substrate region. The substrate is often connected directly to the drain (often considered as a fourth connection). Note that in the figure, the drain connection is shown on the *bottom* of the device rather than on the top. This helps to illustrate where the on-resistance components manifest themselves and is also consistent with the construction of the VD MOSFET.

As we shall later see, there is also a diode connected from the drain to source through this region that is part of the model.

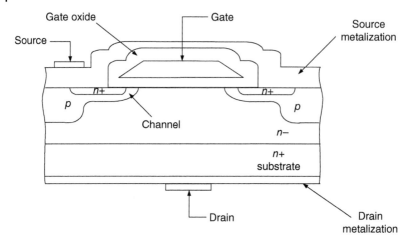

Figure 2.24 Physical structure of discrete power N-channel MOSFET.

The main resistive components that make up the on-resistance in a discrete power MOSFET are shown in Figure 2.25. Many of these contributors are specified by the vendor as one value for power converter applications under specific conditions. The total on-resistance is a compilation of these components in series.

$$R_{DS_{ON}} = R_P + R_S + R_{CH} + R_{ACU} + R_J + R_D + R_{SUB} \tag{2.32}$$

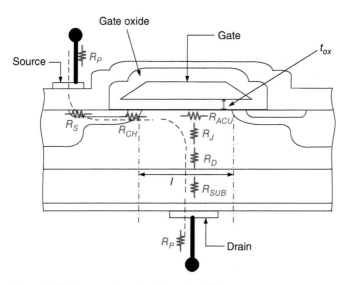

Figure 2.25 Cross-section showing main RDS components.

Note that R_P is shown twice in the figure (one for the drain side and one for the source) though it is typically lumped into one element. R_P is the combination of package, interconnect to silicon, and metallization resistances. For low-voltage, high-current power devices, this may be significant [6]. The other resistance elements are defined herein. R_S is the *source* diffusion resistance. R_D is the *drift* resistance of the extended drain region. R_{CH} is the *channel* resistance of the MOSFET. R_J is the JFET component of the resistance between the two body regions. R_{ACU} is the *accumulation* resistance, which is due to the formation of an accumulation layer in the p-base layer in the center of the trench. And finally, R_{sub} is the *substrate* resistance.

The on-resistance derivations are relatively complex and a detailed study of them is beyond the scope of this text. However, as mentioned in the previous paragraph, most vendors specify the *total* on-resistance under certain operating conditions for the power MOSFET device when they are used for power switching applications. Usually, for low-voltage applications, such as most computer platforms, the channel, accumulation, drift, JFET (sometimes), and parasitic resistances are considered the dominant components [5]. The channel resistance may be estimated from the physics and construction of the device. The following equations are shown for reference only. When analyzing the on-resistance for power loss applications, the vendor's data sheet is always the recommended source.

When the switch is turned on, the drain current flows through the channel region. The channel resistance may be computed from the drain-to-source current, I_{DS} and the voltage across the drain V_{DS}.

$$R_{CH} \cong \frac{V_{DS}}{I_D} = \frac{W}{2\mu_e L C_0 (V_{GS} - V_T)} \qquad V_{GS} - V_T > V_{DS} \qquad (2.33)$$

where W is the channel width, μ_e is the mobility constant, L is the length of the channel, and C_0 is the oxide capacitance per unit area. The gate voltage must be greater than the threshold voltage (and drain to source voltage) in order for Eq. (2.33) to be valid. The channel length L is the length of the transistor as shown in Figure 2.23 into the page.

The accumulation resistance is due to the flow of current spreading into the JFET region which is aided through an accumulation layer that is formed below the oxide gate. For a vertical or VD-MOSFET, this is approximated with the following:

$$R_{ACU} \cong \frac{1}{2\pi \mu_A L C_0 (V_{GS} - V_T)} \qquad (2.34)$$

where μ_A is the mobility constant in the accumulation layer. As with all the formulae, the equation only applies for specific conditions of operation.[4]

---

4 Constants, such as μ_A and others are specific to the device under question which includes cell topology for the actual MOSFET. This is why it is best to use the values given by the vendor for the on-resistances.

The drift resistance is due to the flow of current from JFET region into the drain and substrate region. For reference, the drift on-resistance may be found through the following formula (2.5):

$$R_D \cong \frac{4\rho_D t W_C^2}{\pi a(a + 2t)} \tag{2.35}$$

where a is the radial cross-sectional area of the drift region current flow, t is the thickness of the drift region, ρ is the drift resistivity, and W_C is the cell width.

The parasitic interconnect resistances are dependent on the geometry and construction of the device (as well as the operating conditions). Many of the recent advancements in the discrete power MOSFET over the past 10 years have been due to packaging [4]. This is because advancements in manufacturing techniques for these power devices have driven the internal resistance of the silicon when active to values which are often 10 times lower than the external resistances of the interconnect. Thus, due to temperature effects and other environmental factors, the specifications of the on-resistance are often given with graphs as well as in tables. The reason is the conductivity of the materials being different (such as solder and copper as compared with silicon) and when the values are lumped together, there are usually very limited constraints under which the specifications hold. Moreover, the vendor will usually specify the maximum values for the on-resistance under these constraints (along with the typical values) so that the engineer may compute the boundary conditions for the power loss computation. This has helped designers to build power systems with much smaller footprints as well as with higher power. The shrinkage of the packaging, along with the silicon, has led to more efficient designs because the distances between devices have led to smaller resistance and inductive parasitics.

In today's discrete devices, package considerations and connections to the platform are key to the most efficient operation for the device. At the back of many specifications, highly defined rules for layout and manufacturing are provided to help minimize these parasitics within the board design. Thus, though the PI engineer usually only models the basic behavior of these devices, it is important to note how the assembly into the actual board and platform can affect the over-resistance and thus power loss computation.

The analysis of the static power loss in a power converter is often key to the successful design in a system. The power converter designer is usually the individual responsible for ensuring that their design meets the system specifications and thus will always have to perform a power loss simulation and/or analysis of their design. However, they are not the only designers who need to have a grasp of the impact of power loss in the system. Because PI engineers are often required to have a more active role in the system design, having a grasp of the fundamentals of the power consumption in a power converter helps to give them a broader understanding of the impact they will have on the overall system design. The following example illustrates the use of a vendor data sheet to compute the power loss due to the on-resistance of a given device in an operating environment.

Example 2.2 A PI engineer is trying to analyze the losses during the on-times of an NMOS device to determine the overall efficiency of the converter system. Some key parameters for computing the device's on-resistance are given in Table 2.3. In addition, a data sheet from the discrete power vendor shows a plot of the overall on-resistance as a function of temperature.

Table 2.3 Parameters for computation of on-resistance.

Parameter	Value
R_S, R_{Sub}	$0.5 \times R_D$
W/L	1
μ_e	1e20 C/V s
C_0	1e-17
V_{GS}	4.5 V
V_T	1.5 V
μ_A	1e25
R_J	$0.5 \times R_D$
R_D	$0.3 \times R_{ACU}$

During steady state, the current values, for the peak and trough, are,

$$I_P \cong 6.5 \text{ A}, \quad I_T \cong 3.5 \text{ A} \tag{2.36}$$

Assume that the junction temperature of the device is 80 °C at steady state operation. Also assume the frequency is 1 MHz and the on-time is 60% for this device.

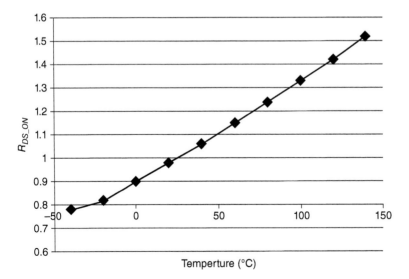

Figure 2.26 Plot of R_{DS_ON}.

First, estimate the total on-resistance of the device from the data above. Second, determine the total loss due to the RMS current through the device during this time. And third, graph the relative contribution of each resistance. Use Eq. (2.32) to sum up the total resistances. Which components dominated the on-resistance calculation?

Solution:
The vendor data sheet shows the on-resistance at a given temperature. At 80 °C, the total on-resistance is 1.24 mΩ from the graph above.

The total loss is due to the RMS current during steady state operation.

$$R_{DS_{ON}} I_{RMS}^2 = (1.24e - 3)\left(\frac{0.4}{3}\right)(6.5^2 + 3.5 \times 6.5 + 3.5^2) = 12.7 \text{ mW}$$

$$(2.37)$$

To break down the contributions, we need to determine the individual on-resistances.Thus, the total of all of the on-resistances should equal this when the computations are complete. It is a simple matter than to plug into the formulae above for the different resistances. The channel resistance is computed directly using the values in Table 2.3.

$$R_{CH} \cong \frac{W}{2\mu_e L C_0 (V_{GS} - V_T)} = \frac{1}{2(1e22 \times 1e - 17)(4.5 - 1.5 \text{ V})} = 167 \text{ μΩ}$$

$$(2.38)$$

We may also compute directly the on-resistance due to the accumulation layer,

$$R_{ACU} \cong \frac{1}{2\pi(1e26 \times 1e - 6 \times 1e - 17)(3)} = 66.3 \text{ μΩ} \qquad (2.39)$$

The other values are now computed relative to each other.

$$R_D \cong 0.3 \times R_{ACU} = 20 \text{ μΩ} \qquad R_J \cong 0.5 \times R_D = 10 \text{ μΩ} \qquad (2.40)$$

And for the rest of the values,

$$R_S, R_{Sub} \cong 0.2 \times R_D = 2 \text{ μΩ} \qquad (2.41)$$

The parasitic resistance is the difference between the total on-resistance and the sum of the others. We may now graph the individual contributions from each of the parasitics.

It is clear from the graph (note the log scale) that the dominant contribution was from the package parasitics. Since the total on-resistance was 1.24 mΩ at this temperature, this would be the one area to focus on to minimize the losses for the design. Though 13 mW or so does not seem like a large number, this is only one of many contributing factors to the loss in a converter, as we shall later see. Note that this loss was just for the NMOS device. The PMOS losses would also need to be computed.

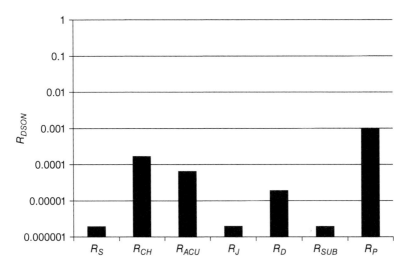

Figure 2.27 Graph of relative R_{DS_ON} values.

We now turn to the dynamic losses in the power MOSFET. The AC losses, as noted earlier, are comprised of the base frequency component and the transition currents and voltages between turn-on and turn-off of the devices. Steady state operation during the switching cycle results in the charge and discharge of the power MOSFET capacitances. The charge stored and emptied results in losses that are dissipated in the lossy networks of the device since little, if any, of the energy in charging and discharging of these capacitances is transferred to the load. The transition losses are due to the imbalance when the upper and lower bridges turn off and on respectively.

As shown in Figures 2.19 and 2.20, the operation of the PMOS and NMOS switching behavior is shown under *ideal* conditions. Because the converter switches at a given frequency under steady state and constant current conduction conditions, the MOSFET also has losses during this time of operation. To examine this behavior better, an electrical model is often helpful. A simplified model is shown in Figure 2.28 for an NMOS device. The passive elements are labeled in the model to help in running a SPICE simulation to represent the correct behavior. There are large libraries in most SPICE packages where an engineer can simply find the correct vendor model and apply it. However, if the device is highly integrated into a PMIC, then this may become a more difficult task. To get an understanding of the behavior, an engineer can start with a fundamental model such as that shown in Figure 2.28.

Starting at the right of the figure, there are two main capacitances near the gate: C_{GS} and C_{GD}. For discrete devices at higher voltages, the gate to source capacitance is often the dominant capacitance that is used in the computation of the switching losses of the device. Often vendors specify the charge on

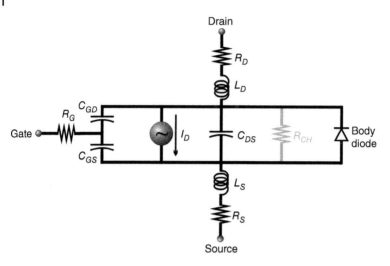

Figure 2.28 Simplified model for NMOS device.

the device due to switching rather than the capacitance [7]. The drain to source capacitance, C_{DS}, may also contribute significantly to the switching losses for lower voltage constructions. This is particularly true for highly integrated devices. A body diode is shown also across this connection which exists in the substrate of the device. In reverse bias, this diode may also break down at higher voltages, and large currents may surge through it (sometimes during transition). This is also specified by the vendor under specific electrical conditions.

The relative position of the capacitances is shown in Figure 2.29. In the data sheets provided by power MOSFET vendors, these capacitances are usually given in terms of other capacitances that may be measured directly on the device itself when characterizing it for operation. The switching losses for a power MOSFET are usually quite complex. Using the second term in Eq. (2.22), we can estimate the losses.

$$P_{NS} \cong f \cdot \left(Q_G \cdot V_G + \frac{Q_{OSS}}{2} \cdot V_{in} + 2 \cdot V_{BD} \cdot I_{out} \cdot t_{off} \right) \qquad (2.42)$$

The first part of the equation under the parentheses is the gate loss due to the switching, where Q_G is the total gate charge when the voltage to the device (V_G) is applied. The total gate charge is normally specified in a table for a given device. It is important to examine the conditions for such parameters because the values are specified under certain constraints.

The second component is the turn-off loss, where Q_{OSS} is the output charge (gate and source shorted) and is usually associated with C_{OSS} where

$$C_{OSS} = C_{GD} + C_{DS} \qquad (2.43)$$

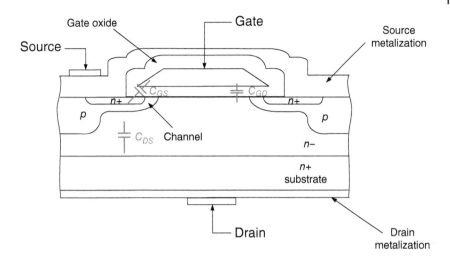

Figure 2.29 General illustration of where key capacitances reside in NMOS.

This gives the indirect relationship between the capacitance and the charge. Q_{OSS} is also specified in a table along with C_{OSS}.

The final portion of the loss to the NMOS device is the body diode losses, where V_{BD} is the body diode voltage. Before the high-side PMOS is turned on, inductor current (I_{out} in this case) is flowing through this body diode for a duration t_{off}, which is the minimum turn-off time of the FET. The relationship between the charge on the device and the capacitances typically involves an integration over a given time period and across the voltages of the device [9].

The PMOS device has similar characteristics and the switching losses may also be estimated from the second part of Eq. (2.21).

$$P_{PS} \cong f \cdot \left(\frac{Q_{SW}}{I_G} \cdot V_{in} \cdot I_D + Q_{RR} \cdot V_{DS} + Q_G \cdot V_G + \frac{Q_{OSS}}{2} \cdot V_{in} \right) \quad (2.44)$$

The first three components are the losses when the upper switch is turned on. Here, Q_{SW} is the total switch charge on the device.

$$Q_{SW} = Q_{GS} + Q_{GD} \quad (2.45)$$

Q_{RR} is the reverse recovery charge on the device. Q_G is the total gate charge on the device which is also specified by the vendor. Q_{OSS} is the small signal output capacitance with the gate and source terminals shorted. This element is part of the last component under the parentheses and is considered part of the turn-off losses of the PMOS. The elements for this portion are the same as the NMOS device discussed above.

Using the vendor's specifications, an engineer can readily estimate the switching power loss for either a P or NMOS device. However, there are

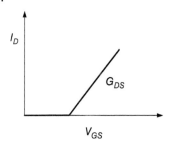

Figure 2.30 Voltage/current relationship (simplified) for MOSFET.

some important details which must be understood before one can compute the losses. First, for discrete MOSFET devices, the switching characteristics between the N and PMOS are very different. Figure 2.30 shows how the gate voltage relates to the transfer characteristics for a MOSFET device. G_{DS} is the trans-conductance for the device. In fact, the charge across the nodes are not constant and are highly dependent upon the gate voltage (as well as other factors). For the PMOS device, the gate voltage goes negative as does the gate current. Figure 2.31 shows the waveforms (inverted for PMOS) illustrating the operation. The gate voltage increases until its threshold voltage is reached (V_{TH}) at time t_1. As the voltage continues to increase (relative to V_{in}), the gate charges increase up to a point where the voltage plateau is reached (t_2).This is the *plateau voltage* where the drain current tends to flatten out. As the gate voltage increases further, the drain current increases but only slightly until the voltage across the drain to source is minimized (t_3), whereas the total charge tends to increase toward t_4 as the gate voltage is increased.

Estimating the power loss in a power converter is just one of the steps toward understanding the losses in the converter and the system. Relative to SPICE modeling, power converter designers usually have very detailed CAD data on the transistors and may determine the power loss over a range of operating conditions. For the PI engineer, just getting a basic computation of the power loss usually suffices which makes the previous formulae applicable for a first principles estimate of the discrete device's dynamic losses.

As shown in Figure 2.28, a simplified model may be created for a device once the values are known. This model, however, is understandably not very useful for determining the exact behavior of a device today. The reason is that

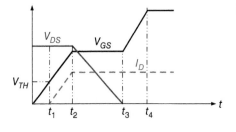

Figure 2.31 Gate and drain waveforms.

vendors typically develop very sophisticated SPICE and CAD models for their MOSFETs which usually meet BSIM (Berkeley Simulation) standards to some level and have very specific parameters in them. When building up a full switching model for a converter, it is usually best to try and get the actual vendor-specific models for the MOSFETs and then run SPICE level tests to ensure they operate correctly. The simplified model (and loss equations) mentioned above can be used as a check to make sure that the switching loss behavior is within the bounds of the expectations of the circuit that it is being used for. If the basic parameter values are applied in the model above and if the results are vastly different when compared with the vendor's model, it may mean that there is something wrong with the model relative to the manufacturer's specifications. It is often a good idea to check any vendor's SPICE model first before it is used for a detailed analysis to ensure data validity. At the end of the chapter are some problems which illustrate how to use a vendor table to estimate the losses in a power converter. In Section 2.3, we discuss inductors in power conversion.

2.3 Inductors

The section on the L–C filter earlier covered how the inductor works as a filter (along with the capacitor) and illustrated the importance of operating the buck regulator within certain limits. For the buck, the inductor is the key passive element, which helps to store energy in the system and is the main device that delivers current into the load. In many designs, however, it is also one of the key components that contribute to the overall power loss which is why it is important to gain a fundamental understanding of its electrical characteristics. The focus will therefore be to develop an understanding of its operation with respect to power loss and the PDN. Finally, the main goal in this chapter will be to develop an inductor model that may be used by the PI engineer for their overall power integrity analysis.

Most buck regulators in high-performance systems today operate at relatively high frequencies in order to deliver the current to the load. This is mainly due to the large dynamic range of the currents in a silicon device as well as due to the proximity of the converter to that load. In most cases, the inductor is placed close to the device, resulting in much lower impedance between it and the silicon. Because of the lack of space, often a smaller and less effective capacitance is the only other available storage element in the path. For this reason, the inductor must deliver current faster than the switching regulator behind it. Moreover, the inductor must not only deliver this current quickly but also do so in an efficient manner.

There are two basic components when considering the loss in an inductor: the windings and core. The windings have two loss elements that require analysis:

DC and AC. The DC loss is a function of the resistance of the metal of the inductor wire itself. The AC loss, if the frequency is high enough, can contribute to the inductor metal or *skin-effect* resistance for the windings which is why the resistances are placed in series and the loss is additive. The core power loss is a function of the inductor material. They are also a function of the time-changing magnetic fields from currents within the core, which means they are part of the overall AC losses.

The basis for the operation of the inductor for power switching applications, and thus for their loss mechanisms, starts with a treatment of field theory. Many of the details of the physics behind this operation, along with the equations, are presented in Chapter 4. A simplified version with respect to the inductor is given here for the reader. There are many approximations associated with the equations and physics and will be noted where applicable [10].

A physical view of a cross-section of a simple inductor is shown in Figure 2.32. For illustration purposes, a *C-Core* based structure is considered. A C-Core is simply what the name describes: the magnetic core material is shaped like a *C* where there is a gap in the material as shown in Figure 2.32. A wire is wrapped around a core of magnetic material from the node of the voltage regulator bridge circuits to the output of where the capacitor is connected. There are many different physical structures for an inductor that are used by vendors besides the C-Core. This particular construction was chosen for simplicity to help in understanding the physics and operation for a discrete device.

The flow of current into the inductor is as indicated through the winding from V_{NODE} to V_{out} when connected to a buck regulator. The windings are

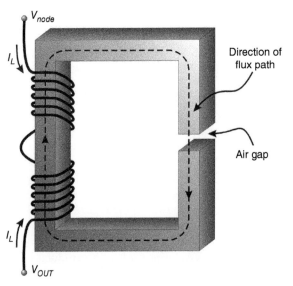

Figure 2.32 Simple inductor representation.

wrapped around the core similar to a solenoid. Using the right-hand rule, we can determine the direction of flux through the core. If one grasps their right hand around one of the windings in the middle with their thumb in the direction of the current flow, the flux path is shown to be in the direction of the arrows in the core region. Note that there is no winding around the air-gap in the C-Core representation. Flux will flow through this region even when gapped. This practice is common for many ferrite-based inductor designs to limit or prevent saturation [8].

When current flows through the windings, a magnetic field is induced within the core region. More magnetic flux is also contained within the core. The magnetic field H and magnetic flux density B are related through the following equation:

$$B(t) = \mu_r \mu_0 H(t) \, W/m^2 \tag{2.46}$$

In general electromagnetics theory, H and B also have a spatial dependence. Note that the spatial dependence here has been removed here for simplified treatment.[5] We assume that the flux is completely contained within the core and is uniform throughout. Also, there is no *leakage* inductance. Leakage inductance is due to the portion of flux that is in the region that resides between the windings (and outside of it in this case) and the magnetic core (which in this case is air). Since there is no flux outside the core, in this analysis, the inductance there will be zero. The magnetic flux density $B(t)$ is directly related to the magnetic field strength $H(t)$ through the permeability of the magnetic material as shown in Eq. (2.46). The magnetic flux density is given the dimensions Webers/meter². The magnetic field $H(t)$ has dimensions Amperes/meter. The permeability constant μ_0 has the following value:

$$\mu_0 = 4\pi \times 10^{-7} \, H/m \tag{2.47}$$

The permeability constant μ_r is related to the material and is dimensionless. For air, it is 1. For ferrite-based materials, it can be in the range of 3000 or even higher. The higher the permeability, the larger the ability to store energy in a given volume of magnetic material. The total flux ψ may be found through integration of Eq. (2.46). Assuming n turns on the core, the total flux is then

$$n\psi(t) = Li(t) = \iint B(t)dA \cong A_c B(t) = nA_c \mu_r \mu_0 H(t) \, W \tag{2.48}$$

This is related to the inductance L as shown. Once again, assuming the magnetic field is constant throughout the core, we may approximate the magnetic field, with n turns around it.

$$H(t) = \frac{ni(t)}{l_m} \, A/m \tag{2.49}$$

---

5 The spatial dependence is removed since it is assumed that the flux and magnetic field is uniform throughout the core.

where $i(t)$ is the current and l_m is the magnetic path length as shown in Figure 2.32. The inductance may therefore be found from Eq. (2.48). Substituting in Eq. (2.49) and solving for L, we get

$$L = \frac{\psi}{i} = \frac{n^2 A_c \mu_r}{l_m} \tag{2.50}$$

This shows that the inductance depends only upon the geometry of the core, the permeability, and the number of windings. It should be noted that this is a special case with the given assumptions and does not hold for all inductor constructions. The reluctance is often determined for many structures as well. The reluctance of the core is a measure of the resistance of the material to induced magnetic fields. The reluctance for the inductor in Figure 2.32 includes the reluctance of the main core plus that of the air gap.

$$\mathfrak{R} = \mathfrak{R}_c + \mathfrak{R}_g = \frac{l_m}{A_c \mu_0 \mu_r} + \frac{l_g}{A_c \mu_0} = \frac{1}{A_c \mu_0} \left(\frac{l_m}{\mu_r} + l_g \right) \tag{2.51}$$

Note that the reluctance is the *inverse* of inductance. The reluctance of an inductor should not be confused with the resistance. The determination of the reluctance for an inductor helps to gain an insight into what the gap should be to prevent or limit saturation (among other important characteristics) since it is often where most of the flux is stored in the inductor.

The effect of gapping a core may be seen in Figure 2.33. Gapping changes the slope of the $B-H$ curve, resulting in an effective permeability that is lower than that when not gapped. Thus, the core can withstand more magnetic flux (impressed magnetic field) inside it before saturating.

We earlier discussed the operation of the filter and buck converter in terms of its voltages and currents. Before we investigate the details of power loss for the inductor, it is important to go over its operation when used in a buck regulator. The operation of the inductor goes as follows:

When the upper switch in a buck regulator is turned on, we know that current flows into the inductor and causes a voltage across it. We recall this from

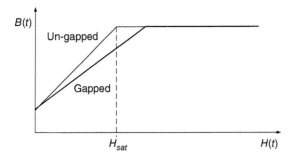

Figure 2.33 Changes in magnetic flux density with gapping and not gapping core.

the previous discussion for the operation of the buck regulator in this chapter where

$$V_{in} - V_{out} = L\frac{di}{dt} = n\frac{d\psi}{dt} \tag{2.52}$$

This falls out directly from Eq. (2.48) when we take the derivative of both sides. When this occurs, the rate of change in current creates a changing magnetic field in the core (and in the region outside the windings, though again we consider only the case where the fields outside the core are essentially zero). This also increases the magnetic flux in the core and the magnetic flux density (B). This is shown in the time-changing fields for H and B in Eq. (2.53).

$$\frac{dH(t)}{dt} = \frac{n}{l_m}\frac{di}{dt}, \quad \frac{dB(t)}{dt} = \frac{n}{A_C}\frac{d\psi}{dt} \tag{2.53}$$

This shows the relationship more geometrically between the magnetic field and the magnetic flux density in Eq. (2.46). We can also see that there is a relationship between the voltage for an inductor and the magnetic flux density from the previous two equations.

$$A_C\frac{dB(t)}{dt} = \Delta V \tag{2.54}$$

This states that the change in magnetic flux is related to the change in voltage across the inductor. For an inductor, there is a hysteretic loop (Figure 2.34) between the fields B and H. This is where the energy is both transferred and lost. As the current flows into the inductor, energy is stored as magnetic flux in the core with impressed magnetic field H. This is shown in the upper curve in the figure. But when the energy is transferred to the load, there is some loss

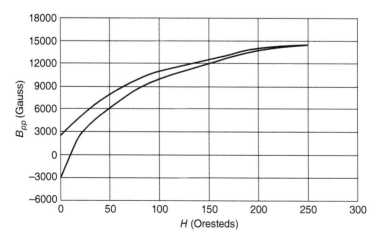

Figure 2.34 Hysteresis loop in *B–H* curve for magnetic core.

within the core which is evident by tracing the lower curve in the figure. This loss in energy may be shown mathematically.

The transfer of energy occurs when the lower switch turns on and the upper switch turns off, decreasing the rate of change of current through the inductor. Using Eqs. (2.52) and (2.53) and substituting, we get

$$\frac{dB(t)}{dt} = \frac{L}{A_C}\frac{di}{dt} \tag{2.55}$$

This now relates the rate of change of *current* through the inductor to the change in magnetic flux in the core.

The energy transferred in one cycle is related to the voltage and current and thus to the magnetic field and the magnetic flux density. This can be seen through integrating the power over one cycle (recall "Energy = Power × Time"). Integrating the power (voltage times the current) and substituting in for Eqs. (2.46) and (2.53), we get

$$E_C = \int_0^T i(t) \cdot v(t)dt = \int_0^T H(t)\frac{l_m}{n} \cdot nA_C\frac{dB(t)}{dt}dt = l_mA_C\int H(t)dB \tag{2.56}$$

This now gives us the total energy in the system over one cycle. The energy lost per cycle is related to the volume of the material as shown in Eq. (2.56) ($l \times A$). We now have the foundation to compute the losses in more detail.

A simplified loss model for an inductor is shown in Figure 2.35. The DC loss is represented by a series resistance as is the AC loss. The core loss of the material is shown shunted across an ideal inductor. Both the AC and core loss vary as a function of frequency. In general, all of the parameters vary as a function of temperature.

The DC component is fairly straightforward to analyze. Many vendors specify the DC resistance or *DCR* for the device under certain operating conditions. Fundamentally, the DC losses are computed from the basic equation:

$$P_{L\_DC} = I_{DC}^2 R_{DC} \tag{2.57}$$

where I_{DC} is the DC or current through the inductor. The resistance for the inductor is determined from the geometry of the construction, which is often

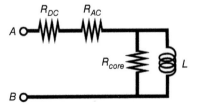

Figure 2.35 Simple model for inductor AC and DC losses.

complex for many discrete designs due to odd shapes that are due to the fabrication process. The basic equation is shown as:

$$R_{DC} = \frac{l}{\sigma A} \tag{2.58}$$

where A is the cross-sectional area of the component, l is the length, and σ is the conductivity of the material. The conductivity is a function of temperature; though for most inductors the winding material of choice is copper, it is often specified over a temperature range.

$$\sigma_T = \sigma(T_0)(1 - kT)\, \text{S/m} \tag{2.59}$$

where T_0 is the conductivity of copper specified at room temperature (typically 20 °C) and k is the percentage change in conductivity for every degree Celsius of increase of T_0. The constant k for copper is ~0.39%. The general relation in Eq. (2.58) holds true if the frequency of operation is below the *skin depth* of the copper conductor. But what if it is not? If the frequency is high enough and the copper thickness is relatively large, then much of the current will crowd in a region, which is near the edge or sides of the conductor. This is important since this will increase the losses for the current path of the inductor. The skin depth of the material is related to the conductivity and frequency of the material,

$$\delta = \frac{1}{\sqrt{\pi \mu f \sigma}} \tag{2.60}$$

where μ is the permeability of the material as described earlier. The skin depth is a measure of the penetration (in this case), of the current in the material. One skin depth means that the material has penetrated ~37% into the material. For example, if a converter is switching at 5 MHz, the skin depth for the conductor would be

$$\delta = \frac{1}{2\pi \sqrt{(1e-7)(5e6)(5.7e7)}} \cong 30\,\mu\text{m} \tag{2.61}$$

Thus, if the thickness of a rectangular piece of copper is greater than 60 μm, then the currents should be within the whole density of this copper region. This is because the currents will crowd toward the surfaces of the conductor as they penetrate, rather than in the middle. For a wide flat wire, this would be predominantly on the two longest sides.

The losses for the wire still have an AC dependence. The power consumed in the wire may be found from the RMS current through the inductor.

$$P_{L_{AC}} = I_{RMS}^2 R_{AC} \tag{2.62}$$

The RMS current may also be found by integrating $i(t)$ through the inductor over a period.

$$I_{RMS} = \sqrt{\frac{1}{T} \int_0^T i(t)^2 dt} = \frac{\Delta I}{\sqrt{3}} \tag{2.63}$$

where ΔI is the change in current over one period or the peak-to-peak current through the inductor.[6]

The AC losses for the core of an inductor are typically more complex and require an in-depth study. The types of magnetic materials that discrete power inductor fabricators typically use are *ferrite*-based materials. Ferrites are ceramics within a group of solid-based magnetic materials that form crystalline structures from a sintering process with other metallic oxides at high temperatures. These *ferri-magnetic* materials have properties where the resistivities of the materials and permeabilities are high relative to other materials. For power applications, they are well suited due to their ease in manufacturing and low cost.

The magnetic properties are often complex and cannot be represented by a single scalar permeability though most suppliers usually give only one value in their specifications. This is because the materials can have strong interactions between their magnetic dipole moments with impressed magnetic fields. Thus, a detailed representation of the permeability requires a tensor matrix in most cases when analyzing the behavior in detail.

Fortunately, for power converter applications, the frequency of operation is relatively low which simplifies the analysis. Moreover, most inductors are *gapped* (as previously discussed) to prevent saturation of the material for low impressed currents and thus an *effective* permeability is often adequate to gain an understanding of the inductance value for the component. Moreover, because the inductors will saturate at a given current and temperature, the inductance is usually specified over an operating range for the designer.

2.3.1 Losses in Power Inductors

This section goes through a fairly stringent analysis of the losses in a magnetic core. It is recommended that the reader review the math and physics in Chapters 4 and 7 to familiarize themselves with methodologies presented herein.

The losses usually encompass two dominant mechanisms: eddy and hysteretic.[7] Eddy current losses are due to the flow of currents that are orthogonal to the direction of flux flowing in the magnetic material. An illustration of how eddy currents flow within a magnetic structure is shown in Figure 2.36. As magnetic flux enters the material, from Lenz's law, currents may flow through the structures that are *induced* from a flux that opposes the flux that was applied. These currents may circulate around the magnetic core as a function

---

6 The RMS current here is derived for IP which is the peak current from zero. Often the RMS current is derived using the change in the ripple or $\Delta I/2$, which results in $I_{RMS} = \Delta I/\sqrt{12}$.

7 For thin-film magnetic structures a third loss mechanism must be considered and is usually called *anomalous* loss and is due, in part, to the physical construction.

Figure 2.36 Illustration of eddy currents flowing in magnetic material.

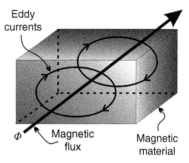

of AC excitation and can represent a significant portion to the loss of a core if not controlled in some fashion.

In film or laminate structures, a dielectric is often placed between layers of magnetic material to limit the flow of these currents, thus reducing the effects of these eddy current losses.

Eddy current losses in magnetic structures, even at lower frequencies, can be quite complex. This is because there are more than one mechanism that can cause eddy currents to flow within a magnetic core structure. For example, for amorphous structures, eddy currents can flow within the domains internal to the magnetic core that are different than the classical eddy current flow through a uniform material. If the structure is isotropic, however, then we can make some reasonable assumptions and determine a fundamental equation for a structure where the currents and considered uniform throughout the material and the structure itself is isotropic.

We may use Maxwell's equations and compute the total power and energy going into and out of a block of material. We start with Ampere's and Faraday's laws. (For a review of these equations, see Chapter 4.)

$$\mathbf{\nabla} \times \vec{\mathbf{H}} = \frac{\partial \vec{\mathbf{D}}}{\partial t} + \vec{\mathbf{J}} \tag{2.64}$$

where H is the magnetic field, D is the electric flux density, and J is the current density. And for Faraday's law we have

$$\nabla \times \vec{\mathbf{E}} = -\frac{\partial \vec{\mathbf{B}}}{\partial t} \tag{2.65}$$

Within the inductor, the displacement current, due to the flow of charge through the dielectric, is very small.

$$\vec{\mathbf{J}}_d = \frac{\partial \vec{\mathbf{D}}}{\partial t} \cong \mathbf{0} \tag{2.66}$$

Thus, we can simplify the first equation above to

$$\mathbf{\nabla} \times \vec{\mathbf{H}} = \vec{\mathbf{J}}_c \tag{2.67}$$

The next step is to take the dot product with H and E respectively of both Eqs. (2.65) and (2.67) and then subtract one from the other.

$$\vec{H} \cdot \nabla \times \vec{E} - \vec{E} \cdot \nabla \times \vec{H} = -\vec{H} \cdot \frac{\partial \vec{B}}{\partial t} - \vec{E} \cdot \vec{J}_c \tag{2.68}$$

The term on the left-hand side may be simplified using a vector identity in the appendix.

$$\nabla \cdot (\vec{E} \times \vec{H}) = -\vec{H} \cdot \frac{\partial \vec{B}}{\partial t} - \vec{E} \cdot \vec{J}_c \tag{2.69}$$

This equation essentially comes from the Poynting's theorem. If we want the average power per cycle, then we can integrate Eq. (2.69) over a period of time.

$$\int_{t_1}^{t_2} \nabla \cdot (\vec{E} \times \vec{H}) dt = -\int_{t_1}^{t_2} \left(\vec{H} \cdot \frac{\partial \vec{B}}{\partial t} + \vec{E} \cdot \vec{J}_c \right) dt \tag{2.70}$$

This equation is a variant of the Poynting's theorem which states the following:

> The rate of energy transferred per unit volume in a volume equals the rate of work done on a distribution of charge and the flux energy leaving that volume.

With respect to the equation, the term on the left-hand side is the total energy now per unit volume that enters a physical structure. If we now integrate over the volume of the space where the fields enter the structure, we will get the total energy going into it.

$$W = \int_{t_1}^{t_2} \left[\int_V \nabla \cdot (\vec{E} \times \vec{H}) dV \right] dt$$
$$= -\int_{t_1}^{t_2} \left[\int_V \left(\vec{H} \cdot \frac{\partial \vec{B}}{\partial t} + \vec{E} \cdot \vec{J}_c \right) dV \right] dt \tag{2.71}$$

Equation (2.71) gives the total energy flowing into the volume on the left-hand side, while the right-hand side is the total energy dissipated plus the total energy stored within a unit of time.

We can use the divergence theorem (Chapter 4) on the left-hand side of the equation to get the total energy flowing into a surface.

$$W = \int_{t_1}^{t_2} \left[\int_S (\vec{E} \times \vec{H}) \cdot \vec{dS} \right] dt = -\int_{t_1}^{t_2} \left[\int_V \left(\vec{H} \cdot \frac{\partial \vec{B}}{\partial t} + \vec{E} \cdot \vec{J}_c \right) dV \right] dt \tag{2.72}$$

Our focus now is on the left-hand side of the equation where we want to determine the total energy lost in the structure as a function of the energy flowing into the system. There are two parts in Eq. (2.72): the first part relates to the hysteretic losses in the system and the second portion to the eddy currents. The second portion we will relate to the eddy currents that are generated in the ferromagnetic material. Here we make the substitution for the electric field where

$$\sigma \vec{E} = \vec{J}_c \tag{2.73}$$

Thus, the second part of Eq. (2.72) becomes

$$-\int_{t_1}^{t_2} \left[\int_V \left(\frac{\vec{J}_c \cdot \vec{J}_c}{\sigma} \right) dV \right] dt \tag{2.74}$$

In this case, we assume that the material is both isotropic and homogeneous. This is a reasonable assumption for most ferrites where the material lattice structures are mainly crystalline and the flux is usually uniform due to the lower frequencies of operation through the material. Also, since $\mathbf{E}$ and $\mathbf{J}$ are in the same direction, the dot product results in the following:

$$\int_{t_1}^{t_2} \left[\int_V \left(\frac{J_c^2}{\sigma} \right) dV \right] dt \tag{2.75}$$

We dropped the negative sign here since it is assumed that this is the energy lost in the system. We now have our formula for the eddy current lost in the system. We now need to solve for the current density in the structure. We refer to Figure 2.37,

The currents flow orthogonal to the direction of flux as shown in the figure per Maxwell's equations. To determine the solution to the current density, we start with Faraday's law and then expand. Taking the curl of both sides of Eq. (2.65), we get

$$\nabla \times \nabla \times \vec{E} = \nabla^2 \vec{E} - \nabla(\nabla \cdot \vec{E}) = -\nabla \times \left(\frac{\partial \vec{B}}{\partial t} \right) = -\mu \frac{\partial}{\partial t} \nabla \times \vec{H} \tag{2.76}$$

Figure 2.37 Eddy currents in block of magnetic material.

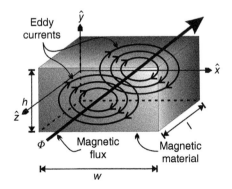

The second term for the divergence goes to zero and we substitute for the right- and left-hand sides for the current density.

$$\nabla^2 \overrightarrow{J}_c = -\mu\sigma \frac{\partial \overrightarrow{J}_c}{\partial t} \tag{2.77}$$

If we assume that J is sinusoidal, then we can simplify the previous expression.

$$\nabla^2 \overrightarrow{J} = -j\omega\mu\sigma \overrightarrow{J} \tag{2.78}$$

where it is now assumed that the current density is a conduction current.

The solution to this type of boundary value problem may be found analytically through the methods outlined in Chapter 7. Here we only state the solution and give the results. The boundary conditions are related to the currents that are flowing throughout the structure in Figure 2.37. In this case, current cannot flow out of the structure; thus, the boundary conditions are

$$J_z = J_y = 0 \tag{2.79}$$

This is because we have assumed that the height of the magnetic block is very small relative to the width in the x direction, and the currents essentially do not flow in the y direction. In the x direction, we have currents flowing in the positive and negative direction.

$$\frac{\partial J_x}{\partial y} = \frac{\partial^2 J_x}{\partial y^2} = 0 = \frac{\partial J_x}{\partial x} = \frac{\partial^2 J_x}{\partial x^2} \tag{2.80}$$

And on the boundary where the initial current flows, we assign the conditions.

$$J_x|_{y=h/2} = J_{x0}, J_x|_{y=-h/2} = -J_{x0} \tag{2.81}$$

The methodology for the solution to this boundary value problem may be found in Chapter 7. We leave it as a problem at the back of that chapter to solve for the reader. For now, we will show the solution

$$|J_x| = \frac{\omega\sigma|\Phi|}{2l} \left(\frac{\sinh^2 \frac{y}{\delta} + \sin^2 \frac{y}{\delta}}{\sinh^2 \frac{h}{2\delta} + \sin^2 \frac{h}{2\delta}} \right)^{\frac{1}{2}} \tag{2.82}$$

Here as in the reference, we have assumed that the equation we are solving is a parabolic partial differential.

$$\frac{d^2 J_x}{dy^2} = j\omega\mu\sigma J_x \tag{2.83}$$

where the substation has been made for the argument on the right-hand side.

$$j\omega\mu\sigma = \frac{(1+j)^2}{\delta^2} = k^2 \tag{2.84}$$

where δ is the skin depth for the material. To find the eddy currents, we integrate Eq. (2.75) over the variable of interest or

$$P_E = \int_V \left(\frac{J_c^2}{\sigma} \right) dV = 2 \int_0^{h/2} \frac{|J_x|^2}{2\sigma} wl \, dy = \frac{\omega^2 \sigma |\Phi|^2 w \delta}{8l} \left(\frac{\sinh \frac{h}{\delta} - \sin \frac{h}{\delta}}{\cosh \frac{h}{2\delta} - \cos \frac{h}{2\delta}} \right)$$

(2.85)

To get the maximum eddy current loss, we now take the limit of the function above

$$P_E = \lim_{\frac{h}{\delta} \to 0} \frac{\omega^2 h \sigma w \delta |\Phi|^2}{8lh} \left(\frac{\sinh \frac{h}{\delta} - \sin \frac{h}{\delta}}{\cosh \frac{h}{\delta} - \cos \frac{h}{\delta}} \right) = \frac{\omega^2 h \sigma w \Phi_{max}^2}{24l}$$

(2.86)

Since the total flux is related to the flux density by BA, we can substitute in to get the average eddy current loss per unit volume

$$\frac{P_E}{V} = \frac{\omega^2 h^2 \sigma B_{max}^2}{24}$$

(2.87)

This formula may be found in a number of references and can be used to estimate the losses in a magnetic structure due to eddy currents. Often the total eddy currents are computed and we get the more familiar formula:

$$P_E = \frac{\omega^2 h^2 \sigma B_{max}^2 V}{24}$$

(2.88)

For rectangular structures with laminations, the power loss may be estimated from the following:

$$P_E = \frac{lwt^3 B^2}{24n^2 \rho}$$

(2.89)

where l, w, and t are related to the geometry of the construction – in this case, it is assumed to be *rectangular* in form (e.g. the length, width, and thickness). The parameters n and ρ are the number of *laminations* and the *conductivity* of the magnetic material, respectively.

Hysteretic losses are due to the energy lost in the loop per cycle of operation. The power lost over this cycle is related to the frequency of operation. This directly falls out of Eq. (2.56).

$$P = fwtl \int H \, dB \quad \text{(over one cycle)}$$

(2.90)

where f is frequency and the other parameters are related to the volume of the material. The hysteretic loss may be estimated for rectangular shapes from the equation

$$P_H = 2fV H_C \Delta B$$

(2.91)

where ΔB is the change in magnetic flux over a range in the hysteretic portion of the curve.

To simplify analyzing magnetic structures, many inductor vendors supply what are termed *Steinmetz* parameters for the losses in the inductor core. The Steinmetz equation is an empirical formula where certain parameters are matched to fit the measured curves for power losses in the core as a function of frequency and impressed magnetic flux B.

$$P_L = kf^\alpha B^\beta \tag{2.92}$$

The parameters k, α, and β are often supplied by the vendor and may be used to compute the core losses in the construction.

Normally, the power designer computes the losses in the inductor based upon the operation of the converter. This usually ends up as part of the overall efficiency computation for the design. When modeling the converter in SPICE, it is often necessary to represent these losses in terms of *resistive* elements. Unfortunately, the losses change as a function of frequency. However, in many ferrite-based inductor solutions, the $B–H$ loop does not change with frequency and thus if the power loss is computed at a given frequency, and the switching frequency is estimated, the losses may also be readily computed. Equation (2.92) for the Steinmetz losses was derived for sinusoidal excitation. An updated variant of the formula was created that may give a more accurate representation of the losses in the core for power converter applications [10]. Once the characteristics of the voltage regulator are understood, this formula may be used to estimate the losses in the inductor.

One of the more common ways to estimate the losses in an inductor is to go back to the data sheet for a specific magnetic material and compute the losses as a function in the change in flux. The change in flux can be estimated using Eq. (2.54).

$$\Delta B = \frac{\Delta V \cdot t_D}{A_C} \tag{2.93}$$

where t_D is the on-time of the converter and ΔV is the voltage across it during this time. ΔB is the *peak-to-peak* magnetic flux density over a switching cycle in the inductor. Figure 2.38 shows an example graph of losses as a function of ΔB in a core material for a power application. The core resistance is then computed using the *RMS* voltage across the inductor. The *RMS* voltage is computed from the square wave voltage across the inductor. It is simple to integrate this under the square root and solve for the *RMS* voltage.

$$V_{RMS} = \sqrt{\frac{1}{T}\int_0^T V_{RMS}^2 dt} = \sqrt{\frac{1}{T}\int_0^{DT} V_{IN}^2 \, dt} = \sqrt{DV_{IN}^2} = V_{IN}\sqrt{D}$$

where D is the duty cycle for the converter. The core resistance is then computed by taking the power loss for the core from the graph and the *RMS* voltage. Note that power loss is now a function of the change in B,

$$R_{AC} = \frac{P(\Delta B)}{V_{RMS}} \tag{2.94}$$

An example illustrates the method for computing the resistances for a given inductor.

Example 2.3   A power integrity engineer is developing a simplified power converter model to analyze the behavior of the converter in the system. A discrete power inductor is part of the model. The converter delivers, during normal operation, 2 A continuous current and operates at 2 MHz in continuous conduction mode (CCM). The converter achieves CCM right when it hits 2 A DC. It runs off of an input voltage (average) of 3.7 V and regulates its output to 1 V. The designer is given the following information about the inductor from the converter designer (Table 2.4).

Estimate the losses for the inductor (windings and core) using the data from Figure 2.38 and the data on the switching converter above. Determine the resistance values for the model in Figure 2.35.

Solution:

Table 2.4 Inductor parameters for loss calculation.

Parameter	Data
A_C	$0.04 \times 0.04 \, \text{mm}^2$
Wire length	$2.5 \, \text{cm}$
Wire cross-section	$0.1 \times 0.1 \, \text{mm}^2$
Temperature of operation	$80 \, ^\circ\text{C}$

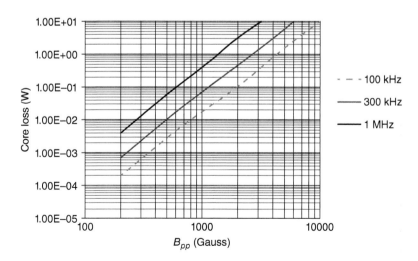

Figure 2.38 Core power loss as function of ΔB.

The first thing to determine is the winding loss. It is best to check if there are any skin-effect losses at the frequency of operation. Because the wire is not at 25 °C any more, we will need to compute the new conductivity first.

$$\sigma_T = 5.7e7(1 - 0.0039 \times 80) = 3.92e7 \ \text{S/m} \tag{2.95}$$

Now we can check the skin depth. The wire is 100 μm thick, so if the skin depth is less than 50 μm, we will need to see if they dominate over the DC losses.

$$\delta = \frac{1}{2\pi\sqrt{(1e-7)(1e6)(3.92e7)}} \cong 80 \ \mu\text{m} \tag{2.96}$$

As can be seen, the skin depth is greater than the thickness of the wire on one side. Thus, the resistance that we need to compute is the DC resistance by itself. Using Eq. (2.58), we get

$$R_{DC} = \frac{l}{\sigma A} = \frac{2.5e-2}{(3.92e7)(100e-6)(100e-6)} = 64 \ \text{m}\Omega \tag{2.97}$$

We now have the DC resistance for the wire based on the operating temperature. The DC current is determined from the operation of the device, which was given to us as 2 A continuous. Thus, using Eq. (2.57).

$$P_{W\_DC} = I_{DC}^2 R_{DC} = (4) \times (0.064) = 255 \ \text{mW} \tag{2.98}$$

The AC ripple current may be found from the equation for the *RMS* current.

$$I_{RMS} = \frac{4}{\sqrt{3}} = 2.31 \ \text{A} \tag{2.99}$$

The AC losses are computed using the DC resistance since this is the value that the inductor AC signal sees.

$$P_{W\_AC} = I_{RMS}^2 R_{DC} = (16/3) \times (0.064) = 340 \ \text{mW} \tag{2.100}$$

The core losses are now computed from the graph shown in Figure 2.38. To determine the loss, we need to estimate the change in flux. Using Eq. (2.100), we may compute this directly.

$$\Delta B = \frac{\Delta V \cdot t_D}{A_C} = \frac{(3.7)\left[\left(\frac{1}{3.7}\right)(1e-6)\right]}{(0.04e-3)(0.04e-3)} = 625 \ \text{G} \tag{2.101}$$

Examining the graph, we find this is approximately 100 mW. To get the core resistance, we can compute the *RMS* voltage.

$$V_{RMS} = V_{in}\sqrt{D} = 3.7 \times \sqrt{\frac{1}{3.7}} = 1.92 \ \text{V}$$

Back-calculating for the resistance we get

$$R_{AC} = \frac{P(\Delta B)}{V_{RMS}} = \frac{0.1 \ \text{W}}{1.92 \ \text{V}} = 52 \ \text{m}\Omega \tag{2.102}$$

There is now enough information to build the loss model in SPICE and run simulations for the inductor. Notice that we could have also estimated the number of windings on the core since the length of the wire was given. However, since we were not given the routing to the board connections, this would have resulted in only an approximation.

2.4 Controllers

Earlier in this chapter, we discussed some of the key components that make up the buck regulator. These elements included specifically the switches, filter, and the inductor. In this section, we discuss some of the basics with respect to the controller. If we consider the other elements as the instruments in an orchestra, the controller would be akin to the musicians and the conductor directing those mechanisms.

In most of our discussions concerning these other components thus far, we have focused primarily on their impact on power loss; however, this is only one facet to the controller design and its operation. Though the controller power loss usually impacts only the overall efficiency for a buck regulator when the currents are very low, it is the key to managing the response and operation for the converter itself. The regulator's response is very critical for dynamic operation within a system and thus has a large impact on the efficacy of the power integrity design. The droop behavior, under heavy *continuous* load conditions (such as a step response), is one of the most important metrics that a power integrity engineer investigates to ensure the efficacy of the power distribution for a given power rail in the design. It is therefore imperative that the PI engineer responsible for developing the PDN network has a basic understanding of the feedback of the buck regulator and its control.

One of the goals here is to develop an overall model for a simple buck converter that can be used for analyzing the system response with or without a PDN in place. Though it is not normally required to build power integrity models with active converter models, having the understanding of how the converter may affect the power delivery can be important to many PI problems.

Power converter modeling usually requires a background in control systems as well as an understanding of compensation methodologies and the componentry, which make up the controller circuits. Many excellent texts on power conversion include methods on how to design and compensate for the perturbations in a system [11]. However, even a cursory discussion of this subject is well beyond the scope of this text. Nonetheless, an overview of the basic model and an introductory understanding of the fundamentals behind such a model creation are warranted. Thus, a brief discussion and example of a buck converter model will be presented here.

Power integrity engineers are quick to ask why such a model would be necessary to predict the behavior of the system. Since the goal is to understand the time-domain response of the system and the voltage regulator's response is usually much slower than the network's response, what is the value in having a model that emulates the voltage regulator? The answer lies in understanding the type of behavior a power integrity engineer is interested in and the *time window* in which that behavior is relevant. In the past, most engineers who were interested in PDN behavior simply focused on the passive elements and performed both time- and frequency-domain analyses on the network alone. It was common when building these frequency-dependent sweeps and time-domain step responses to use simple R–L networks to model the regulator and even leave the source as ideal and/or open loop. This was because there was often sufficient energy stored in the network and power filter to reduce the regulator droop to an acceptable level. However, in many of today's systems, the need to reduce capacitance and inductance (basic energy storage) while increasing performance has resulted in two basic realities: First, the response of the regulator must in fact increase to keep the droop in check, and second, the dominant droop in the system is often due (at least in part) to the regulator and its filter. This means that to recover from this droop event requires the regulator to respond within a shorter period of time and that minimizing the effects of the regulator lag (response time), in transferring energy, is essential. It also means that the PI engineer must develop their PDN with the power converter filters and response as part of the overall power delivery, rather than simply decouple the system after the fact. It is for these reasons that a deeper understanding of the workings of the feedback system in a buck regulator is an important part of the power integrity engineer's education.

There are a number of different control topologies for a feedback system within a buck converter. The two most common are voltage mode (VM) and current mode (CM). Voltage mode essentially means that the feedback control is based on monitoring the output voltage of the converter and then responding to a change either higher or lower than the voltage set point. A current mode system responds to the change in current in a similar fashion and also has a voltage feedback loop. Inherently, a current mode system is slightly more complex since it requires the system to detect a change in current, which makes the sensing slightly more challenging. The advantage is that CM systems are nearly always faster than VM-based systems since the current sense circuits are often faster than the voltage sense circuits in the feedback. Moreover, the output filter is typically sized smaller which speeds up the response of the system. In this section, to illustrate the methodology, a voltage mode feedback system will be created.

For many engineers, SPICE-level circuit models for converters on their platform boards are developed by vendors and are made available to the engineers developing the systems. In fact, sometimes the design of those converters is performed in house and the engineers themselves have very useful models that can help in predicting system level power integrity behavior in the time domain. However, for the power integrity engineer, there is usually little to any documentation that goes with these models and without the understanding of their basic inner workings, the model is usually treated like a black box and the engineer is left to trust that the model accurately predicts the system behavior. This is where it is important to have a basic understanding of how these models are constructed and function so when analyzing a system droop problem, the PI engineer can begin to diagnose its cause more easily and determine if the results actually make sense.

2.4.1 A Simple Feedback System

In the design of a voltage regulator, the controller plays the role of ensuring that the output voltage is maintained to the specific value (or values) expected or programmed. The feedback system adjusts or compensates for deviations in the system to ensure this occurs. If there were no perturbations in the system, the voltage would remain the same as the reference it was compared to. However, in real systems, there are often significant deviations that cause the voltage to vary over its operation which require the converter to compensate for these variations and readjust the voltage back to its initial set point.

For the following overview, we have made an effort to keep the nomenclature consistent with the references cited so that the reader has some continuity with respect to these. This is important since if the student wishes for a deeper dive into the subject matter, the nomenclature here will be relatively consistent to aid in that process. Thus, the descriptions should therefore seem familiar to the reader.

Figure 2.39 shows the key elements of the buck converter along with the sources, which dominate the effects of the output of the voltage in a system. The input voltage $V_{in}(t)$ in a computer platform for a DC–DC converter can vary due to many factors, some of them from the switching behavior of the converter itself. The load current $i_{load}(t)$ (as most power integrity engineers are familiar with) usually has the largest effect on the perturbation of the output voltage, and thus the preponderance of engineers spend most of their time focused on this behavior. The control signal $d(t)$ helps to alter the effects on $V_{out}(t)$ and thus is also an input function of the output voltage. This is the duty cycle of the system,

Figure 2.39 Elements which affect output voltage.

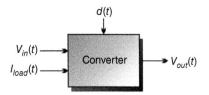

which varies as a function of time. As the input voltage and output current also vary, perturbing the output voltage, the control signal adjusts the converter to correct for the desired output voltage for the system.

If a significant impedance over the frequency range of the system existed between the output of the voltage regulator and the load (with the PDN in between), the power integrity engineer would likely have less interest in the design of the converter than there would be today. The reason is that *temporally*, due to the time lag dependence in the system, the response of the regulator would essentially be governed by the perturbations at the filter node and the compensation thereof, and thus there would only be a need to respond to changes at the output of the filter rather than at the load. This implies that if significant load deviations occurred at the source, then the burden would fall upon the PDN design and the PI engineer to ensure that the perturbations were bounded to an acceptable level and to a much lesser degree the design of the converter.

In today's more complex system designs, it is getting more difficult to distinguish between what the PDN addresses and what the voltage regulator is able to compensate for. This is, in part, due to *both* the speed of the regulator and the impedance between its output and the load. Thus, as the impedance decreases over a given frequency range and the speed of the regulator increases, more of the affects due to these voltage deviations fall within the bandwidth of the regulator. Figure 2.40 illustrates how the bandwidth of the converter is required to increase as the impedance decreases. The axis for the VR bandwidth is relative and is shown in log scale.[8] As the frequency of

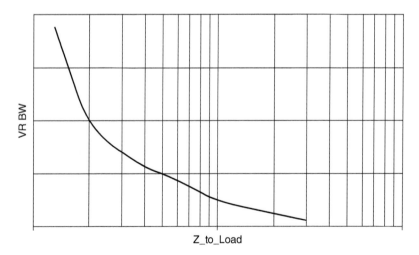

Figure 2.40 Illustration of VR bandwidth as function of impedance to load.

---

8 The bandwidth range may vary for different design spaces which is why the axis' are somewhat arbitrary and shown without numbers.

the load increases, depending upon the PDN design, the power converters' bandwidth may be more than adequate to meet the slower voltage changes in the system. However, if the impedance between the source and the load decreases significantly, while the PDN remains the same, then the burden will eventually need to fall on the voltage regulator to respond to a larger portion of the deviations. Understanding this interaction between the voltage regulator and PDN is essential to ensure the efficacy of the power delivery path. Because of this, the power integrity engineer must play a greater role in understanding how the voltage regulator affects the power delivery system.

Today, in many power delivery systems, the output sense point for the voltage is typically placed as close to the load point as possible. This usually means it is positioned somewhere near the silicon load or even as part of the silicon power distribution interconnection. Thus, the converter feedback signal has the potential to see significant high-frequency voltage variations when the load changes. However, the feedback is usually always filtered somewhat to limit the bandwidth of the noise in order to allow the converter to adjust within its loop response.

A simple block diagram, added to the block of Figure 2.39, of a feedback system is shown in Figure 2.41. The signal $d(t)$ is adjusted according to the feedback from the output voltage as shown in the figure. A gain block for the sensor is fed back and compared against a reference voltage, which determines the output voltage set point. The difference between the reference voltage and the feedback signal is the error signal in the voltage.

$$v_e(t) = v_{ref} - v_{fb}(t) \tag{2.103}$$

This error signal is then fed into a compensator gain signal which adjusts the voltage back to its correct point while driven into the PWM (pulse-width modulator) block which adjusts the duty cycle to the proper ratio to make sure that the voltage is correct. This signal now controls the power stage, which determines the on-time of the upper switch for a fixed frequency design.

The design of the full feedback and control block can be very complex. In fact, depending upon the topology of the feedback system, multiple poles and zeros

Figure 2.41 Feedback added to basic perturbation and control block.

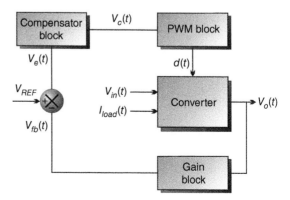

may be required to stabilize the system properly. Thus, the closed loop transfer function may end up being difficult to construct and analyze once all of the components of the system are added. This is why the study of control systems for power conversion design requires significant time and rigor to ensure that a power regulator remains stable during the full range of variability in a system. As mentioned previously, there are many excellent texts that describe the theory behind a good generic control system and some references are provided at the end of this chapter for the reader to examine. However, even for simple buck regulator designs, the control system behind it may be extremely complex. Therefore, in general, the detailed design of a feedback system for even a simple buck regulator is beyond this text. However, it is important for the foundation of the power integrity engineer to have a basic understanding of how this system operates; this will become even more important for future chapters when the analysis of a full power delivery system is discussed. Therefore, in Section 2.4.2, we show a design example for a control system for a single-phase buck regulator. The discussion is therefore brief compared to other treatments and is only intended to give the reader a starting point for how to develop a proper closed-loop converter model. The reader is encouraged to review the references in the bibliography at the end of this chapter for more information.

2.4.2 Generalized Controller Feedback Design Setup

In this generalized setup, we will assume that the converter is a single-phase buck regulator. To simplify the design, we will also assume that the input voltage is nearly constant and has little to no perturbations on it. For now, we will focus on the design of a simple feedback system for a voltage mode system. In Section 2.4.3, we will develop a SPICE-level model that may be used for a full time-domain simulation with a PDN attached to it at some point. We follow similar methods to those illustrated in the given references [12] and try to maintain consistency in many of the diagrams and methods. The main deviations will be in the areas related to analyzing the PDN for power integrity applications.

The basic diagram for a general design is shown in Figure 2.42.

The only difference between Figures 2.41 and 2.42 is that a few of the circuit elements for the buck have been added to illustrate how these components will fit within the overall SPICE model at a later time. In addition, to simplify the design, we replace the bottom switch with a diode so that there is only one control point into the model. We could have just as easily put another FET here and then inverted the control signal to drive it. In the full SPICE model, we would have used an upper and lower FET switch.

We can turn the diagram in Figure 2.42 into a control system block for analysis purposes. (We will later turn this back into a circuit block in Section 2.4.3 when we develop the equivalent SPICE model.) The generalized control block

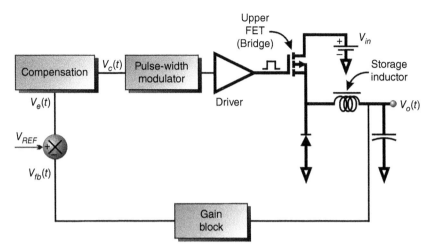

Figure 2.42 Simple buck with feedback system.

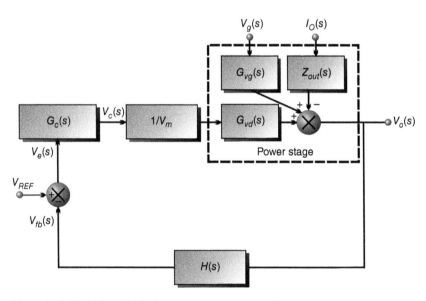

Figure 2.43 Equivalent block diagram to converter model.

diagram for the converter system is shown in Figure 2.43. We are mainly interested in the small signal state representation for the frequency domain behavior and thus the signals are represented in the complex s-plane now rather than in the time domain. Note that for these signals, we have replaced the time-dependent variables with the frequency-dependent variables to create a small signal AC model.

Replacing the circuits with equivalent gain blocks, we can create the open- and closed-loop transfer functions for our model. Moving from the output voltage $v_o(s)$ through the feedback loop, we can build the entire closed loop transfer functions for the system. The feedback, $v_{fb}(s)$, is the voltage after the gain block, or

$$\hat{v}_{fb}(s) = H(s)\hat{v}(s) \tag{2.104}$$

Note the accent symbol on top of the voltages to represent the small signal approximation for the signals. The error signal is now equivalent to the reference signal minus the feedback voltage.

$$\hat{v}_e(s) = \hat{v}_{ref}(s) - H(s)\hat{v}(s) \tag{2.105}$$

Continuing on through the loop, we can build the equation for the gain of the system. The compensation gain block is multiplied by the error signal to get $v_c(s)$, which in turn is multiplied by the PWM block ($1/V_m$). This signal is then multiplied by the gain block for the control signal $d(s)$. The generalized version of this includes the input voltage variations, which we will assume has small variations. The equation then becomes

$$\hat{v}_o = \frac{G_c G_{vd}}{V_m}(\hat{v}_{ref} - H\hat{v}) + G_{vg}\hat{v}_g - i_{load}Z_{out} \tag{2.106}$$

where we have dropped the "s" notation for simplicity. Rearranging the equation and solving for v, we arrive at the expected result.

$$\hat{v}_o = \hat{v}_{ref}\frac{G_c G_{vd}/V_m}{1 + HG_c G_{vd}/V_m} + \hat{v}_g\frac{G_{vg}}{1 + HG_c G_{vd}/V_m} - i_{load}\frac{Z_{out}}{1 + HG_c G_{vd}/V_m} \tag{2.107}$$

The loop gain of the system is found by inspection

$$T(s) = H(s)G_c(s)G_{vd}(s)/V_m \tag{2.108}$$

This is the product of gains around the forward and feedback paths. Our objective now is to determine these system gain- functions and then compensate the system correctly within our model. Once the system is compensated, we can look at the closed-loop stability of the design.

2.4.3 Buck Regulator Design Example

Given that our end goal here is to develop a model for a practical single-phase buck regulator, it is easiest to show how to compensate for a voltage regulator model using an example. In this case, we will be designing a control feedback system for a buck regulator for a mobile-based application and thus the currents will be lower than those for larger systems. The starting point will be the specifications for the regulator that give us the basic parameters of the

Table 2.5 Parameters for controller design.

Parameter	Value	Definition
R_{load}	0.5 Ω	Load resistance for steady state analysis
C_{out}	300 µF	Output filter capacitance
L_{out}	2 µH	Output filter inductor
F_{SW}	800 kHz	Converter switching frequency
v_{ref}	1.25 V	Input reference voltage
V_m	1.5 V	Peak PWM triangle voltage
V_{out}	1 V	Steady state output voltage
V_{in}	4.5 V	Steady state input voltage
R_D	$0.3 \times R_{ACU}$	

design. Table 2.5 shows the important parameters we will need to develop our model.

Our first objective is to draw a circuit that represents the parameters in the table and shows where the unknowns are. Figure 2.44 illustrates the model with the appropriate areas highlighted on where we need to find solutions.

The model represents most of the key parameters that we will need to build a closed-loop feedback system. We can compute the proper values for the rest of components and then determine the compensation network necessary for the stability of the system under the specified perturbations in the system. At a later time, we will tie this together with the PDN for the overall system to determine how the circuit responds to these perturbations.

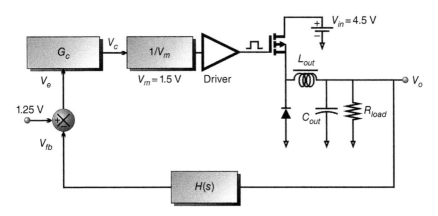

Figure 2.44 Model with key elements added.

The load resistance will represent the steady state current at the given output voltage. In this case, the output current is computed to be

$$\frac{V}{R_{load}} = \frac{1}{0.5} = 2\,\text{A} \tag{2.109}$$

The LC filter parameters are given for the regulator as well. These will allow us to compute the output impedance of the system. Often the LC filter is sized to help meet both the stability of the converter as well as for the required energy storage in the system. The switching frequency and PWM peak voltage allow us to compute and design the PWM circuit. The sensor gain is determined from the ratio of the steady state output voltage to the reference voltage. This assumes that the error signal is small which means that the loop gain will be large implying that we will be left with a negligible error.

$$v_e \cong 0 \rightarrow \frac{v_{ref}}{v_o} \cong \frac{1.25\,\text{V}}{1} = 1.25 = H \tag{2.110}$$

This may be used as our sensor gain function. The open-loop transfer functions for the buck regulator are given below without proof. Reference [13] shows the derivations for both. Eventually, we will be filling out the block diagram in Figure 2.43 so that we may have a fully compensated feedback controller for the system.

$$G_{vd}(s) = \frac{V}{D} \frac{1}{1 + s\frac{L}{R} + s^2 LC} \tag{2.111}$$

Equation (2.111) is the open-loop transfer function for the controller. Here V is the fixed output voltage and D is the fixed duty cycle, while the LRC values are those given in the table for the output filter and load. The function is already in a form that will allow us to manipulate and plot relatively easily the gain of the two pole system,

$$G_{vd}(s) = G_0 \frac{1}{1 + s\frac{1}{Q_0 \omega_0} + \left(\frac{s}{\omega_0}\right)^2} \tag{2.112}$$

where Q_0 is the quality factor of the circuit, which we discussed in Section 2.2.1. The resonant angular frequency, ω_0, can also be found by inspection of Eq. (2.112). The magnitude of the function can be found by taking the square root of the imaginary plus the real portion

$$\|G_{vd}(j\omega)\| = G_{d0} \frac{1}{\sqrt{\left(1 - \left(\frac{\omega}{\omega_0}\right)^2\right)^2 + \frac{1}{Q^2}\left(\frac{\omega}{\omega_0}\right)^2}} \tag{2.113}$$

This is the math that we will need to develop a Bode plot of the gain for the system to check its stability. The phase angle can be found by taking the inverse tangent of the real over the imaginary portion of Eq. (2.112)

$$\angle G_{vd}(j\omega) = -\arctan\left[\frac{\frac{1}{Q}\left(\frac{\omega}{\omega_0}\right)}{1-\left(\frac{\omega}{\omega_0}\right)^2}\right] \tag{2.114}$$

The other two transfer functions have the same poles as G_{vd}

$$G_{vg}(s) = G_{g0}\frac{1}{1+s\frac{1}{Q_0\omega_0}+\left(\frac{s}{\omega_0}\right)^2} \tag{2.115}$$

The output impedance is solved by looking back at each element in parallel.

$$Z_o(s) = \frac{sL}{1+s\frac{L}{R}+s^2LC} \tag{2.116}$$

The goal of this exercise is to determine the compensation for the model in Figure 2.44 where we first find the loop gain of the system $T(s)$. We can see right away that we can substitute in for $G_{vd}(s)$ and the other gains to get $T(s)$.

$$T(s) = \frac{G_c(s)H(s)}{V_m}\frac{V}{D}\left(\frac{1}{1+s\frac{1}{Q_0\omega_0}+\left(\frac{s}{\omega_0}\right)^2}\right) \tag{2.117}$$

To first find out where the system is, we may plot the uncompensated loop gain for the system. Let us assume that $G_c(s) = 1$. Consequently, Eq. (2.117) reduces to

$$T_{uc}(s) = \frac{H}{V_m}\frac{V}{D}\left(\frac{1}{1+s\frac{1}{Q_0\omega_0}+\left(\frac{s}{\omega_0}\right)^2}\right) \tag{2.118}$$

We may plug in for the given quantities and plot the uncompensated transfer function. Plotting for the uncompensated loop gain we get the results shown in Figure 2.45.

As shown in Figure 2.46, the unity gain crossover frequency is approximately above 1 kHz. This is too low for our design. Moreover, there is less than 5° of phase margin per the figure. We would like the crossover frequency to be in the 40 kHz range and have sufficient phase margin that the system is both stable and has sufficient dampening.

As a general rule, we would like to obtain at least 45° of phase margin to ensure a stable design. However, to make sure that our overall system has sufficient margin, we will be increasing this to 55° of phase margin. The figure illustrates where the phase of the system crosses over and where the gain is at unity, or 0 dB. We would like to push the frequency response out while increasing the phase margin in the system. The phase margin is found by adding 180°

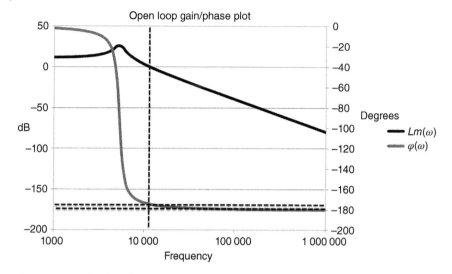

Figure 2.45 Bode plot of uncompensated gain function.

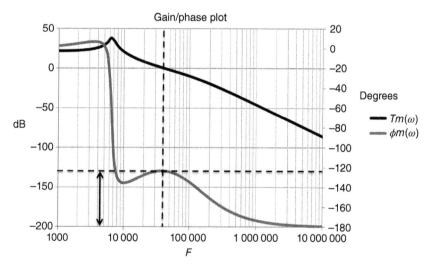

Figure 2.46 Bode plot for compensated VR controller.

to the phase plot and then examining the difference when the gain function crosses zero on the Bode plot.

$$\text{Phase margin} = 180 + \angle G_{vd}(j\omega) \tag{2.119}$$

There are various compensation methods, which may be used for a voltage mode system. Voltage regulator designers often talk of compensation methodologies such as *lead* and *lag* systems. A proportional plus differential, or PD, compensator is called a *lead* compensator. The reason is because the network adds phase lead to the system. By increasing the phase margin while increasing the crossover frequency, this type of network is well suited for our purposes. Often designers add a proportional plus integration or PI control to the network as well, resulting in what is called a PID controller. For our purposes, we will focus on designing a PD controller for our voltage regulator system.

The general compensation network is shown in Eq. (2.120).

$$G_c(s) = G_{c0}\left(\frac{1 + \frac{s}{\omega_z}}{1 + \frac{s}{\omega_p}}\right) \qquad (2.120)$$

The pole and the zero, along with the static gain G_{c0}, are the values we need to calculate. We have specified that our crossover frequency is now 40 kHz and our desired phase margin is 55°. The network above will add a slope to the final Bode plot for the gain function, resulting in shifting out the gain. The first equation we need is for the gain function G_{c0}.

$$G_{c0} = \sqrt{\frac{f_z}{f_p}} \qquad (2.121)$$

This now gives us one of the variables. The next equations are for the pole and zero frequencies which are found from the following:

$$f_z = f_c\sqrt{\frac{1 - \sin(\theta)}{1 + \sin(\theta)}}, \quad f_p = f_c\sqrt{\frac{1 + \sin(\theta)}{1 - \sin(\theta)}} \qquad (2.122)$$

where θ is the new phase margin that we are interested in. Plugging for the gain, and new pole and zero values we get

$$G_{c0} = 3.2, \quad f_z = 12.6\,\text{kHz}, \quad f_p = 127\,\text{kHz} \qquad (2.123)$$

The new plot is shown below. Note that the crossover frequency is essentially at 40 kHz per our design. The plot also shows that we have less phase margin at the lower frequencies. For some designs it may be necessary to compensate for the low-frequency phase margin. Also note that the gain has increased slightly at the lower frequencies. This, again, may be undesirable depending upon the low-frequency excitation of the system. For example, if a certain activity on this rail has a stimulus that excites the load at or near the Q of the circuit, it may be desirable to shift this resonant point to the right or left of the curve and determine a better operation for the converter.

Section 2.5 on controllers shows the development of a SPICE model from the data that was generated earlier. The system is not quite ready for a fully developed PDN model with the controller example, but a SPICE model for the converter portion can be constructed so that a full system can be linked together at a later time. In the Section 2.5, the transfer functions for the converter are utilized so that a working SPICE model may be used in a power integrity simulation.

2.5 Integration of Closed Loop Model into SPICE

There are numerous techniques and methods on how to develop a buck converter model in SPICE. However, most of them are for designers interested in emulating the behavior of their voltage regulator design with respect to its overall response and mode switch behavior. Simply put, these approaches are for power conversion design and not specifically for power integrity. Nonetheless, an accurate SPICE model, even a complex one, can be applicable when analyzing PI problems. In this case, at least for the power integrity engineer, the internal workings of the voltage regulator is less interesting than how it responds to perturbations at the load output. For this reason, choosing a simpler model that is more conducive to what the PI engineer requires is often a better approach.

Modeling the losses and behavior of the power switches along with the passive elements is certainly within the purview of the power integrity engineer's domain. And, when building a model, it is important to take certain circuit characteristics into account. However, we will leave the power loss analysis using such models as an example in the appendix for the reader to follow for a later time. In this section, the focus will be on emulating a design similar to the one developed in the previous section. Therefore, some simplifications to the elements in the model will be presented.

The objective will be to start by developing the circuits for an open-loop model and then close the loop as specified previously. After building the simpler model, we will add some levels of complexity to the circuit and then resimulate.

Figure 2.47 shows the first simplifications for the model. Notice that the upper MOSFET has been replaced with a controlled switch resistor circuit. This simplifies getting the model running without having to worry about the effects of the MOSFET within the circuit. The resistor is intended to represent at least some of the on-resistance of the MOSFET when it is turned on. Usually this is a fairly low value as seen in the previous part of the chapter. However, it will also help in understanding the characteristics of the circuit when a more representative device is added back. Also, notice that a lower NMOS device has not been added back into the circuit as yet and a diode is in its place. This will allow a single control signal point into the power stage rather than have to time two signals at once. The other parts of the output stage and impedance have not changed.

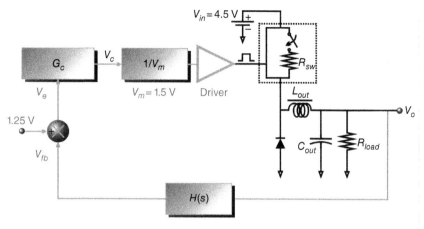

Figure 2.47 Initial model setup for SPICE.

If we run the circuit shown in Figure 2.47, without the other components in place, we can see the behavior of the circuit to make sure it is operating correctly. The output of the LTSPICE simulation shows the voltage at the load and the current into the capacitor of the circuit (Figure 2.48).

There are a few interesting things to note from the results. First, the output voltage overshoot is quite large. The voltage deviates from the mean (1 V) to nearly 1.8 V as shown. In a proper closed-loop system, we would expect this overshoot to be minimized, even during ramp-up. However, the steady-state voltage is dependent upon a few factors. First, if the finite impedance of the inductor is small, the voltage drop across it would also be small. However, at 1 MHz operation, the impedance is significant enough that there is a noticeable drop across it at steady state, which causes the output voltage to not match

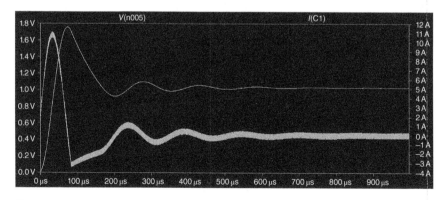

Figure 2.48 Simulation output from open-loop of circuit.

the set point of the duty cycle perfectly. We can see this by computing the magnitude of the impedance of the inductor at 1 MHz.

$$|Z_L| = |j\omega L| = \omega L = 2\pi f L \cong 12.6\,\Omega \tag{2.124}$$

Thus, for a current operating at a similar frequency, the drop would be significant. Moreover, it is evident that the current waveform rise has a reasonably high-frequency content. At DC, we expect the voltage drop to be essentially zero since we have not added a series resistance yet. However, the RMS ripple current through the inductor will still have an effect, which will result in some voltage drop across the inductor. We may temporarily change this by adjusting the pulse width for the drive signal into the switched resistor to compensate for the difference in voltage.

The third item that is important to note is the current into the capacitor, which is also on the graph. Note that this is initially very high. It is common to have large in-rush currents into the output capacitor when the converter is first turned on. Since the capacitor is ideal, we would expect this to happen. Unfortunately, adding a simple series resistor, or ESR, to the capacitor will not reduce it. This is because the ESR is typically quite small (5–10 mΩ in this case) in most designs. Thus, we will need to rely on the startup control of voltage regulator to fix this. Often a startup circuit limits the duty cycle of the PWM control during startup to limit the overshoot and inrush current into the capacitor.

The next part of the circuit that needs to be built is the PWM generator. The PWM circuit consists of a triangular wave generator and a comparator. The comparator takes the input from the triangular wave generator and compares it against the signal from the compensation network.

The idea behind it is to create a triangular wave ramp signal (positive side of op-amp) that is periodic at the frequency of operation of the voltage regulator. The signal is compared against the input signal, which is the compensated output of the reference signal. When the triangle wave voltage exceeds the reference voltage on the negative input, the comparator output generates a signal at the output node. This simple circuit creates the PWM square wave signal that is necessary for driving the controlled resistor circuit. In a real system, the voltage threshold for the PWM generator will need to be shifted up to account for the input and output voltage when an actual MOSFET device is added back into the circuit. Thus, we will add a DC voltage offset for the input to ensure this signal change through the appropriate swing later. The representation is shown in Figure 2.49.

The PWM signal output should be a square-wave function, and the output voltage should ramp and stabilize after some period of time as shown in Figure 2.50. The voltage overshoots because there is no feedback in the system and because there is no direct control on the PWM signal to slowly ramp up the current into the output capacitor. Since the goal is not to realize the entire

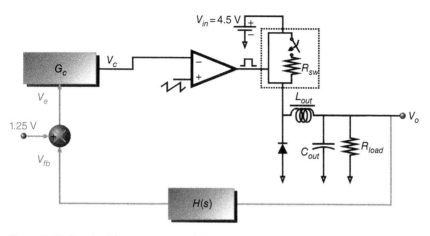

Figure 2.49 Circuit with comparator added.

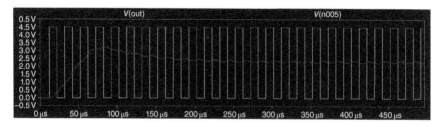

Figure 2.50 PWM open-loop drive signal and output voltage.

buck converter operation (startup, establishing initial conditions, etc.), usually allowing the converter to stabilize to a steady state condition is satisfactory.

There are some important factors to note when building the comparator circuit in SPICE, specifically because op-amps in general are *not* comparators. Though they both have inverting and non-inverting input symbols and a single output (single-ended versions), they behave quite differently. Thus, when building this circuit in SPICE, the op-amp requires some configuring. To generate the proper square wave, the op-amp's gain, gain-bandwidth, and slew rate will need to be adjusted. The slew rate will determine how fast the comparator will slew once the triangular wave ramp crosses over from the reference. Typically, the slew rate in the SPICE model is given in volts/μs. Thus, if the full swing of the triangular wave signal is from 0 to 4.5 V (to the input voltage rail), and if the ramp rate needs to be in 1 ns, then the slew rate should be

$$\text{Slew rate} = \frac{V}{s} = \frac{4.5\,\text{V}}{0.001} = \frac{4.5\,\text{V}}{\text{ns}} = 4.5e-9\,\text{V/s} \tag{2.125}$$

Setting the slew rate to this value ensures that the comparator output slew rate will be fast enough. Another important metric is the GBW or gain-bandwidth product. This is multiplication of the open-loop gain and the frequency at which it is measured. Often this value may be set to something reasonably high (10^8) to ensure that the GBW does not affect the output of the comparator.

Finally, the open-loop gain (Avol) of the op-amp itself needs to be determined. Normally, this is set to a high value as well (5×10^4) to ensure that the signal amplification is high once the transition occurs.

After changing the values for the generic op-amp in the LTSPICE library, the square wave output from the comparator circuit may be examined. The probe output is shown in Figure 2.51. Sometimes it is good to place Zener diodes at the output to help drive the voltages to their proper states.

As stated previously, the duty cycle will need to be adjusted to account for the load impedance effects. Thus, for the time being, a correction is added to the DC value of the feedback into the comparator for the PWM which will set the duty cycle and thus the output voltage. A plot of the output voltage with the PWM added is shown in Figure 2.52.

The output voltage settles as expected, but there is still a large overshoot at startup. Once the compensation network is added, this should change. The important thing to note is that the PWM circuit is operating as expected. If this were not the case, a designer would need to spend the appropriate time debugging the circuit to ensure it was operational before moving onto the last part of the model development.

The final part of the circuit is to add in the compensation network, the reference, and the feedback transfer function H. The feedback is a simple resistor divider that represents the ratio of the output voltage to the reference voltage

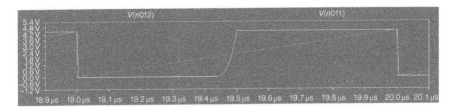

Figure 2.51 LTSPICE output ramp with square wave output over one cycle.

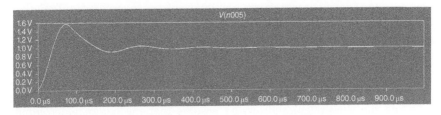

Figure 2.52 Output voltage plot from LTSPICE circuit with PWM driver.

for the proper gain. As analyzed in Section 2.4.3, if the output voltage is to be set to 1 V and the reference is 1.25 V, then the transfer function sets V_{fb} to the proper value so that the signal is compared against the reference. In this case, the gain function would be adjusted from 1.25 V. However, since the objective is to set the output voltage to 1 V, this already presupposes that there will be some gain in the system. It is more practical here to simply add the divider network and thus shift the gain to below 1 rather than add an amplifier to the circuit. This signal then goes to the negative feedback of the amplifier. The reference voltage is then the input to the positive side of the amplifier. The compensation network has both a pole and a zero in the feedback path. The pole is represented by an RC network with R_4 and C_2 and the zero with R_3 and C_1. The gain function may be found by recalling that we can solve for the voltages as currents as shown in Figure 2.53.

The gain function is the ratio of the output to the input voltage. We assume that the impedance into the op-amp is very high; thus, we essentially get the result in Eq. (2.126).

$$i_2 - i_1 = i_a \cong 0 \tag{2.126}$$

Since this is the case, the currents are by definition equal which means that we can solve for the voltages in the same way or

$$\frac{V_i}{Z_2} = \frac{V_O}{Z_1} \rightarrow \frac{V_O}{V_i} = \frac{Z_1}{Z_2} \tag{2.127}$$

The lead compensation network has two networks where a capacitor is in parallel with a resistor as shown in Figure 2.54 . Replacing the impedances with these values, we get the following:

$$\frac{Z_1}{Z_2} = \frac{R_3 \parallel C_1}{R_4 \parallel C_2} = \frac{\frac{R_3 / \frac{1}{sC_1}}{R_3 + \frac{1}{sC_1}}}{\frac{R_4 / \frac{1}{sC_2}}{R_4 + \frac{1}{sC_2}}} = \frac{R_3}{R_4} \frac{1 + sC_2R_4}{1 + sC_1R_3} \tag{2.128}$$

This is in the form of Eq. (2.120), where

$$G_0 = \frac{R_1}{R_2}, \quad f_z = \frac{1}{2\pi C_2 R_4}, \quad f_p = \frac{1}{2\pi C_1 R_3} \tag{2.129}$$

Figure 2.53 Solution to gain function.

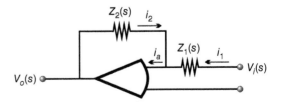

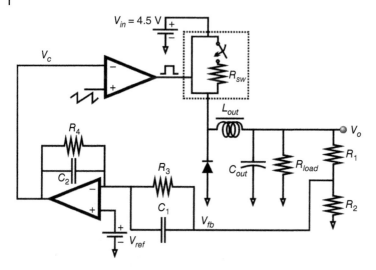

Figure 2.54 Full model with compensation network and PWM.

The divider network of R_1 and R_2 is the transfer function for H. Thus, we now have the complete system which is shown in Figure 2.54.

The reference voltage is now the voltage relative to the new divider network; that is, at a nominal voltage of 1 V at the output, we may divide down the output voltage and feed this value into the compensator to compare with the new reference. The choice of the resistor values for R_1 and R_2 are somewhat arbitrary. In an actual system, one would choose these values to be reasonably large to limit the current from the output to ground. Also, it would be appropriate to have the divided voltage reasonably high relative to the output so that a practical reference could be realized with actual circuitry. A reasonable voltage for the reference is 0.75 V. Thus, if we choose R_1 to be $10\,\mathrm{k\Omega}$, the value for R_2 may be computed.

$$R_2 = \frac{V_n R_1}{V_o - V_n} = \frac{0.75 \times 10\,\mathrm{k\Omega}}{0.25} = 30\,\mathrm{k\Omega} \tag{2.130}$$

where V_n is the node voltage of the divider network. In this case it is V_{fb}.

The frequencies for the compensation network were determined in Section 2.4.3. The objective now is to solve for those frequencies with practical resistor and capacitor values. Again, the objective is to choose values that result in realizable circuit implementations. It should be noted that this is not absolutely necessary; however, as the model becomes more complex, having values that do not work appropriately with devices (such as MOSFET models) when they are added makes constructing the final model more difficult.

Given the parameters, we can now build and run the SPICE model.

The frequencies for the compensation network are given in Eq. (2.123). We may choose a value for one of the components and then solve for the other. A small high-frequency cap (described in Chapter 3) value for a multilayer ceramic is 0.01 μF. We can use this value for the 0 and a 0.001 μF for the pole. Thus, solving for the resistor values gives

$$R_3 = \frac{1}{2\pi f_z C_1} = \frac{1}{2\pi(12.6 \times 10^3)(0.01 \times 10^{-6})} = 1.26 \text{ k}\Omega \qquad (2.131)$$

Similarly for R_4 we get

$$R_4 = \frac{1}{2\pi f_p C_2} = \frac{1}{2\pi(127 \times 10^3)(0.001 \times 10^{-6})} = 1.25 \text{ k}\Omega \qquad (2.132)$$

We note that the values are very close to each other. Thus, we can use the same resistor for both the pole and the zero here.

Adding in the compensation network to the circuit that has already been developed and running a simulation for a transient response (step current), the output voltage is as shown in Figure 2.55. As can be seen in the figure, the droop behavior is captured. The step current is relatively large (4 A) to illustrate the response. In a real system, a droop this large would likely not be acceptable. This PDN behavior is addressed in later chapters, however. Note that there is some low-frequency ringing, but the output voltage dampens due to the lead compensator. It can also be seen that the voltage is still stable even with a large current step added. A more sophisticated feedback system may be constructed to dampen the behavior further in this design. For illustration purposes, it is clear that the circuit operates as expected.

The SPICE model that was created here was for a voltage mode design topology. A current mode system would be better if a faster response were required. There are many examples both on the web, in texts, and in papers that illustrate how to develop a working model for a buck regulator. The reader can choose from a number of references to follow the different methods outlined. The work we have developed in this chapter will be used in later chapters for simulations, with PDNs added to determine the overall behavior of the system.

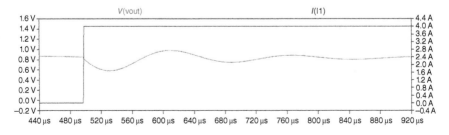

Figure 2.55 Time-domain simulation of integrated model in LTSPICE.

2.6 Short Discussion on System Considerations for Power Conversion Integration

Chapter 3 discusses some of the salient system considerations while integrating the active and passive power components into a platform. Although this subject warrants a thorough discussion when developing and analyzing power converter systems in detail, only a cursory overview will be given here.

As discussed in this chapter, it is apparent that there are multiple components, both active and passive, within the power converter itself that transfer current from one part of the circuit to the other. Moreover, the current, in many cases, can be substantial and the potential to generate noise in the system is therefore high. Noise can manifest itself in a number of ways and can affect other components that are sensitive to it. There are two components to this noise that we are interested in: susceptibility and emissions. Susceptibility is the unwanted coupling of noise in a system to couple into another. Emissions are noise that may couple to the environment causing violations to specific electromagnetic compatibility standards in the industry set by the FCC. Emissions are discussed in more detail in Chapter 3.

Noise is one of the significant system considerations that power integrity engineers must be aware of when examining the power delivery system as a whole. Though usually an electromagnetic interference (EMI) engineer spends the bulk of their time determining the effects to the outside world and therefore examines the power converter for the source of this noise, the PI engineer must also understand how this affects the other parts of the system. For example, if the power converter generates unwanted high-frequency noise in addition to an efficient DC current, the PI engineer can aid in squelching this noise through decoupling of the PDN or other parts of the power converter circuit. Moreover, placement and routing of the traces properly can also help in reducing these effects.

Many other silicon devices are susceptible to unwanted noise and various frequencies. Depending upon the platform, devices which have radio structures, processors, and high gain, low noise amplifiers may degrade or not function properly if this noise is not reduced to an acceptable level. Thus, it is critical that the power integrity engineer be well versed in the system design.

Layout, route, and noise mitigation techniques for high-performance power converters are a study in itself. Many power conversion texts often dedicate entire chapters in this discussion. Here, we will focus on the elements which are within purview of the power integrity engineer.

One of the key sources of noise generated from a regulator is the high-frequency switching noise and the voltage ripple. Both are periodic with the clock circuit that generates the PWM pulse in the system. The high-frequency noise is due to the fast di/dt of the bridges when they switch from one state to the next. The ripple is due to the noise on the filter and reduces the square wave signal to a DC signal as was seen earlier in this chapter. Because both sources generate relatively large currents, it is critical that good layout and noise mitigation techniques be used to limit the currents from escaping throughout the system.

2.7   Advanced Topics in Power Conversion

This final section is on advanced topics today in power conversion as it relates to power integrity. As discussed earlier in this chapter, there is a change that is occurring in the industry with respect to power conversion that could have a profound effect on power integrity as the state of the art progresses. One of these changes is the advancement in on-silicon voltage regulation [18]. The progression of power conversion from the board to the package to the silicon is a transition that has the potential to revolutionize the way developers design their silicon and integrate their platforms. For the PI engineer, it will change the way the PDN is viewed and how their analytical techniques are applied.

Figure 2.56 illustrates how this world may change from the view most systems are developed today to one that we can envision in the future. The power converters today typically are on a circuit board and deliver power to multiple rails on one or more devices. However, as silicon becomes more integrated and specialized, the number of rails begins to increase, and the need for many smaller converters, close to their loads, becomes more of a necessity. This is shown in the figure on the right.

This important change will likely revolutionize the way engineers view power integrity moving forward. The reason is that once the VR is colocated with the load, the PDN is no longer comprised of large capacitive elements that are distributed on the motherboard. It is now a culmination of integrated capacitive elements on the die itself and some high-frequency components on the package. The voltage regulator, in turn, is now a high-frequency multiphase converter capable of switching at frequencies that could likely exceed 100 MHz with bandwidths beyond 10 MHz [15]. This means the voltage regulator will be designed to respond to changes in the load and deliver current that only high-frequency capacitors are capable of today.

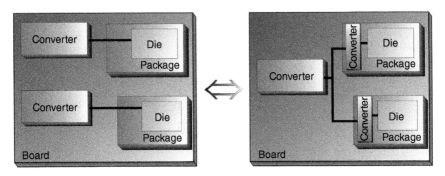

Figure 2.56 Migration of motherboard voltage regulation to on-package and on-die.

The main reason though for this change will likely be for cost and size reduction. An integrated voltage regulator will need to be extremely small, and to do that the components which make up the regulator itself must also be small. The silicon bridges are shrinking generation on generation due to logic process shrinkage, and because the input voltage will be stepped down (likely) from a higher voltage, the size of the bridges will also shrink. Thus, the size of the silicon portion of the design should be within the technology limits necessary to deliver the power to the load.

The LC filter, however, is another matter. In order to shrink this, the frequency of operation must increase proportionately. This is why each phase of the design may need to switch at very high frequencies to drive the size of the inductor and capacitance down. Once they are small enough, these passive elements may reside on silicon as well [18].

One technology which is helping to drive this is the invention of on-silicon coupled inductors. There have been numerous developments in the past decade in this area mainly in the development of magnetic materials as well as in the voltage regulation itself. The key is the coupled inductor itself. Figure 2.57 shows a typical coupled inductor.

The nodes A–A' are driven by the voltage regulator bridges and nodes B–B' are connected to the output capacitor filter for the voltage regulator. When constructed on-silicon, the inductors are extremely small, meaning that the amount of inductance that each one has is small as well. Because of this, the regulator must switch at a higher frequency. To combat this, engineers try to use materials with high permeability to all for storing energy in a small region. However, as in all magnetic materials for such power conversion applications, the materials themselves are susceptible to saturation which reduces their effectiveness for storing energy. By coupling the inductors, engineers can mitigate much of these effects by cancelling out the fields and thus reducing what is called the magnetizing currents within the structure. The magnetizing current is the instantaneous difference between the current in one inductor and the other at any time. By minimizing this magnetizing current, for a given structure and material, much of the saturation effects can be reduced, allowing for use of materials that are conducive for on-silicon processing.

However, with the use of such materials, come other problems. When switching at these higher frequencies, the materials are also more lossy than at lower switching frequencies. The model shown in Figure 2.57 now changes somewhat due to the addition of these parasitics. This is shown in Figure 2.58.

The additional parasitics have losses, which must be controlled as well. The AC losses usually dominate which are mainly in the core of the material.

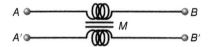

Figure 2.57 Coupled inductor without parasitics.

Figure 2.58 Coupled inductor with parasitics added.

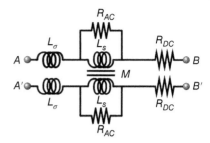

Most thin-film inductor-based materials suffer from both eddy current and hysteretic losses as discussed earlier in this chapter for standard inductors. However, in the case of a discrete non-coupled system, the inductor is typically *gapped* which alters how the energy is stored in the inductor itself. By gapping the system, much of the energy is stored in the gapped area rather than in the core which helps to minimize saturation. Unfortunately, as we have seen, this increases the physical size of the inductor. For many external applications (on board or package), the combination of large size, larger inductance, and slower switching frequency results in a design that meets the engineer's requirements. When the task is to mount the inductors on-die, the challenges are a whole other matter.

To keep the structure small, instead of using individual inductors for the output energy storage, many opt to use coupled inductors. By coupling the inductors, engineers are able to achieve two critical objectives: (i) the physical size is reduced and (ii) control of saturation. The saturation control as mentioned earlier is achieved by controlling the differences in current between one inductor and the one it is coupled to. This can be seen by looking at Figure 2.59.

The Kirchhoff equations for the top inductor in the coupled system are as follows:

$$\Delta V = V_{in} - V_{out} = L_1 \frac{di_1}{dt} - M \frac{di_2}{dt} \tag{2.133}$$

Figure 2.59 Buck regulator in polar form.

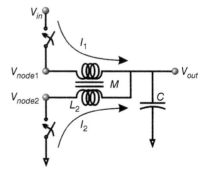

For the first phase and

$$\Delta V = V_{out} = L_2\frac{di_2}{dt} - M\frac{di_1}{dt} \tag{2.134}$$

In terms of average currents and voltages, the equations become

$$\Delta V = V_{in} - V_{out} = L_1\frac{\Delta i_1}{DT} - M\frac{\Delta i_2}{DT} \tag{2.135}$$

and,

$$\Delta V = V_{out} = L_2\frac{\Delta i_2}{(1 - D)T} - M\frac{\Delta i_1}{(1 - D)T} \tag{2.136}$$

We note that the change in voltage for the upper vs. the lower phase is due to different times in the switching cycles of the power FETs. If the duty cycle is 50%, these times would be equal. When one examines the equations closely, it will be evident that if L and M are very close to each other, the change in current could be very high. This is one reason why some amount of *leakage* current from leakage inductance is usually desired in the construction of the coupled system to bound the ripple current which can result in undesirable ripple voltage.

The current through both inductors resemble that in Figure 2.60. The difference between the currents at any instant in time is called the magnetizing current. This is the AC peak to peak that must be controlled to ensure that the coupled system does not saturate.

The flow of current through the coupled system is fundamental to its operation. First, it is the *net* flux that contributes to the saturation effects in the magnetic core. These effects are complex and involve quantum mechanics theory to describe at the atomic level of the material since the behavior is highly dependent upon the chemical makeup of the structure as well as the geometry and excitation. However, a general description here helps to gain insight into how the flow of flux governs the saturation and energy storage within a magnetic core.

In general for magnetic materials, we may represent the atoms in a lattice structure for a magnetic material as an electron orbiting a positively charge nuclei. The orbiting electron creates a differential current loop as shown in Figure 2.61.

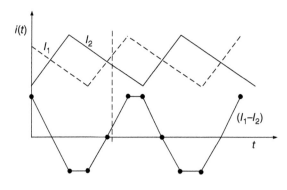

Figure 2.60 Current waveforms for coupled system.

Figure 2.61 Magnetic dipole representation.

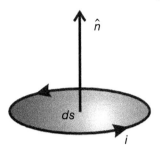

The electron orbiting the nucleus will have an angular momentum which may be represented by the following formula:

$$dm = \vec{n}\, i\, ds \qquad (2.137)$$

The total magnetic moment is the vector sum of all of the magnetic dipole moments in the magnetic structure or

$$m = N\vec{n}\, i\, ds \ \mathrm{A\ m^2} \qquad (2.138)$$

where N is the total number of loops in the structure. For a given general structure, the magnetic dipole moments are randomly oriented inside of a slab of magnetic material. This is shown in Figure 2.62.

In the absence of an applied magnetic field, the magnetic dipoles will be oriented such that there will be no net magnetic dipole moment. When a magnetic field is applied, the dipoles will tend to align in the direction of the magnetic field as shown in Figure 2.63. This results in a net magnetic polarization vector M or

$$M = \lim_{\Delta V \to 0} \left(\frac{1}{\Delta V} m \right) \ \mathrm{A/m} \qquad (2.139)$$

where V is the total volume of the magnetic slab.

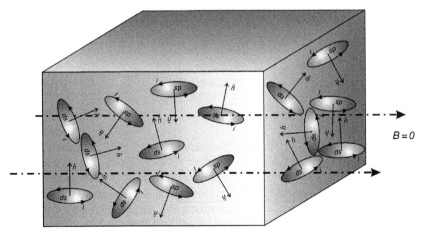

Figure 2.62 Magnetic slab with no applied magnetic field, random magnetic dipoles.

Going back now to overall discussion of why coupled inductors are used, a fundamental issue with on-die voltage regulation is the efficiency of the converter. Because the AC losses are higher due to the switching frequencies, it is necessary to make the passive and active elements as efficient as possible. Moreover, since the regulator will now be in series with another input regulator, as described earlier in this chapter, the overall efficiency, when active, will be lower. This could be a limitation in moving the state of the art toward on-die regulation. Another issue with this approach is the encroachment of the voltage regulator with other silicon and its thermal impact. By moving the VR onto or near the same silicon that it is powering, the thermal density will go up, further reducing the efficiency of the power conversion. These issues and others could delay the advancement of this technology into products for some time.

However, as the silicon loads start to become more segmented and condensed, there will be a need to manage them to make the system more efficient. And when on-silicon voltage regulation becomes a reality, the impact to the power integrity engineer will be many.

The first impact, as mentioned previously, is the change in how the system will be modeled. It will no longer be adequate to model the system with passive networks since the voltage regulator will have a critical impact to the efficacy of the power delivery path. Moreover, since the input rail will likely have more energy storage than the output rail, this network will need to part of the model, complicating the overall power delivery system. Basically, the modeling effort will comprise of not just the passive portions of the system but also the active elements as well. The system simulation methodologies will need to change to accommodate since the active modes of the voltage regulator will now come into play in the quality of the delivery path.

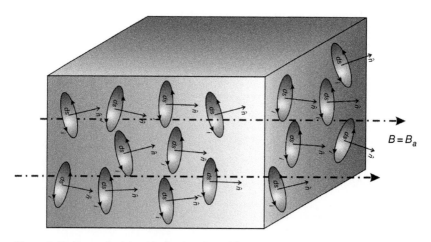

Figure 2.63 Magnetic slab with dipoles aligned from applied magnetic field.

In addition, the voltages for each device will continue to shrink, causing less margin in the tolerances in the system. This will continue to compete with the increasing need to shrink the size of the filter elements and the power silicon. The encroachment of the power delivery onto the functional silicon will be an ongoing battle as power becomes the crucial metric that engineers strive toward lowering with ever-increasing performance and silicon function density.

2.8 Summary

Having a background in power conversion is a critical toolset toward understanding the overall effects in a power integrity analysis. This is truer today than it was 5–10 years ago due to the advancements in delivering power to complex high-performance silicon loads. Whether the power delivery system is *distributed* or *centralized*, the development of the power conversion mechanism will have an impact on the efficacy of the overall power delivery path. Moreover, as part of this understanding, power integrity engineers need to understand the *losses* in a system, which necessitates having a fundamental knowledge of the power converter's efficiency as part of their learning.

With respect to computer-based systems, a key to this education in power conversion is to have a good understanding of the *buck converter* and its workings. The buck converter is one of the central topologies in such platforms (along with the linear regulator). The buck converter essentially is comprised of two switches (the bridge), a controller, and an LC filter. The LC filter does more than filter out the high-frequency switching components from the bridge. It also stores the energy before delivering power to the load. Focusing on the *energy storage* element portion rather than the Q of the filter is key toward gaining an understanding in the overall losses and efficiency of the system.

The bridges, both PMOS (high side) and NMOS (low side), play a critical role in steering the current to the filter and load. A VDMOSFET is a common device used in discrete power switch devices. Basic operation with respect to its on-resistance and internal passive capacitances is important when determining the overall efficiency of the converter. Most devices are *enhancement-mode*-based, meaning their mobility is enhanced through controlling of the gate voltage. Once a basic understanding of the operation and its key elements is obtained, a model may be built that can be used for simulation in SPICE and other programs.

Because the inductor is such an integral part to the buck, having a foundation in its properties is also critical. A basic model that comprises the AC and DC resistances as well as the inductance helps in building a model that can predict both its losses and operation in a buck converter design. Even the AC losses in discrete power inductors are governed by complex physics which include *eddy current* and *hysteretic loss* mechanisms. However, by making some good simplifications, the discrete power inductor may be well understood through basic mathematical formulations at least to a first order.

The controller is the portion of the buck where all the intelligence is. It is also the key to the response of the voltage regulator. Because the impedance between the load and source is continually shrinking, particularly in the higher frequency bands, it is critical to involve the converter in the overall power integrity analysis. Because of this, having a background in how the controller operates is necessary. Building a controller model, whether voltage mode or current mode based, is complex and a detailed study is beyond the scope of this text. However, an overview of a simple version of a control system along with its compensation network helps to understand how to build an actual model for simulation purposes. Once a mathematical control block model is built, a developer may design a buck regulator design in either SPICE or MATLAB Simulink (or possibly some other program).

Problems

2.1 An engineer is working on plotting the efficiency for a buck regulator and he gets the following values from the power converter designer in Table 2.6.

Table 2.6 Data for efficiency plot.

Input current (A)	Output current (A)
0.003	0.02
0.013	0.08
0.033	0.2
0.064	0.4
0.092	0.6
0.152	1
0.224	1.5
0.296	2
0.512	3.5
0.727	5
1.091	7.5
1.446	10
1.818	12.5
2.195	15
2.970	20
3.75	25

If the input voltage is 5 V, plot the efficiency as a function of output current in log scale. Find the maximum efficiency point on the curve. What is it?

2.2 Using Eq. (2.12), and given the same values for the filter in Figure 2.8, and its parasitics $R_{pl} = 1\,\text{m}\Omega$, $R_{pc} = 1\,\text{m}\Omega$, and $L_{pc} = 200\,\text{pH}$, plot the transfer function for the filter over the frequency range of 10 kHz–10 MHz. What has changed from the curve shown in Figure 2.11? Is this what was expected?

2.3 Given the values in the previous problem, add a series resistance to the inductor that changes as a function of frequency at 10 dB/dec from 1 MHz at 500 mΩ to 10 MHz. Plot the new transfer function and the old one and compare the results.

2.4 Derive the equation for the RMS current through a MOSFET as shown in Eq. (2.24).

2.5 Sketch the current and voltage for the PMOS device similar to that in Figure 2.21. Compare your results to the figure to ensure it makes sense.

2.6 Compute the on-resistance for the MOSFET device using the values in Table 2.7 as well as the plot in Figure 2.26. Compare the histogram to the one in Figure 2.27.

Table 2.7 Parameters for computation of on-resistance.

Parameter	Value
R_S, R_{Sub}	$0.45 \times R_D$
W/L	1.2
μ_e	1.7e20 C/V s
C_0	0.3e-17
V_{GS}	3.7 V
V_T	1.32 V
μ_A	0.8e25
R_J	$0.23 \times R_D$
R_D	$0.1 \times R_{ACU}$

2.7 Compute the AC losses for a PMOS device given the following parameters for an NMOS.

Assume the duty cycle is 35% for the operation in the buck converter.

$$P_{PS} \cong f \cdot \left(\frac{Q_{SW}}{I_G} \cdot V_{IN} \cdot I_D + Q_{RR} \cdot V_{DS} + Q_G \cdot V_G + \frac{Q_{OSS}}{2} \cdot V_{IN} \right)$$

$$(2.140)$$

2.8 Compute the AC losses for a PMOS device with following parameters as they relate to the values in Table 2.8.

Table 2.8 Parameters for computation of AC losses in NMOS.

Parameter	Value
f	1 MHz
Q_G	15 nC
V_G	1.5 V
Q_{OSS}	10 nC
V_{in}	4.5 V
V_{BD}	0.75 V
I_{out}	2 A

Use the same values in the previous problem to compute the losses (Table 2.9). How do the results compare with the NMOS in the previous problem? Which one is larger?

Table 2.9 Parameters for computation of AC losses in PMOS.

Parameter	Value
f	1 MHz
Q_{SW}	$0.3 \times Q_G$
I_G	$0.1\% \, I_D$
I_D	3 A
Q_{RR}	370 nC
V_{DS}	0.2 V

2.9 Compute the reluctance of a C-Core inductor given an air gap of $2\,\mu m$. Assume the same cross-sectional area as in Example 2.3. If the effective permeability of the material is 1200 and the magnetic path length is 4 mm, what is the total reluctance of the core?

2.10 Determine the open loop transfer function ($T(s)$) of the block diagram. Use the equation for the new transfer function $K(s)$ as shown (Figure 2.64). How many poles and zeros does it have?

$$K(s) = K_0 \left(\frac{\frac{s}{\omega_{z1}}}{1 + \frac{s}{\omega_{p1}}} \right) \tag{2.141}$$

2.11 Construct the SPICE circuit as shown in Figure 2.55. Add an AC source voltage in series between the output of the compensator and PWM comparator. Sweep the signal from 1 kHz to 5 MHz. The ratio of the input to output voltages on each side of the AC source is the open-loop transfer function $T(s)$. Plot the gain of the system.

2.12 Run the SPICE circuit from the previous problem and increase the step current from 1 to 5 A with increasing steps of 1 A apiece. Determine the changes in the droop voltages at the output. Now perform the same analysis with the capacitance doubled. What is the result? Is there additional ringing in the system? Did the capacitor help? Why?

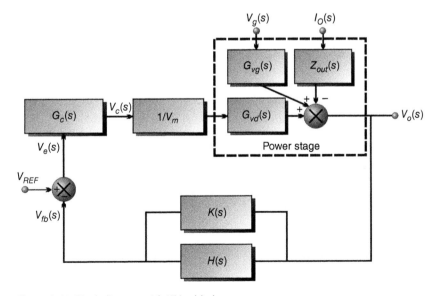

Figure 2.64 Block diagram with $K(s)$ added.

Bibliography

1 Embedded Computing Design (2013). The future of power management in the mobile computing market. http://embedded-computing.com/articles/the-mobile-computing-market/ (accessed October 2016).

2 Fitzgerald, A.E., Higginbotham, D., and Grabel, A. (1981). *Basic Electrical Engineering*, 5e. McGraw-Hill.

3 Kumar, P. (2014). DC-DC converters applications and design trends. *Electronics Maker Magazine* (January 2014). http://electronicsmaker.com/em/admin/pdf/free/DC-DC.pdf (accessed October 2016).

4 Lidow, A. (2015). A new generation of power semiconductor packaging paves the way for higher efficiency power conversion. In: *IWIPP*. Chicago, IL: IEEE, Inc.

5 Biliga, A. (2008). *Fundamentals of Power Semiconductor Devices*. Springer.

6 Chen, Y., Cheng, X., Liu, Y. et al. (2007). Modeling and analysis of metal interconnect resistance of power MOSFET's with ultra low on-resistance. In: *Power Semiconductors and IC's*. Naples: IEEE, Inc.

7 Infineon (2016). 20V N-channel power MOSFET specification. http://www.infineon.com/cms/en/product/channel.html?channel=5546d4624e765da5014e8cf9b4a06b2f (accessed October 2016).

8 Kassakian, J.G., Schlecht, M.F., and Verghese, G.C. (1991). *Principles of Power Electronics*. Prentice Hall.

9 Divakar, B.P. and Ioinovici, A. (1996). Zero-voltage-transition converter with low conduction losses operating at constant switching frequency. In: *Power Electronics Specialists Conference*. Baveno: IEEE, Inc.

10 Schade, N.J. (2010). Simplifying the calculations of core losses in composite inductors. In: *Electronic Products Magazine*. ASPENCORE. https://www.electronicproducts.com/Passive_Components/Magnetics_Inductors_Transformers/Simplifying_the_calculation_of_core_losses_in_composite_inductors.aspx (accessed October 2016).

11 Erickson, R.W. and Maksimovic, D. (2000). *Fundamentals of Power Electronics*, 2e. Springer US.

12 Forsyth, A.J. (1998). Modelling and control of DC-DC converters, Tutorial. http://www2.ulpgc.es/hege/almacen/download/40/40986/00730829.pdf (accessed October 2016).

13 Lee, Y.S., Wang, S.J., and Hui, S.Y.R. (1997). Modeling, analysis, and application of buck converters in discontinuous-input-voltage mode operation. *IEEE Transactions on Power Electronics* 12 (2): 350–360.

14 Baliga, B.J. (2010). *Advanced Power MOSFET Concepts*. Springer.

15 Zumel, P., Garcia, O., Cobos, J., and Uceda, J. (2005). Tight magnetic coupling in multiphase interleaved converters based on simple transformers. In: *IEEE Applied Power Electronics Conference*, vol. 1. Austin, TX: IEEE, Inc.

16 Lee, M., Chen, D., Liu, C.-W. et al. (2006). Comparisons of three control schemes for adaptive voltage position (AVP) droop for VRMs applications. In: *Power Electronics and Motion Control Conference*. Portoroz: IEEE, Inc.

17 Xu, P., Xu, P., Yang, B., and Lee, F. (2002). Investigation of candidate topologies for 12 V VRM. In: *APEC*. Dallas, TX: IEEE, Inc.

18 Harris, P. and DiBene, J.T. II, (2004). Integrated magnetic buck converter with magnetically coupled synchronously rectified Mosfet gate drive. US Patent 6,754,086.

19 DiBene, J.T. II, Morrow, P., Park, C.-M. et al. (2010). A 400 Amp fully integrated silicon voltage regulator with in-die magnetically coupled embedded inductors. In: *IEEE APEC Conference*. Palm Springs, CA: IEEE, Inc.

20 Vafaie, M., Adib, E., and Farzanehfard, H. (2012). A self powered gate drive circuit for tapped inductor buck converter. In: *PEDSTC*. Tehran: IEEE, Inc.

21 Kingston, J. (2002). Application of a passive lossless snubber to a tapped inductor buck DC/DC converter. In: *PEMD*. Santa Fe, NM: IEEE, Inc.

3

Platform Technologies and System Considerations

In the previous chapter, the subject of power conversion was addressed as an important factor to power integrity. In this chapter, the discussion continues with respect to some of the system-level components and issues, which are also critical to the study of power integrity. These components are what make up the power delivery network from the source to the load. Specifically, they are the capacitive, inductive, and resistive elements, which comprise the power distribution network (PDN) of the system. What is more important though is how these components interact with the voltage regulator (VR) and the other parts of the system.

Though the elements themselves are fairly easy to describe electrically, they differ considerably in physical makeup as the electrical path egresses from the motherboard onto the package and finally to the silicon. Thus, it is important to describe the characteristics of each element electrically as the path traverses from the power source to the silicon.

When describing such elements, it is often easier for engineers to focus only on the capacitors in the network. This is because the capacitors are the main source of decoupling and high-frequency charge storage for the system. The inductive elements in the path are strictly for interconnection and thus have little to no energy storage capability. Thus, these elements are usually considered the *parasitics* of the system PDN. In most cases, they are a detriment to the efficacy of the electrical behavior for the power distribution. However, it is critical that they be well understood, and a discussion of these electrical elements within the construct of the PDN is an important part toward understanding the full PDN. The resistive elements are also part of the parasitics in the path, and it is important to describe these correctly as well.

The second main topic of this chapter involves the interaction of these components, as well as the voltage regulator, to examine the behavior of the system. The low-frequency operation of the system is an important aspect to the interaction between the power source and the load and in many ways is the first step toward analyzing the power integrity for a particular PDN. Part of this introduction is the operation of the load line of the system and the guardband effects that are associated with it. The load line is a critical system aspect with respect to the interaction between the power converter and the load behavior. How the errors accumulate in a system is important to the operation of the silicon under the constraints of the platform and its safe operating regions. Because these error bands accumulate over time, temperature, and normal operation, it is critical that the power integrity engineer has a background in understanding them.

The first few sections in this chapter will focus on the elements that make up the source, PDN, and the load, and then the discussion will turn toward the interaction of them within the constructs of the system operation.

The focus will be on the physical structures and how they manifest themselves electrically. The extraction of these electrical circuit elements is discussed in later chapters. The following sections discuss some of the key concepts around how the PDN is constructed from these elements and then how the system operates together with the PDN and the voltage regulator. This includes representations for the source and the load. These representations are used conceptually so that an introduction to the load line may be given. The rest of the chapter discusses noise in the system and how this affects the other components, in particular, layout considerations and coupling mechanisms.

3.1 Physical Elements

Chapter 2 showed the power source as a structure that supplies power to the load. The voltage regulator was broken down to essentially its elements, which comprised the controller, bridge, drivers, and the LC filter. Between the filter and the load is the PDN. This is the central point of the power integrity engineer's focus. The PDN is a conglomeration of capacitors bridged together with interconnections that create a path between the source and the load. Figure 3.1 shows an example path electrically and physically.

Each structure can be represented essentially as an impedance or a network of impedances. The sections which are in series are the parasitic interconnects of the PDN and are represented as combinations of resistances and inductances.[1] The shunt or parallel sections are a combination of parasitic connections but also make up the capacitors of the PDN. The capacitors are more complex and will be described in some detail toward the end of this section. Note that the circles around each section show an overlap between them.

Starting from the left in the figure, the power converter shown in black encompasses the filter and the active elements. Z_1 and Z_2 may be represented by the inductor and capacitor of the filter respectively along with their parasitics. Z_3 and Z_4 are the next stage portions of the PDN which are typically

---

1 As the distances between the traces and or planes become small, the capacitance may play a role in the PDN parasitics; however, usually these are negligible until the path egresses into the silicon where the plane capacitance may need to be considered.

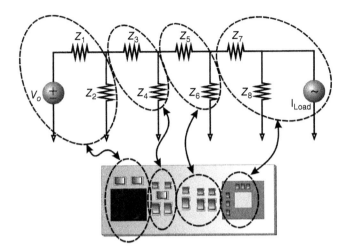

Figure 3.1 Simple electrical/physical representation of path from source to load.

higher frequency capacitors intended to dampen some of the mid-frequency effects from the load as well as to further reduce the voltage ripple in the system. The next elements, Z_5 and Z_6, are decoupling elements and their parasitics located adjacent to the package, and are usually small capacitors intended to continue to minimize high-frequency noise. And finally, Z_7 and Z_8 combine with the silicon load and its decoupling to deliver charge as fast as possible when the load device switches.

As in the interconnect of a MOSFET device, the resistance of the series and shunt connections may be represented electrically as resistive circuits that change as a function of frequency and temperature. The frequency dependence becomes more important as the spectral content of the power signal increases in frequency while the PDN path gets closer to the load. More often than not, however, when analyzing the resistance portion of the traces on the motherboard, the frequency content is low enough that the current is either uniform or nearly uniform throughout the cross-section of the trace. However, this may not be true for the current through the *plane* path. Often the distribution of current on the plane, from one point to another, is smaller than width of the plane between them, and resistance does not flow uniformly throughout the width of the trace. In part this current distribution is because of the distance between the source and sink of the current is smaller than the width. It is then important to analyze the *effective* resistance of the trace in question.

If the current is uniform throughout the conductor, the equation for the resistance is straightforward. The resistance as a *circuit* element that is temperature dependent may be represented by Eq. (3.1).

$$R_t = R_0(1 + kT) \tag{3.1}$$

where R_t is the effective trace resistance and is a function of temperature T. This is the static DC resistance of the trace. As discussed in Chapter 2, the resistance of a physical block may be computed if the current is uniform throughout, that is, if the current density in the cross-section is assumed uniform.

In Chapter 2, the resistance of a core winding for a discrete inductor was examined. The principles are the same here though the focus will be on interconnection systems for the PDN and its components rather than a storage element for a power converter.

The current density J_{xy} in the case of Figure 3.2 is assumed to be constant. Thus, to get the total current going through the cross-section, it is necessary to integrate over the area

$$I = \int_0^h \int_0^w J_{xy} \, dx \, dy = \int_0^h \int_0^w \sigma E_{xy} \, dx \, dy = \frac{Ewh}{\rho} = \sigma E A \tag{3.2}$$

where ρ is the *resistivity* of the material and σ is the *conductivity*. Note the substitution for the current density J with the σE term where E is the electric field. Since the conductivity here is a constant, so will be the electric field. The voltage across the conductor may be found by integrating over the line from one end of the conductor to the other, or, in this case, by integrating over the electric field *inside* the conductor.

$$V_2 - V_1 = \int_0^l E \, dz = El \tag{3.3}$$

where we assume that V_1 is zero. The resistance is now the ratio of the voltage to the current or

$$R = \frac{V}{I} = \frac{El}{\sigma E A} = \frac{l}{\sigma A} \tag{3.4}$$

This is the special case for the resistance of a conductor where the current density is uniform throughout the cross-section. The resistance for a section where the current is uniform throughout will be dependent upon the length, cross-sectional area, and conductivity. The conductivity of copper at 20 °C, for

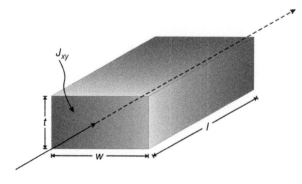

Figure 3.2 Current into block of metal with uniform current density.

the alloys that are used in printed circuit boards and other electronic systems [1], is a well-known quantity as we saw in Chapter 2.

$$\sigma_c(T_0) \cong 5.7 \times 10^7 \frac{S}{m} \tag{3.5}$$

where T_0 is room temperature or 20 °C. The constant, σ, is called the *constitutive* parameter for conductivity, in this case, the conductivity of the material. Copper conductivity changes at a rate of approximately 0.39% per degree Celsius. This means that the resistance will increase as the temperature increases. For example, the change in conductivity from 25 to 100 °C would result in a change in conductivity of

$$\sigma(T_{100}) = \sigma(T_0)[1 + .0039 \cdot (100 - 25)°C] = 7.37 \times 10^7 \frac{S}{m} \tag{3.6}$$

The resistance would therefore increase 29% over a 75 °C rise.

A conductor is also subject to frequency-dependent behavior. At higher frequencies the skin depth of a material comes into play. This is where the current begins to crowd in a region that is smaller than the conductor's cross-section. Thus, the resistance will start to increase at a given frequency for a particular conductive path.

The skin depth is related to the propagation of a wave through a medium. In the case of a signal that propagates through a copper trace through a power plane, the signal will be of fairly low frequency. However, there are high-frequency components that *may* be generated due to the load changing. The skin depth is the amount the wave has dissipated by ~37%, or e^{-1}, as the signal propagates into the medium. The equation is a function of the attenuation constant for a signal. If the medium is a good conductor, the skin depth of the material may be found from Ref. [2].

$$\delta = \sqrt{\frac{2}{\omega\mu\sigma}} = \sqrt{\frac{2}{2\pi f\mu\sigma}} = \frac{1}{\sqrt{\pi\mu f\sigma}} \tag{3.7}$$

This is the general expression for the skin depth in a material. To determine the resistance of a block of copper with a frequency-dependent signal, it is best to examine the physical structure along with the skin depth and then compute. The skin depth, in general, is not an absolute value for the current density. This equation only gives the value at ~63% of the current density in the cross-section of the conductor. However, for purposes of estimating the resistance, or loss in a PDN conductor segment, it is more than adequate.

Figure 3.3 shows a cross-section of a segment of copper similar to Figure 3.2 except that well-developed skin effect is assumed. For now, the nonuniformity of the current density from one point to the next is ignored, and it is assumed that the current is uniform across those two points. If the cross-section of the block is 40 μm and the frequency of the signal is approximately 100 MHz, the

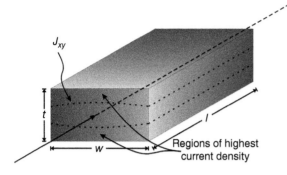

Figure 3.3 Region showing skin depth of material.

skin depth and the resistance may be computed. Let us assume the given parameters for the block below:

$$l = 2\,\text{mm}, \quad h = 40\,\mu\text{m}, \quad w = 100\,\mu\text{m} \tag{3.8}$$

To compute the resistance, we will need to calculate the skin depth. Given the parameters above, the skin depth will be

$$\delta = \frac{1}{\sqrt{\pi\mu f\sigma}} = \frac{1}{2\pi\sqrt{(1e8)(5.7)}} = \frac{1e-4}{2\pi\sqrt{5.7}} \cong 6.7\,\mu\text{m} \tag{3.9}$$

This is the skin depth on *both* sides of the conductor. Thus, the resistance may be computed given the other geometries of the construction.

$$R_{AC} = \frac{l}{\sigma A} = \frac{2e-3}{(5.7)(1e7)(2 \times 6.7e-6)(1e-4)} = \frac{1}{(5.7)(6.7)} = 26\,\text{m}\Omega \tag{3.10}$$

It is interesting to compare this AC value with the DC value where the current was essentially uniform throughout the plane. Computing the DC value we arrive at the solution,

$$R_{DC} = 8.7\,\text{m}\Omega \tag{3.11}$$

As expected, the result is nearly three times smaller than the AC resistance computed for the same block. However, the frequency here is relatively high for most of the signals than one would expect to traverse within a PDN; near the load, this higher frequency would actually be expected. This is because the current switching behavior of a microprocessor or any other computing device can generate very high switching currents which can create relatively fast di/dt, resulting in signals that may propagate through the nearer portions of the PDN at high frequencies.

The next question is how this relates to one of the physical structures in Figure 3.1. This can best be answered through a simple analysis. Before examining this, however, it is best to discuss the effects due to nonuniform current densities. Most layout engineers spend a significant amount of time optimizing the layout in the motherboard, package, and silicon planes to ensure that the area and stackup of the structure are minimized. Either directly or indirectly, this has an impact on the cost of the system and thus is a critical factor in the overall design of the platform. This is why routing the PDN becomes a challenge. The objective is to achieve a low impedance path while controlling the dimensions of the interconnect. Therefore, it is not uncommon to see reasonably narrow planes routed between capacitive structures from the output of the voltage regulator to the load. These traces have to carry both the AC and DC current, and even with the use of numerical extraction tools to determine the effects, it is often difficult to find out where the highest current densities are.

As discussed earlier, there are two important components to consider: the DC and the AC portions of the distribution. However, as the currents egress from one section to the next, it is not uncommon for the current densities to be very focused in regions that are higher than desired. To see this, one can analyze a structure from two points in a plan and then trace the currents between them. Figure 3.4 shows the source and sink of currents between two points in a plane. It is important to remember that this is positive going current and that there is a return current associated with this one. For now, the analysis will be focused on the one directed as shown.

As before, the current will follow a function, but now it is no longer constant. The current density function is in the z–x direction. Normally the function is not known. However, in this case, we can assume a certain linear function to find the total current through a section of the plane. In Chapter 4, the closed form solution for the resistance between two circular points will be analyzed as well.

First, let u's assume that the current density is linear between the two points but varies more in the z direction than in the x direction. The function for the current density in the plane is then

$$J_{xz} = J_0 \frac{(\vec{i}\,x + \vec{j}\,z)}{(x + 2z)} \tag{3.12}$$

Figure 3.4 Current density between two points in plane.

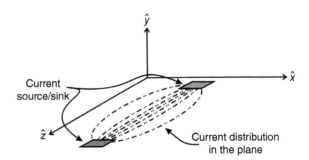

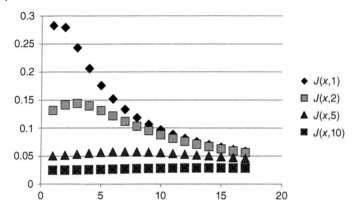

Figure 3.5 Plot of magnitude of current density for values of z.

The current density is a vector quantity and has magnitude and direction in the x–z plane. It is assumed that the current density is uniform in the y direction; that is, if one were to slice the plane in the x–y direction, the current density would be the same in the y direction. The plot of the current density magnitude for the function above for various distances along the z direction is shown in Figure 3.5.

The current density magnitude of the function gives an indication of the decrease in the function as the dimensions move outward across the x–y plane. It is possible to view this by imagining that these are the lines of current density *flux* across the plane region. The decrease from left to right would be expected if the sources were smaller than the plane width, which is nearly always the case.

The total current may be found by integrating over the x and z regions or

$$\iint J_{xz} \, dx \, dz = \iint J_0 \frac{(\vec{i} x + \vec{j} z)}{(x^2 + 4z^2)} \, dx \, dz \tag{3.13}$$

The integration in x may be done by separating it into simpler components[2]. These are indefinite integrals, so the integrations will have constants when the final integrals are done. The two integrals in x will be

$$\int \frac{x}{(x^2 + 4z^2)} \, dx, \quad \int \frac{z}{(x^2 + 4z^2)} \, dx \tag{3.14}$$

The first integral may be computed using a direct substitution.

$$\int \frac{x}{(x^2 + 4z^2)} \, dx = \int \frac{u}{2u^2} \, du = \int \frac{1}{2u} \, du = \frac{1}{2} \ln(x^2 + 4z^2) + C1 \tag{3.15}$$

---

2 Many of these integrals may be looked up on the web, but it is instructive to go through the analysis to gain an insight into the process as well.

The second integration is slightly more complicated and involves a trigonometric substitution.

$$\int \frac{z}{(x^2 + 4z^2)} \, dx = \frac{z}{4z^2} \int \frac{1}{\left(\frac{x^2}{4z^2} + 1\right)} \, dx \tag{3.16}$$

We can now substitute for the numerator function.

$$\frac{x^2}{4z^2} = \tan^2 \theta$$

Substituting in for the above values results in the new integral

$$\frac{1}{4z} \int \frac{2z\sec^2\theta}{(\tan^2\theta + 1)} \, d\theta = \frac{1}{2} \int \frac{\frac{1}{\cos^2\theta}}{\left(\frac{\sin^2\theta}{\cos^2\theta} + 1\right)} \, d\theta = \frac{1}{2} \int \frac{1}{(1)} \, d\theta \tag{3.17}$$

Integration gives the simple result as follows:

$$\theta = \frac{1}{2} \arctan\left(\frac{x}{2z}\right) + C2 \tag{3.18}$$

Thus, we now have the integral in terms of z to perform the final integration

$$\frac{1}{2} \int \left[\ln(x^2 + 4z^2) + \arctan\left(\frac{x}{2z}\right) + C_x\right] dz \tag{3.19}$$

where C' is just the constant of integration that falls out as a combination of the two constants C1 and C2.

The integration in z is obviously a two-part operation. The first integral may be solved with a combination of integration by parts and a trigonometric substitution.

$$S_1 = \frac{1}{2}[\ln(x^2 + 4z^2)]dz = z \ln(x^2 + 4z^2) - 2\left[z - \frac{x}{2}\arctan\left(\frac{2z}{x}\right)\right] + C_{z1} \tag{3.20}$$

The second integral may be also solved through integration by parts but is a little more involved. The solutions for both of these are left as an exercise at the end of this chapter.

$$S_2 = \frac{1}{2} \int \arctan\left(\frac{x}{2z}\right) dz = \frac{z}{2}\arctan\left(\frac{x}{2z}\right) + \frac{1}{4}\arctan\left(\frac{x}{z}\right) + C_{z2} \tag{3.21}$$

The total current is now

$$I = J_0[S_1 + S_2 + C_x z + C_z] \tag{3.22}$$

The total current I is now described in functional form. This can also be plotted to determine where the current is flowing as a magnitude through the plane. The current through the region of interest is now a function of x and z and may be plotted to get an insight into how this current flows from one point to the other (see exercise at the end of the chapter).

The next component in Figure 3.1 is the inductance of the path. For plane structures the PI engineer is most interested in the external inductance of the plane. The external inductance is a function of the magnetic fields created by the currents through the plane and contributes to the overall impedance of that path, along with the resistance. Here we are most interested in the magnetic fields due to the conduction current in the plane itself. The internal inductance is due to the internal fields in the metal itself. Though these fields exist, their contribution to the overall inductance is small compared to the external inductance.

For a plane structure where the width and length of the plane are relatively long compared with the distance between them, the fields dominate between the two plane pairs rather than on the outside of the plane. This is, in part, due to the electric potential between them, which increases the charge density on the sides where the two conductors face each other.

The external inductance between a plane pair may be estimated with the following simple formula:

$$L_p = \frac{\mu_0 A}{l} \tag{3.23}$$

where l is the length, μ_0 is the permittivity in air, and A is the cross-sectional area between the two planes where the magnetic flux flows. The formula may be easily derived by looking at Figure 3.6. The two planes are assumed to be very thin so that the current flows uniformly throughout the structure. The total current may be found by integrating around the path of the conductor.

$$I = \oint H_x \, dx \tag{3.24}$$

The integration in the y dimension is assumed small; since there exists (essentially) no fields outside of the region between the two conductors, this contribution may be ignored. Thus, the magnetic field is then

$$H_x \cong \frac{I}{w} \tag{3.25}$$

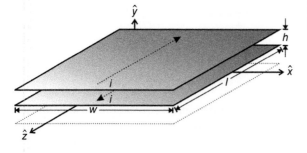

Figure 3.6 Two copper planes in space.

The total flux between the two structures may be found by integrating over the area between the two planes. In this case, it is the area through which the flux flows between them.

$$\psi = L_p I = \int_0^l \int_0^h \mu_0 H_x \, dy \, dz = \frac{\mu_0 I l h}{w} \rightarrow L_p = \frac{\mu_0 A}{w}, \quad h \ll l, w$$

This only holds true if the separation between the planes is much less than the width and the length as indicated. The result in the above equation is very useful for computing the inductance between two plane pairs, or to get an estimate of what the inductance may be in a structure where current is flowing from one segment in a PDN to another. This formula breaks down, even when the height is small relative to the other parameters, if the current is not considered uniform between two segments. This was seen in the previous analysis for the resistance. The fields are only considered constant if the current is uniform. If the current flows nonuniformly throughout the planes, a different analysis will be necessary.

If the flow of current is between two vias, then the inductance will be due to the currents in those two vias between the plane pairs. Figure 3.7 shows the structure. This is similar to the via connections routing to the contacts of capacitors where the connections are to adjacent power and ground planes.

The magnetic field egresses radially outward between the two vias. The total current is found from the integration around the loop.

$$I = \int_0^{2\pi} H_\rho \rho \, d\phi \tag{3.26}$$

This yields the magnetic field over the region of interest.

$$H_\rho = \frac{I}{2\pi\rho} \tag{3.27}$$

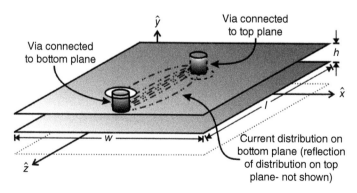

Figure 3.7 Vias in planes with current flowing in each.

The inductance is now found by integrating the magnetic flux density in the region between the two vias.

$$\psi = LI = \int_0^h \int_{\rho 1}^s B_\rho \, d\rho \, dz$$

$$= \int_0^h \int_{\rho 1}^s \frac{I}{2\pi\rho} \, d\rho \, dz \rightarrow L = \frac{\mu_0 h}{2\pi} \ln\left(\frac{s}{\rho 1}\right), \quad \rho 1 \ll s \tag{3.28}$$

The inductance may be plotted for various separation distances and radii. Figure 3.8 shows this plot. Intuitively, as the separation between the vias increases, one would expect the inductance to increase. However, the rate of increase in inductance is small relative to the separation. This is because the current flows radially outward and the magnetic field, as shown by Eq. (3.27), decreases as the inverse of the radius.

When laying out capacitive structures in a PDN, it is important to minimize the inductance between the structures. There is a methodology introduced by Reuhli [3] called the partial element equivalent circuit method that breaks down the circuit representation of the inductance (or other circuit element) in, say, a loop into partial element components that represent the whole loop. For lumped element modeling, or conceptualization of a sub-element loop in a PDN construction, it is often advantageous to visualize the path in this unique manner. However, with regard to a loop inductance between, say, the interconnect between two capacitors in a plane, understanding this inductive path is important to the construction of the PDN as a whole.

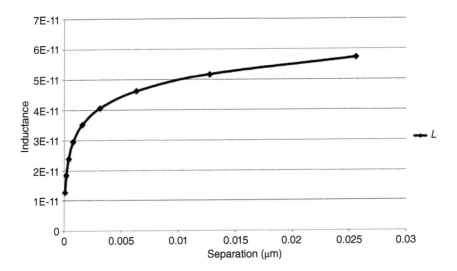

Figure 3.8 Plot of inductance as a function of separation between vias.

The inductance between two vias in a plane was analyzed above. However, what about the entire path where current also flows on the planes? The question is, is there a contribution to the inductance from these currents in the loop and how does this manifest itself in the building of a circuit that might be used, for example, in a Simulation Program with Integrated Circuit Emphasis SPICE model of the PDN?

To answer this, it is best to go back to the physical construction of the plane/via pairs in terms of what they are. Figure 3.7 shows this in more detail. The vias are usually connected to different planes because one represents the positive voltage side of the plane, while the other the return path. In a multi-layer printed circuit board, it is often a challenge in today's systems to have the power and ground planes adjacent to each other. However, for this analysis, it will be advantageous to place them next to each other. Currents will flow therefore in the loop as shown between the two planes. One can imagine that both DC and AC currents will flow depending upon the excitation at the load. The current path therefore comprises the vias on either side of the interconnection and the upper and lower planes between. From an inductance point of view, engineers usually represent this as one complete inductance. However, using *partial* inductance methods, this inductance may be thought of as multiple *partial* inductances [1]. Figure 3.9 shows a representation of these partial inductances. Since there are four distinct portions here, it is easy to visualize the partial inductances as four separate circuit elements. However, as shown in the figure, there are also mutual elements which couple between them.

The inductance that we are interested in is a result of the total flux contained within the surface bounded by the vias and planes. The regions outside of this loop, for example, above and below the planes, have no flux since it is assumed that the planes are wider and longer than the separation between them by a

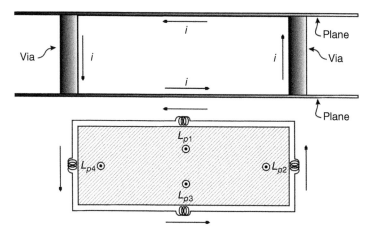

Figure 3.9 Currents through vias/planes and corresponding partial inductances.

significant amount. The next question one might ask is what about the regions outside of the two vias? In this case there is flux from the currents on either side. However, if one integrates over these regions, it can be shown that the flux cancels (especially if the currents are equal and opposite), resulting in no net flux from either region [4].

Mathematically, the concept of partial inductance starts with the total inductance of a loop. This inductance is a function of the total net flux in this loop and the current creating it,

$$L_{lp} = \frac{\Psi_{lp}}{I} = \frac{1}{I} \int \vec{B} \cdot d\vec{S} \tag{3.29}$$

where L_{lp} is the loop inductance. The integration is over the surface S. To illustrate this concept, it is best to introduce the *magnetic vector potential* (or sometimes termed magnetic potential). The curl of the magnetic vector potential is the magnetic flux density vector,

$$\vec{B} = \nabla \times \vec{A} \tag{3.30}$$

Using Stokes' theorem, we can substitute for the integral in Eq. (3.29).

$$L_{lp} = \frac{1}{I} \oint \vec{A} \cdot d\vec{l} \tag{3.31}$$

The magnetic vector potential direction will be orthogonal to the direction of the magnetic flux. From Figure 3.9, it is evident that the magnetic flux vector direction is into the page. Thus, the magnetic vector potential will be in the direction parallel to the page and across the surface enclosed by the four wires that make up the contour. Because the resulting integral is now a *contour* integral around the enclosed surface, we can break up the integral as a function of sections of the contour around that path. For Figure 3.9, this comprises the four sections or,

$$L_{lp} = \frac{1}{I} \int_{C1} \vec{A} \cdot d\vec{l} + \frac{1}{I} \int_{C2} \vec{A} \cdot d\vec{l} + \frac{1}{I} \int_{C3} \vec{A} \cdot d\vec{l} + \frac{1}{I} \int_{C4} \vec{A} \cdot d\vec{l} \tag{3.32}$$

Thus, each section can now be broken into individual partial inductances. The self-partial inductances are easily identified as the individual inductance of each section of the loop or,

$$L_i = \frac{1}{I_i} \int_{Ci} \vec{A} \cdot d\vec{l} \tag{3.33}$$

While the mutual partial inductances are due to the coupling between the elements,

$$M_{ij} = \frac{1}{I_i} \int_{Cj} \vec{A} \cdot d\vec{l} \tag{3.34}$$

As an example, let us examine two wires in parallel and determine both the loop inductance and the partial inductance of each wire.

Example 3.1 Compute the inductance between two wires that are very long with a relatively small wire diameter on each.

Solution:
The structure is shown in Figure 3.10. The current in each wire creates a net flux through the region between them. Note the dots of the direction of flux per the right-hand rule coming out of the region. The solution to the entire inductance for the loop may be computed using the basic formula for the self-inductance of the two wires. However, let us use the magnetic vector potential to compute the self-partial inductance of one wire and then add the other partial inductance to it to get the total self-inductance for the loop.

The total self-inductance of the loop will be approximately the sum of the two partial inductances (neglecting the partial inductances from the ends or any fringing effects)

$$L_{loop} = L_{p1} + L_{p2} \tag{3.35}$$

Thus, we are going to compute the self-partial inductance of L_{p1} which is only on one side. We are neglecting the inductance due to the ends as we will also assume that the separation is small relative to the length ($s \ll l$).

The magnetic vector potential may be found from Eq. (3.36) [4],[3]

$$A(\hat{r}) = \frac{\mu_0}{4\pi} \int_V \frac{J(\hat{r}')}{|\hat{r} - \hat{r}'|} d^3 r' \tag{3.36}$$

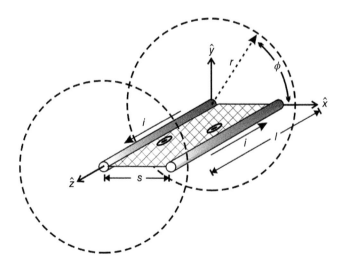

Figure 3.10 Two parallel wires in space.

---

3 This is shown without proof here. The equation falls out directly from the Biot–Savart law.

The vector $\boldsymbol{J}$ is the current density in the wire (A/m³). The distance r' is the length from the origin to the differential element of current. In this case $r' = 0$ since it lies along the z axis. The distance r is to the observation point.

The integral is over the volume that contains the current; thus, the integration will sum the total current in the wire. Integrating around the wire yields the magnetic vector potential in the z direction.

$$A_z = \frac{\mu_0 I}{4\pi r} \tag{3.37}$$

Because the integration is for the *current* only, it is a simple result. The net partial inductance is now the integral over the length of the finite wire on the side that we have chosen for L_{p1}.

$$L_{p1} = \frac{1}{I_p} \int_{C1} A_z \cdot dl = \frac{\mu_0 l}{4\pi} \ln\left(\frac{s}{r_1}\right) \tag{3.38}$$

The total inductance is now twice the inductance in Eq. (3.38), which is the solution for the total loop inductance between two wires.

$$L_{loop} = \frac{\mu_0 l}{2\pi} \ln\left(\frac{s}{r_1}\right) \tag{3.39}$$

We now turn to the subject of capacitance; in particular, the capacitance between two parallel planes. The capacitance between the planes in the PDN – interconnect capacitance specifically – is typically very small and has little to no effect on the contribution to the PDN. Normally, in a PDN extraction, this capacitance is ignored since the dielectric constant of the motherboard is quite low and the separation is large relative to what the other dimensions would need to be to acquire a significant capacitive structure. The capacitance of two planes separated by a distance s where we can make some reasonable assumptions is easily derived from first principles (ignoring fringing effects). Figure 3.11 shows an isometric of two planes where the electric field is in the direction of the y axis.

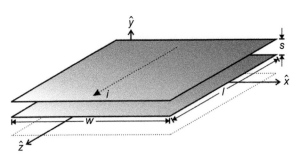

Figure 3.11 Two parallel planes.

Assume a charge Q exists on the lower plate such that the electric field is in the positive y direction. Then, the voltage across the two plates may be computed using Gauss's law.

$$V = \int_0^s E_y \cdot dl = E_y s \tag{3.40}$$

The total charge enclosed by the surface that circumscribes the region in the x–z plane is found from the electric flux vector $\boldsymbol{D}$, where the vector is in the same direction as the electric field.

$$Q = \oiint \boldsymbol{D}_y \cdot dx\, dy = \int_0^w \int_0^l \varepsilon E_y dx\, dy = A\varepsilon E_y \tag{3.41}$$

where $A = lw$. The capacitance is now the ratio of the total charge to the voltage or

$$C = \frac{Q}{V} = \frac{A\varepsilon E_y}{E_y s} = \frac{\varepsilon A}{s} \tag{3.42}$$

For most plane structures this value is typically very small. As an example, we use a typical board separation of 8 mils (0.2 mm). A reasonable region between two capacitors in a PDN reveals dimensions between the two devices (the area A) as the following:

$$l = 2\,\text{mm}, \quad w = 1\,\text{mm}$$

Thus, the capacitance of this structure, not including fringing and limited to this region, would be

$$C = \frac{\varepsilon A}{s} = \frac{(8.854 \times 10^{-12})(1 \times 2)(10^{-6})}{(200)(10^{-6})} \cong 90\,\text{fF} \tag{3.43}$$

In general, capacitance in most regions of the PDN would not show much affect until the values reached into the 10 nF or more. Thus, this is the reason PI engineers tend to ignore these affects.

The next section focuses on the electrical characteristics of actual capacitors that are used for the construction of the PDN. These are the workhorses of the power delivery path and are designed to maintain the efficacy of the delivery of current from the source to load under all of the operating conditions for the life of a product.

3.1.1 Capacitors for PDN Applications

The main device that is used for PDN and power conversion filtering is the capacitor. In general, the capacitor is probably the most heavily used component in an electronic system and therefore one of the most important. This device is used not only for power decoupling but also for noise suppression, electromagnetic interference (EMI) mitigation, energy storage, and general circuit design, to name a few of the applications. Unfortunately, for such a critical and heavily used device, in a number of ways, the device is often not well understood for many of its applications.

Though in concept the capacitor at first glance appears simple, it can be rather complex when analyzing it as part of a larger system. This is because its circuit quality can change substantially (depending upon the technology of the capacitor) over the life of the device and under various use cases. This is why it is important to understand the capacitor in some detail before using it within a computer or an electronic platform.

There are numerous types of capacitors that are useful for computer systems, and a thorough study of every one of them is well beyond the scope of this text. However, one of the more popular and common devices that are applicable for motherboard-level PDNs and power filtering is the multilayer ceramic chip capacitor (MLCC). The ceramic chip capacitor in today's systems spans a wide variety of sizes and applications, and there is a plethora of devices available to the engineer.

The MLCC device is comprised of many layers of ceramic dielectric with inter-stitched electrode layers that are connected to opposing terminals. A section view of an MLCC is shown in Figure 3.12. The device connects to the circuit board at the ends through a soldered connection which in turn connects to the power layers or traces on the board.

There are many classes for MLCCs that today fall under the IEC (International Electro-technical Commission). The standards are categorized into three classes where the most used device (in general) for electronic systems is class 2. Class 2 devices are typically made with a dielectric that has a higher permittivity and thus have a higher capacitance per unit volume. The material is usually a variant of barium titanate ($BaTiO_3$) with various additives. The permittivity may be as high as 14 K. Within this class, the devices are characterized by a letter/number/letter code combination, which designates the lower temperature

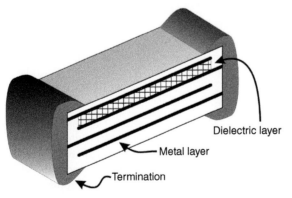

Figure 3.12 MLCC construction.

Dielectric layer

Metal layer

Termination

by the first letter, the upper temperature by the number, and the change in capacitance due to temperature by the last letter respectively [5]. Though there are a number of combinations available to the designer as products, there are typically three types of devices that are relatively common for electronic systems today: X7R, Y5V, and Z5U.[4] A Class 1 device, NP0, is also seen in electronic systems and is used for applications that require better temperature stability. The typical characteristics for each capacitor type are shown in Table 3.1.

Depending upon the type of capacitor, it can vary over a wide range of environmental conditions. Moreover, the electrical characteristics of ceramic capacitors may also change during operation. There are two parameters that are important to keep in mind when analyzing a PDN with MLCCs: (i) the variances of the electrical characteristics of the device as a function of frequency and (ii) operating temperature dependencies. A basic electrical model for the capacitor is shown in Figure 3.13. This simple model is often used for determination of the *impedance* for a typical capacitor. In most cases, this model is more than adequate to predict the frequency behavior of the device. However, this circuit only models the electrical behavior at any given temperature or time. The capacitor also has a *frequency*-dependent behavior for the ESR or, the *effective series resistance* of the capacitor that also changes. Depending upon the material properties, the ESR of an MLCC can vary widely over its frequency range of operation.

Table 3.1 Typical capacitor limits for some common MLCC devices.

Dielectric type	Lower temp. limit (°C)	Upper temp. limit (°C)	%Cap change over temp.
X7R	−55	+125	±15
Y5V	−30	+85	+22/−82
Z5U	+10	+85	+22/−56

Figure 3.13 Simple RLC model for a capacitor.

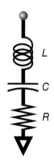

---

4 There are many variances to the above such as Z7U. The variances are usually related to different dielectrics.

The basic equation for the impedance follows directly from the circuit diagram.

$$Z(j\omega) = \frac{1}{j\omega C} + j\omega L + R_{ESR} \tag{3.44}$$

The resonant point may be found by differentiating the previous equation with respect to ω, and then setting the result equal to zero.

$$\frac{\partial Z(j\omega)}{\partial \omega} = 0 = \frac{-1}{j\omega^2 C} + jL \rightarrow \omega_0 = \frac{1}{\sqrt{LC}} \tag{3.45}$$

The result is not surprising given that this is the resonant frequency for an LC circuit. Plugging this back into the equation yields

$$Z(j\omega_0) = \frac{\sqrt{LC}}{jC} + \frac{jL}{\sqrt{LC}} + R_{ESR} = -j\sqrt{\frac{L}{C}} + j\sqrt{\frac{L}{C}} + R_{ESR} = R_{ESR} \tag{3.46}$$

This simply says that the lowest resonant point for a capacitor is the ESR. However, unfortunately, this is only true if the ESR has no frequency-dependent behavior. The fact remains, that it *does* and the equation must be modified to include this. Thus, our new equation shows this behavior.

$$Z(j\omega) = \frac{1}{j\omega C} + j\omega L + R_{ESR}(j\omega) \tag{3.47}$$

Differentiating with respect to the angular frequency and setting to zero gives us the relationship between the *frequency*-dependent ESR and the other components in the impedance profile.

$$\frac{\partial Z(j\omega)}{\partial \omega} = 0 = j\left(\frac{1}{\omega^2 C} - L\right) + \frac{\partial R(j\omega)}{\partial \omega} \tag{3.48}$$

Because the behavior of the capacitor is such that the ESR is now frequency dependent, a new minima may be established depending upon the type of capacitor and its material properties.

It is clear that the ESR is an important quantity when examining the characteristics of a capacitor. It becomes even more important when examining the capacitor for its loss characteristics within a PDN or power supply filter. This is when the frequency-dependent behavior of the resistance becomes more interesting.

Figure 3.14 shows the normalized (to 1 MHz) frequency dependence for a common MLCC device for both the impedance and the ESR of the device.

The curve in black shows the change in impedance as a function of frequency. The curve in gray shows the ESR as a function of frequency. Note that the ESR in this case does not minimize at the resonant point for the standard impedance model. This is possible because the ESR may have a *different* frequency dependence than the overall electrical behavior for the capacitor. Most capacitors have a resonance point called the *series resonant frequency* (SRF) for the capacitor. This is where the impedance for the device is lowest and is an *indication*

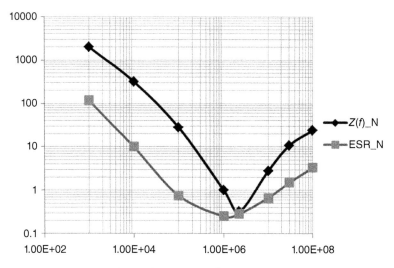

Figure 3.14 Normalized impedance (1 MHz) for an MLCC capacitor.

of the ESR at that given frequency. Typically, an engineer wants to place the device and to operate it at or near its resonance frequency to maximize the energy transfer and minimize its losses. The ESR is important for applications where current ripple (such as in a voltage regulator) is present. The ripple in a system translates directly to loss in the ESR of the capacitor. The graph shows that for minimizing losses, it may be better to operate this device at a lower frequency. Once again, it is important to attempt to operate a capacitor at the point where the ESR is lowest to minimize these losses.

The other change that can occur in a system is the variance of the device due to temperature. This is mainly a function of the change in dielectric constant for the given material as the system varies in temperature. In many electronic systems, from battery powered to server platforms, the ambient temperature may vary over a wide range. In fact, depending upon where the system is operated over the globe, the temperature in the platform could easily vary over 100 °C. The reason is the actual board and device-level temperatures can change due to internal self-heating and environmental conditions from thermal gradients that occur within the platform. These temperature changes are common for electronic systems and vary from device to device and platform to platform depending upon the cooling solution and environment. Suffice it say, however, ensuring the efficacy of the electrical system due to temperature variability is one of the more critical items that a system designer must consider.

Figure 3.15 shows the change as a function of temperature for the capacitance. It is not uncommon for the capacitor to vary electrically throughout the operation of an electronic system. This is why engineers must design their systems such that the capacitor variances are taken into account during

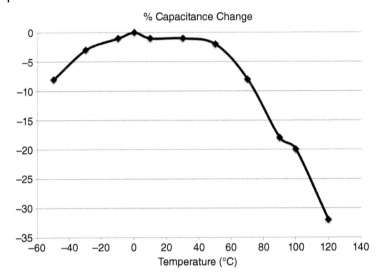

Figure 3.15 Example capacitance change as a function of temperature for X7U dielectric.

normal operation. A good example of this is the filter for a buck regulator. As we saw in the previous chapter, the stability of the VR is dependent upon the filter impedance at the output. If the variance due to the filter was significant, the loop stability could be compromised. The figure, however, shows only the change in a capacitor for variations during *normal* operation. But what about the changes over the life of the device?

It is not uncommon for certain types of MLCCs to vary by more than 100% over the life of the device in a platform. This is mainly due to the change in leakage current in the dielectric and construction over many cycles of impressed voltage [6]. Eventually, the device can break down if the excitation voltage goes beyond the operational limits for the capacitor. Over the past decade or more, the dielectric and electrode thicknesses for the parallel plate MLCC have gone down to the point where the number of layers are in the thousands and the dielectric layers are into the nanometer range [5]. Along with these trends, the grain size for the materials has also been decreasing to keep pace. However, as the distances between plates decrease, the opportunity for voltage breakdown across a given plate pair can also increase, leading to degradation in the capacitor over time and sometimes failure. It is therefore critical when designing a PDN structure to ensure that the boundaries of the capacitor are well understood.

An example of this change for a Z5U capacitor is shown in Figure 3.16. Note that for this device the performance degrades over time due to aging. It should also be noted that the degradation in performance can take years (and often does!), which is indicative of the MLCC reliability in general. Thus, as long as the

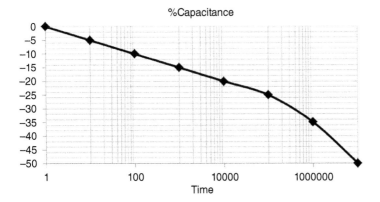

Figure 3.16 Percent capacitance change example for Z5U device.

capacitor performance is maintained over the expected life of the platform, the change may be manageable. However, if the environment is stringent enough (voltage and temperature) relative to the devices operational boundaries, the degradation may be accelerated.

One of the key parameters that alter performance of a capacitor (in this case its capacitance) is the device's polarization. For a parallel plate capacitor, like in a generic MLCC, the capacitance is proportional to the change in the polarization vector P.

$$C = \left(\epsilon_0 + \frac{\partial P}{\partial E} \right) \frac{A}{h} \tag{3.49}$$

The polarization vector is in the same direction as the applied electric field E and is related to the electric flux density D.

$$D = (\epsilon_0 E_a + P) \tag{3.50}$$

where E_a is the applied electric field. Polarization in a capacitor occurs when an electric field is applied (voltage) across the plates. The dipoles inside of the dielectric become "polarized" and line up in the direction of the electric field. When a net polarization still exists within the dielectric when the electric field changes direction, a net remanence in the polarization may exist, which results in a hysteretic effect for ferro-electric-based materials such as barium titanate [6]. Thus, when the capacitor is discharged, a hysteresis can exist, resulting in a curve that traces a different path as illustrated in Figure 3.17. The polarization vector has a nonlinear relationship to the applied electric field. The polarization may be approximated for a barium-titanate-based capacitor using the following formula [6]g

$$P = N\hat{p} \left[\coth \left(\frac{\hat{p}E}{kT} \right) - \frac{kT}{\hat{p}E} \right] \tag{3.51}$$

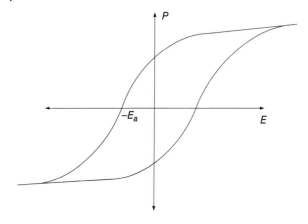

Figure 3.17 Polarization vector relationship to electric field in parallel plate capacitor.

where N is the domain density for the dielectric, $\hat{p}'$ is the average dipole moment, E is the electric field strength, k is Boltzmann's constant, and T is the temperature in Kelvin. A simple plot of the function in Eq. (3.51) for a non-polarized capacitor is shown in Figure 3.18 illustrating the effect without polarization (charge curve with depolarization) normalized to an electric field excitation. If the device shows not net polarization, it will follow the curve as shown in the figure.

Thus, as the electric field (or voltage applied) across the capacitor changes, the slope changes which in turn effects the capacitance. This in turn implies that the capacitance at any one time is dependent on the previous state of the device.

What is important to note is that the capacitance may change depending upon a number of factors including degradation over time. The polarization

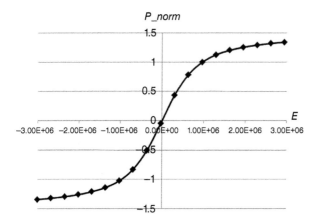

Figure 3.18 Polarization of capacitor as a function of electric field.

vector for a parallel plate capacitor will change over the life of the product due to the applied electric field or voltage on the device. It is a strong function of the AC signal duty cycle and amplitude and breakdown of the dielectric can occur over time. Figure 3.19 shows the change in capacitance over time due to two different applied AC voltages as a function of temperature.

It is thus not unusual to see the capacitance change as a function of applied voltage, and when designing a PDN, this must be considered when analyzing the system or simulating it to find the droop or other characteristics in the power path. As seen in Figure 3.19, the effective capacitance may also change due to applied voltage. Moreover, over time, the capacitance may also trend downward to due degradation in the construction as indicated in Figure 3.16.

The issue here is that it is difficult, if not impossible, to predict the failure of a device using a typical model. This makes it even more difficult to predict the degradation of the device over time in order to predict the behavior of the droop on the PDN. As we shall see in later sections, this becomes a problem because statistically the lowest droop in the system, depending upon the frequency of excitation, is not necessarily when the capacitance has degraded to its lowest point. What also makes this more challenging is that the capacitors in the PDN tend to degrade at different rates within a platform. This is because some devices are located closer to heat sources, while others may also tend to see higher AC voltages.

This begs the question of what are the correct values to use when analyzing a capacitor within the context of PDN simulation? To answer this, it is best to go back to the manufacturer's data sheets and the use case for the capacitor.

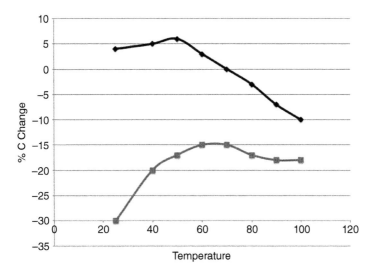

Figure 3.19 Change in capacitance for two AC voltages (gray is 50 × smaller voltage than black).

If the capacitor is not stressed relative to its voltage excitation and if the average thermal gradients are reasonably well known in the system, then an engineer can map this behavior back to the data sheet specifications for the device.

There are two important points to focus on when analyzing a capacitor within a given platform PDN system. The first one is the start-of-life (*SOL*) value for the capacitor. This gives the initial expected value for the device given its starting tolerance band (ToB). The second is the end-of-life (EOL) value. Both are important since they give upper and lower bounds to the device. Using these two data points, an engineer may evaluate the capacitors' performance over the life of the product when doing an analysis of the system PDN.

Section 3.2 discusses the interaction between the system and the PDN and how the power delivery path is affected by the load and the power source.

3.2 Power Delivery System Interaction

As we have seen in the previous sections, the components which make up the PDN interact with the power supply in a very specific way in today's high-performance electronic systems. It is no longer sufficient for the voltage in a regulator to be set to a single value and have the system respond that is acceptable for the system or the load. Thus, additional controls and operational modes have been added to the voltage regulator and the power management control systems to ensure that the power source operates to maximize the efficiency and performance of the product. In this section, we discuss a number of these important system-level functions with the expectation of understanding the behavior of the PDN and power source as a single structure. Specifically, the discussion centers around power load-line basics. Voltage regulator load line and its interaction with the system is critical since the load line is programmed to interact directly with the load to ensure an optimally operating system.

3.2.1 Power Load Line Fundamentals

Due to the recent advancements in high-performance silicon, engineers have been looking for ways to optimize the performance of their devices while maintaining or reducing the overall power. The continued improvement to CMOS silicon processes has pushed the envelope of what computational density within a given processing unit is capable of. Though the overall power consumption on a per processing unit (for a fixed number of gates, per se) basis appears to continue to shrink, architects and designers are finding more ways to increase the processing capability of SoCs and other advanced silicon with the goal of giving users a better experience in the end product.

The issue, of course, is that while the performance has increased, system-level design has had to adapt to keep up. This is where the power delivery has become such an important element to the overall performance. Though voltages are not shrinking at the same rates they once did, the dynamic behavior of the load has certainly increased along with its power density. This has made the design of the PDN and the power system a crucial part to the overall system.

Fundamentally, there are two potential issues or concerns a designer must contend with as it relates to the voltage that is controlled when supplying power to the silicon load under all operational conditions: the *minimum* regulated voltage and the *maximum* regulated voltage. Most designers focus on the minimum voltage that is required for a system and under what conditions that voltage needs to be maintained. If a processing unit only has a single operational clock frequency, a single voltage is normally assigned to it. Typically, that frequency is the maximum frequency a processing unit can maintain over the range of the fixed operating voltage, load changes, temperature, and so forth in order to maintain the data integrity of the system. The data integrity is dependent upon the timing relationship between the clock and data as the signals propagate through the functional unit (Figure 3.20). For a given voltage set point, there are tolerances in the system that must be maintained to ensure that under all conditions for a given piece of silicon, the efficacy of the data is maintained. Otherwise, the processing unit might not work correctly, or the system would require additional logic and other data correction mechanisms to maintain the efficacy of the data, which would slow down the system and compromise the performance.

However, silicon processing speeds are highly dependent upon the voltage that is supplied to the functional blocks, and this direct relationship helps designers to run workloads through the functional units at different clock rates, depending upon the requirements of that workload. These different frequencies, or workload *states*, are common for microprocessors and other computational units and are dependent upon the voltage and frequency relationship that is maintained for a given performance state. In the microprocessor world,

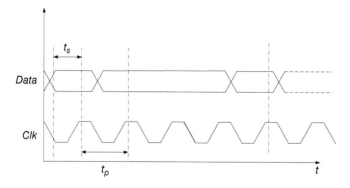

Figure 3.20 Clock to data relationship for logic signals in a functional unit block.

this relationship within a core or a microprocessor is commonly called a processor state or *p-state* for short [7]. For a given processing unit, there may be multiple *p-states* associated with it. Each *p-state* has a maximum frequency of operation and minimum voltage that must be maintained throughout. Figure 3.21 illustrates how *p-states* may be associated with the voltage and frequency of the system.

The *p-state* frequencies are shown normalized as are their corresponding voltages. As the voltages decrease, the frequency of operation also decreases as expected. For a given processing system, this is important because the power consumption in each *p-state* is not equivalent. Moreover, as the voltages increase, the active power consumption goes up as the square of the voltage.[5] This might be tenable for many platforms, except for the fact that the leakage power in the unit goes up as much as the cube of the voltage [8]. Because of the large power consumption at the high end of operational frequency curve, architects and designers often try to minimize the activity of the processing in this regime to ensure that the power does not exceed the limits of the system. However, this can be a difficult proposition since many workloads require the highest frequency and thus the work needs to be performed very quickly in order to improve the end customers' experience. This is why many processing systems have heuristics and other optimization systems in place to try and minimize the times the processing unit is in its highest *p-states*.

A detailed review of how a processor operates with different work- loads – even with respect to its power consumption – is well beyond the scope of this text. Nonetheless, it is important here to examine how power is controlled and delivered at the system level to satisfy the requirements of the

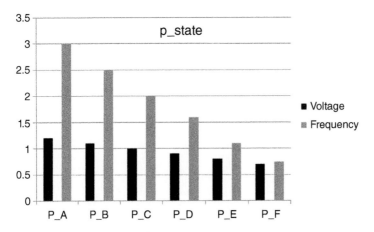

Figure 3.21 *p-State* voltage and frequency (normalized).

---

5 This assumes everything else is constant in the system, including the frequency.

processing unit. In addition, for the power integrity engineer, it is important part of the education of how to deliver a quality power signal from the voltage regulator to the load. The control of the voltage between the voltage regulator and the operation of the digital processing unit is an integral mechanism that must operate coherently in order to preserve the data integrity of the system. Furthermore, the mechanism must deliver power in the most efficient manner in order to optimally compute the data within the constrained environment the silicon is placed. This is why, in part, the *load line* was invented. The load line, in theory, is a static operational mode with respect to the clock frequencies that propagate internal to the system. However, it has a dynamic component to it that cannot be ignored. The load line, as its name implies, is a ratio of the voltage in the system to the current being delivered. Both variables are considered *static* as compared to the clock frequencies of the functional load. The definition of "static" here is relative of course. As mentioned earlier in this section, the load line was originally intended to help ensure the reliability of the system (maximum voltage) while maintaining an optimal voltage being delivered to the load [9]. The reason for the consideration for the reliability was due to the voltage exceeding the reliability limits for the silicon on a periodic basis. If the voltage continued to exceed these limits, the device could break down and fail.

Figure 3.22 shows a somewhat idealized load line. The slope is truly the ratio of the voltage to the current as the voltage regulator adjusts the system to the current as mentioned. Note that there is a maximum and minimum current associated with the load line. This occurs at the ends of the load line where the system starts adjusting the voltage up or down. Since the load line is associated with a device's *activity*, there is always an offset to the current at I_{min} (meaning the current will not be zero). However, the voltage does not have the same limits as the current. This is because there are *guardbands* to both the minimum and maximum voltages around the load line as we shall see.

As the current in the system increases, the voltage also decreases. One might think that this would be a consequence of simply adjusting the voltages through

Figure 3.22 Idealized load line.

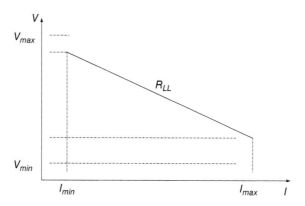

different *p-states*. However, this is not the case and is *not* how the load line was intended to function. Each *p-state* has its own load line associated with it that makes for a somewhat more complicated design of the power converter but aids in the reduction of power consumption at the system level. Since the slope of the load line is dependent upon both the voltage and the current, adjusting it to meet the proper operational limits of the load is critical for the design of the feedback system in the voltage regulator.

As an example of how the load line works for a given load, let us assume that a processing unit has been placed in the *p-state* P_C. Per our simple illustration in Figure 3.21, this would mean the normalized voltage and frequency pair would then be $(V_C, F_C) = (1, 2)$. With respect to the upper voltage limits, the maximum voltage should be lower than that in the highest *p-state* simply because, given a similar slope to the other load lines (which may or may not be the case), the reliability limit for the silicon should be within a safe operating region. This means that there is some *headroom* for this upper voltage. However, what about the lower voltage limit and overall power consumption? Depending upon the workload activity for this state, the power could be larger than desired. Thus, it is possible that the designers may wish to change the *slope* of the load line to minimize loss in the system. Since for this *p-state* the maximum current should be less than that in the higher *p-states* (we will discuss this more later), one would expect the power to be less. However, this is where the system design comes into play. If the workload activity tends to be highly active in this *p-state*, and the overall consumption is dominated here as well, then the architects would be expected to try and optimize around this operational point to improve the power consumption. Thus, depending upon the dynamic behavior of the overall system, the goal would be to make the load line as shallow as possible to minimize loss.

Unfortunately, this is not always the best option for the system design. A shallow load line has impacts beyond the potential savings in power. Before embarking on a discussion of power savings, however, let us discuss how the load line is created.

Figure 3.23 shows the distribution path from the output of the regulator to the load. Note that there is a resistance in the path due to the plane resistance in the PDN. As we have seen earlier, this plane resistance varies and is dependent upon a number of factors including temperature.

Thus, there is a *tolerance band* associated with this resistance, which will be discussed in Section 3.2.1.1. For a given *p-state*, we would expect (to a first order) that the load-line slope is associated with this resistance and that the drop is due only to the DC voltage difference in the path.

$$R_{path} = R_{LL} = \frac{V_{LL}}{I_L} \tag{3.52}$$

Thus, Eq. (3.52) implies that the load line is equivalent to the resistance of the path and the ratio of the load-line voltage to the load current. In addition, the

Figure 3.23 DC path from voltage regulator to load.

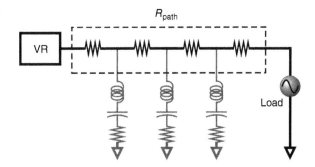

R_{path}

VR

Load

load-line voltage is dependent upon the preset *p-state* and the load current. Thus, functionally, this relationship can be defined explicitly now since the load-line voltage has multiple dependencies, or with regard to the *p-state* voltage.

$$R_{LL} = \frac{V_{LL}(P_S, I_L)}{I_L} \tag{3.53}$$

Basically Eq. (3.53) states that V_{LL} is the *dependent* variable and I_L is the *independent* variable. Thus, once a given *p-state* (P_S) is set, the voltage will adjust to the static change in load current.

Since the voltage at the load is what is critical to the operation of the functional unit, the voltage set at the regulator, given a nonzero resistance in the path, will always be set higher than the voltage found at the load. This *set* voltage is typically called the *voltage identification* or *VID* voltage of the regulator. The VID voltage usually has a tolerance associated with it along with an offset that is related to the drop across the resistance of the system. Depending upon how the path is designed, the controller will preprogram a fixed offset in the voltage regulator as a starting point and the load-line slope which will adjust the output voltage relative to the load through the feedback system while monitoring the static current changes,

$$V_{VID} = V_{FU} + R_{LL}I_L \tag{3.54}$$

where V_{FU} is the functional unit voltage, or the voltage that is measured at the load.

The question arises now as to why the load line changes with each *p-state* setting. It seems at first glance that it would be easier to simply maintain a constant programmable load line rather than change it as a function of the *p-state* of the system. In many platforms with VID controlled regulators, this is indeed the case. However, to minimize power loss in the system, it is more advantageous to adjust the load line when the *p-state* is also changed. Figure 3.24 and the following discussion illustrate why.

Since each *p-state* has associated with it a given voltage and frequency pair, this implies a given power consumption at the load that is essentially fixed for

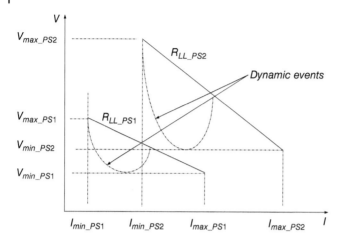

Figure 3.24 Load-line differentiation with *p-states*.

that *p-state*. The power in a functional processing unit may be estimated by the following simplified equation:

$$P_{PS} = I_{PS}V + P_{lk\_PS} = (A_f C_D V F_{PS})V + P_{lk\_PS} \tag{3.55}$$

where P_{PS} is the power consumed in the functional unit for a given *p-state*. The variables here warrant some explanation.

The term in the braces on the right-hand side of the equation is the *current* at the load. When multiplied by the voltage, it gives the total power in the functional unit that is being drawn. This is the *active* power in the functional unit. The second part of the equation is the *leakage* power in the device.

The first term within the braces for the current is called the activity factor A_f. The activity factor is the percent of the functional block that is active during a given workload. This is essentially,[6] and statistically, the number of gates/transistors that are active when the functional block is operating and doing work. C_D is the dynamic capacitance or *switched* capacitance for that unit when it is active. It is derived from the fact that a transistor's active power generally is dominated by its gate or *switched capacitance* when the device is operating at a given frequency. It is an aggregate or lumped quantity for the functional unit. The voltage is the actual instantaneous voltage during the active workload cycle. F_{PS} is the frequency of the active *p-state* during this cycle.

The leakage current is also proportional to the voltage and increases as a power of α.

$$P_{lk\_PS} = kV^\alpha \quad 2 \leq \alpha \leq 3 \tag{3.56}$$

---

6 The activity factor is an estimated average quantity based on what workload is running at the time. Normally, an activity factor is assigned to the block for certain software benchmarks for processors as well.

where α for advanced CMOS-logic-based silicon today ranges between 2 and 3 as shown. Thus, it is easily seen that the power will increase substantially as the frequency also increases since both the active power and frequency increases with every *p-state* change. If the *p-state* changes are linear, that is, for a given change in *p-state* there is a fixed corresponding change to the voltage and frequency, then it is easy to see the increase in the active current and power.

$$(P_{PS} + \Delta P)_{Active} = (A_f C_D)(V + \Delta V)^2 (F_{PS} + \Delta F) \tag{3.57}$$

This means that the incremental power increases as the *square* of the voltage and *linearly* with frequency. One can see the change in active power as a ratio of a linear increase in the increase of each *p-state* change. Assuming a constant activity factor and switched capacitance the ratio becomes

$$\frac{P_{n+1}}{P_n} = \frac{(A_f C_D)(V + \Delta V)^2 (F_{PS} + \Delta F)}{(A_f C_D)(V)^2 (F_{PS})}$$

$$= \left(1 + \frac{2\Delta V}{V} + \left(\frac{\Delta V}{V}\right)^2\right)\left(1 + \frac{\Delta F}{F_{PS}}\right) \tag{3.58}$$

For the graph in Figure 3.25, the increase in power and current as a ratio of the smallest *p-state* overall is quite evident. The increases are clearly exponential for the active power. This justifies engineers and architects doing their best to reduce the power consumed in the highest *p-states* if possible.

As was stated earlier, this also assumes the activity factor and switching capacitance remain constant, which is more often *not* the case as they are highly dependent upon the workload activity. Because of this, the current step changes are *bounded* for each *p-state*. This brings us back to Figure 3.24.

For the first and lowest *p-state*, the current is limited by the operation of the system which means that a dynamic step change will result in a limited

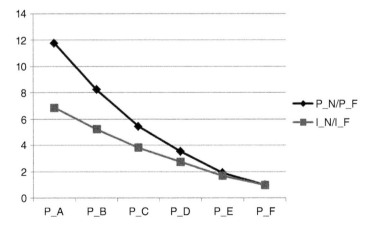

Figure 3.25 Graph of power and currents as a ratio of the lowest *p-state*.

droop voltage for given step change. This may be seen dynamically in the figure. Because of this, the *p-state* load line may have a slope that results in the minimum voltage point aligning to the V_{LL_min} as shown in the figure. When the *p-state* is increased, the voltage is shifted up but so is the slope of the load line. This is because the current step may increase (all else being equal) and the load-line slope may also need to increase. This is a subtle, but important change to the *p-state* load line because designers are required to minimize the losses across the entire operational range while at the same time provide the critical performance for the functional unit during its operational state.

Note that in the figure, the *p-state* dynamic event never exceeds the minimum load-line voltage. This is because if it did, then additional voltage guardband would need to be added to the system to guarantee operation across all the changes in the system. If the reverse were true, that is, if a dynamic event was *less* than the load line minimum voltage, then the slope of the load line would be steeper than required, and there would be additional losses in the system that would not be necessary. It is now clear that a droop event plays an important role in the design of the load line and the power losses overall. This means from a system design point that the load line and PDN *must* be designed as a single unit and an integral part of the power delivery system since their interaction will affect the power consumption of the system, potentially, in a negative manner if they are not considered concurrently.

It should also be noted that the slope of the load line is also intended to compensate for the droop across the load event; that is, by definition, if a droop event occurs at a low or a high current, the PDN will filter it sufficiently that it will never drop below the load line minimum voltage as shown in Figure 3.24. Thus, the minimum point in the voltage load line is not intended to be fixed at one particular load current, but for all of them.

Another important point about the load line is that though it can guard against the droop events in the system, the actual response of the voltage regulator is not designed in most cases to mitigate the dynamic voltage changes. This is the job of the filtering in the PDN system. The major job that the load line does is that in working with the PDN system, it gives the voltage point at any given static load a starting point for the droop event such that the droop never exceeds the minimum voltage set by the minimum voltage (highest current) in the load line. This means that if the load line and PDN are designed together, that the system should work at its optimum point to guarantee system operation while minimizing power delivered to the functional unit.

As a final discussion point, it should be noted that there is a maximum acceptable slope associated with the load line given the minimum acceptable voltage for operation and the maximum current for the highest *p-state*. As was mentioned earlier, the device's voltage is also bounded on the upper end due to reliability constraints. Thus, when a current unload event occurs, the voltage must also not exceed this V_{max} point or else the device could get damaged. It is therefore in the best interest of the systems, power integrity, and functional units' designers to develop the full loop design from the load through the PDN and back into the voltage regulator to ensure that the efficacy of the power delivery is designed to optimize for the entire required operating range for the silicon.

In the Section 3.2.1.1, we turn toward understanding how the load line varies as a function of the different variables in the system. Specifically, we discuss the tolerance bands associated with the load line.

3.2.1.1   Tolerance Band

In the Section 3.2.1, the focus was mainly on the load line as a fixed slope corresponding to a given *p-state*. In reality, the load line has a *tolerance* associated with it called the *tolerance band* or ToB. The ToB is the variation in the load line due to the elements which create it, namely, the circuits which actively adjust the load line. Since the load line is essentially controlled from measurements made in the power converter for the voltage and current in the system, it is expected that the circuits and their tolerances can also vary which will affect the load line.

There are various methods that may be applied by voltage regulator designers, which can measure the voltage and current in a system voltage rail. For this discussion, it is not truly relevant to discuss these methods in detail. However, the variations due to these methods are important relative to the variations that a designer might see in the load-line ToB. In general, we can address two important aspects to the ToB with regard to the load-line variations. The first is the *fixed offset* tolerance which may be applied to the low current starting point for the load line. The second is the variation in the slope of the load line, which is essentially the *drift* of the load line over the load current range.

The fixed offset error in the ToB occurs (mainly) due to an error in the starting offset voltage relative to the low-load static load-line current. Figure 3.26 illustrates this point in the upper right portion of the curve. The error may be due to a number of factors including the temperature in the system. Though this has been stated a number of times, it should always be noted for these discussions that the tolerance band and load line itself are with respect to signals that are considered largely static with regard to the variations that may manifest in the system. Thus, the upper and lower starting points for the voltage assigned to the low-load load-line current has a fixed *error* tolerance band where the voltage is typically bounded across the expected ideal load-line slope. In the figure, it should be noted that the error is shown *imbalanced*. This was done intentionally, since the ToB here is not necessarily symmetric about the expected load-line resistance. The reason is that though circuit errors may occur on the high or low side of the current and voltage measurement methodology, the overall error may be offset due to a number of factors. Often this error is associated with one or more of the op-amps in

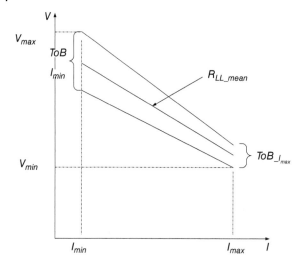

Figure 3.26 Tolerance band in load line.

the system having a gain error that is asymmetric with respect to the input signal [10].

As the two bounded load lines (upper and lower) slope toward the higher currents, it may be seen that neither follow the same slope as the *mean* expected load line. In fact, the ToB error becomes smaller as it approaches the maximum load-line current. This is because, at least for the current measurement, the accuracy in the measurement improves due to the larger relative current signal that is measured and thus the error band decreases. At the end of each load line, the ToB is smallest and usually results in the smallest error.

Because the ToB range varies over the load, it will also vary as a function of which *p-state* is active. Thus, for every fixed load-line set, there will be a tolerance band associated with it as well. To ensure the proper operation of the silicon, it is critical to take into account the variations due to the load line with the ToB and adjust the voltage upward slightly to account for the error. This fixed offset is often backed into the accumulation error for the entire voltage stackup.

It should not be surprising that the error in the load line varies as a function of the system electrical and environmental conditions. The tolerance band for the system comprises a plethora of errors that can accumulate over the range of current. Many of the components that create the tolerance band in the load line are due to the errors in the voltage regulator feedback system. However, not all of the sources in the errors are caused by this. Some of the errors are due to output filter tolerance and the PDN itself. Thus, it is important to understand the origin of these errors as a PDN designer, though many of them are out of the control of the power integrity engineer.

As described in the previous sections with respect to the physical elements that make up the VR output filter and PDN, there are variations in the passive elements in the power delivery path that must be considered when examining the sources of error in the tolerance band of the load line. However, there are also many active components that add to the accumulation in the error stack. Table 3.2 is a list of the main components that accumulate the tolerance band errors in the load line. The ones that are associated with the active components and sources are noted along with the ones associated with the passive elements.

The accumulated error that is due to the manufacturing errors is illustrated in Eq. (3.59) [11].

$$
V_{TB\_mf} = \sqrt{ \begin{array}{l} (V_{VID} \cdot N_V)^2 + (V_{AV})^2 \cdot \left(N_{CE}^2 + \frac{N_{ESR}^2}{n_{ph}} \right) \\ + (V_{AVdyn})^2 \cdot \left(\frac{N_L^2 + N_C^2}{n_{ph}} \right) \end{array} } \tag{3.59}
$$

The equation is a generalized one and is intended to give a statistical view of the accumulated error in the tolerance band for the manufacturing errors overall. The V_{VID} signal is a digital signal that is sent from the load to the voltage regulator. The signal is strictly a digital representation of the analog voltage that the load wished the VR to adjust to at the die. There is typically a feedback sense from this load back into the VR control system to adjust the voltage to the correct point. This signal is called *remote sense* in the field of power systems. This signal is filtered to prevent the amplifiers from responding to high-frequency noise. Thus, the response to the change in the voltage is usually relatively slow. Once the signal is brought internal to the voltage regulator, it is often divided down through sense resistors prior to amplification. The resistors are usually within the silicon and have their own errors. The signal then goes through some voltage amplifier which accumulates additional errors. The digital signal is then converted to an analog signal through a digital to analog converter (DAC). The DAC conversion has an error associated with it that is a percentage of the V_{VID} signal which is the N_V number in the equation. This portion of the error is considered statistical and is associated with the manufacturing ToB since the

Table 3.2 Sources of ToB errors.

ToB component	Active/ passive	Symbol and source	Fixed/ statistical
Voltage identification digital signal	Active	VID – Load	Statistical
Current measure	Active	NCE	Statistical
Voltage measure	Active	Vavp	Statistical
Output capacitor	Passive	NESR	Statistical
Phases in VR	Active	NPH	Statistical
Output inductor	Passive	NL	Statistical

variation is often due to trimming of the reference for the DAC as well as the amplifier and internal circuit resistors in the path that add to the accumulation in the errors.

Figure 3.27 illustrates from a circuit perspective where some of these errors may come from. Note that this is not representative of all designs. Each VR designer may choose to sense the currents and voltages through different methods. For example, the figure shows that current sensing is done through a mirrored upper PMOS transistor rather than through the inductor. Thus, the error accumulation will strictly be different between them (not necessarily worse or better).

It should also be noted that the statistical error accumulation also changes as a function of which mode the voltage regulator is in. For example, at very low loads, as discussed in Chapter 2, the VR may be in what is termed PFM or *pulse frequency modulation* mode. In this mode, the ripple is typically much higher than in pulse-width modulator PWM or CCM (continuous conduction mode). The ripple adds to the overall error in the load line because the signal that is fed back to the voltage sense circuitry is now varying over a wider range which increases the overall error in the ToB. Consequently, the ripple is represented in Eq. (3.60). The total error in the tolerance band is the sum of the errors due to manufacturing in the ToB components, the voltage ripple, and the thermal drift in voltage sense point. Often engineers want to assign the last two components

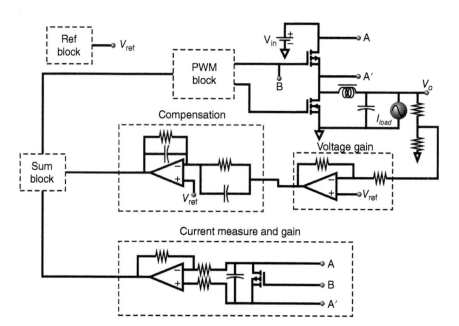

Figure 3.27 Simple circuit representation of blocks in accumulation of ToB errors.

to the voltage guardband rather than the tolerance band. However, one can see this might over-constrain the system design because, as we shall see in the next section, voltage guardband is a *fixed* offset across the load and for a given *p-state*. Moreover, many of the voltage guardband errors are due to system-level variations rather than just the voltage regulator. The load-line errors, on the other hand, vary across the load and are mainly due to issues surrounding the voltage regulator design.

$$ TOB = TOB_{mfg} + V_{ripple} + V_{TT} \tag{3.60} $$

The voltage ripple, as an example, varies as a function of load and modal operation of the voltage regulator. If the current load is very high (statically), then the voltage ripple may be lower than when in a low load operation due to the mode the voltage regulator may be in. In addition, this usually means for a multiphase VR design, the phasor relationship of the currents from each phase will be designed to minimize the ripple, resulting in a lower tolerance band as well. Variations occur when the VR mode switches. When the current is reduced to a lower value from a high current load, the modal change will occur on the slope of the load line [likely] where the current has settled to a lower value. This then would be on one of two outer curves in the ToB as shown in Figure 3.26.

The V_{TT} or thermal drift voltage is the last component in the overall ToB accumulation error. At first, it appears this is strictly a variation during run-time due to thermal variations in the feedback circuits that measure the voltages and currents, specifically, centered around the reference for the main circuits. However, the term is called a *drift* for a reason. Thermal drift may also occur over the life of the product as a result of one or more of the components experiencing physical changes. Thus, there are two components to the thermal drift that need to be considered: drift over life due to thermals and drift due to variations of temperature when operational.

We should make a final note about Eq. (3.60). The manufacturing portion is typically a statistical variation over the life of the product, which is why it is under a square-root operator derived from the *sum of squares* function. The voltage ripple and drift are *fixed* offsets that are due to the operation of the system. Though they are affected by the long-term reliability of the VR and the system, there is really no statistical method to guarantee where the ripple and drift will be at run-time which is why they must accumulate directly as a fixed voltage error with respect to the overall ToB.

3.2.1.2 Voltage Guardband

Up to this point, the discussion has been centered around the load line and its tolerance. Most of the errors that accumulate with regard to the load line are driven by the voltage regulator design, its internal components, and the filter. If these were the only variations in the system, we would stop there and know what the actual voltage set point would need to be for a given voltage rail with the tolerance band of the system. However, there are fixed and statistical offset elements in the voltage that are independent of the load line that contribute to the final value set by the load which is intended to guarantee operation over the range of the *p-state* at any instant in run-time. This additional contribution is typically called the *voltage guardband.*

This guardband is exactly what its name implies. It is an offset that is applied to the preset *p-state* to guarantee that the voltage is at least at the point where the functional unit will operate correctly, under all the changes in the system and over the life of the product. If there were no errors in the system at any instant in time and the voltage set point was measured and set perfectly to the exact correct voltage, then there would be no need for a voltage guardband. However, there are components which exist in an electronic system that will vary over time while in normal operation. These components can cause the actual voltage (relative to a hypothetical exact fixed value) to vary relative to where the load wishes the voltage to be set at. Equation (3.61) shows a representation of how voltage guardband may be assessed.

$$V_{GB} = \sqrt{\sum_i (V(i)_{PR})^2 + (V(i)_{VAT})^2 + (V(i)_{LO})^2 + (V(i)_{TS})^2} \qquad (3.61)$$

Equation (3.61) implies that there are errors across the process (PR), VID offset/aging and temperature (VAT), localization offsets (LO), and tester (TS). Moreover, there is a summation under the root, which means there may be more than one of each item. This is important since there are variances to the guardband within a region (such as process) that may drift at a different rate and magnitude.

The equation is shown as a statistical distribution; though depending upon how the system is studied, a power analyst may choose to represent one or more of these as simply part of the fixed voltage offset. These differences will be discussed shortly.

First, let us examine each term in the equation independently. There are two components to the process variability. First, for a given piece of silicon and for the functional units within that device, the frequency of operation may vary simply due to manufacturing process distribution at the time the device is manufactured and tested. When a device is placed within a *category* or a particular *group* in which they are produced and tested, the device will fall within certain operation boundaries that allow the manufacturer to ensure it will operate correctly within certain bounds based upon how it performed under the given test conditions [11]. The process of categorizing the device is called *binning*. Essentially, as devices are produced and tested, they fall within a preset range of operation. Usually this is associated with a maximum operating frequency and voltage. If the device is considered *fast*, it goes into a higher bin; if it is *slow*, it goes into a lower bin. Devices such as microprocessors are often categorized this way. Depending upon which bin the device was tested to and placed in, there may be a voltage offset from the nominal voltage associated with that binned part. For a given frequency of operation the guardband in the voltage is determined around the binning process based on how much from the ideal (tester reference) that voltage needs to be set to guarantee the correct operation. This is the first part of the process variability.

The second part has to do with aging. As a silicon device ages, the effective speed of the device may slow down. This can occur due to a number of factors [12]. To compensate and maintain timing, the system will be required to increase the voltage relative to the initial process voltage setting. However, rather than adjust to the changes over time, which are very difficult to predict, in most cases, test engineers specify a fixed offset voltage for the device over the life of the product (a fixed increase at the load). Because the variation due to aging is based on a statistical variability, the voltage guardband error falls within this statistical range for the device's test conditions and process control. A typical voltage stackup from the minimum preset *ideal* voltage is shown in Figure 3.28 along with other guardband components. The stackup is shown in a linear fashion; that is, the components add up directly which is often the case. However, depending upon how the power analyst chooses to do this, some or all of the components may be dependent upon each other. Also, since the stackup is based upon a statistical analysis, or in this case root-sum-of-squares (RSS), depending upon how many standard deviations from the nominal the analyst is after, the guardband can vary quite a bit.

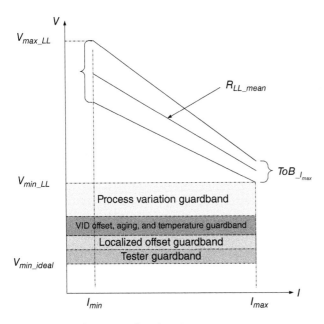

Figure 3.28 Voltage guardband stackup.

The next set of statistical variances are related to the voltage regulator. One might ask why this error was not assigned to the load-line *ToB* rather than affixed to something we call the *guardband*. As in the silicon load, there are components in the VR power and control silicon that vary over the life of the product. Often these errors are grouped together as in this case since they are a culmination of changes to the VR itself. These errors have a direct effect on the offset voltage in the system. The errors are due to aging, temperature, and VID offset in the circuits related to the set point of the voltage. The aging and temperature errors are readily understandable to some extent. This is because much of this error here is due to the *reference* in the controller for the VR. Because the reference is often used across the entire device, any error in this component will affect most if not all of the analog voltage regulator circuits. With respect to the voltage guardband, this will affect the *fixed* static voltage for the device. This is why temperature and aging are essentially grouped together since the main source of the error is the reference, which is affected by both. The reason why it is not included in the load line is because it affects the stackup in a fixed fashion, and though it will have an effect on the load line ToB, the change in the reference will tend to *shift* the load line one direction or the other but should not contribute to the difference in the ToB during runtime.[7] The VID offset, however, is another matter.

The VID offset error should not be confused with the actual offset voltage that the VID signal may incorporate into the digital bit pattern that is sent to the VR to be translated to an analog signal for the feedback of the VR. Depending upon the load and the resistance in the system (PDN static path), the controller in the silicon may adjust a fixed voltage offset to the signal that is sent to the VR. This is NOT the same as the statistical error as shown in Figure 3.28. The actual digital VID reference is the value that the load actually desires. There is an analog reference signal that the VID signal is compared against. This analog reference is the signal that was discussed in Chapter 2 with respect to comparing the feedback signal to the voltage reference. The VID signal, as explained earlier, is a digital representation of the voltage that the load wants the VR to adjust to. There are a number of components in the conversion process that can alter the analog signal relative to the digital signal. But the signal that is sent to the VR is determined by the load, for a load-line-based feedback system, and normally comes from a lookup table that was preprogrammed at the manufacturing test. Since the load does not adjust the voltage itself, it is free to tell the VR to move the voltage to its ideal fixed point. However, while the VR converts the signal, there is a *digital* error relative to an ideal value. This is culmination of discretization of the analog signal, which results in an aliasing error ($\pm$ LSB

---

7 Strictly speaking this is not always the case. The variation in the ToB may come about due to a drift in the reference relative to the other components, which use this reference. However, this variation is considered quite small in comparison to the other ToB components.

or *least significant bit*). Moreover, the DAC can convert the signal with a small error, resulting in a slight offset relative to the ideal. This total conversion error is often termed the *VID* offset voltage because the result is a slight offset from the ideal digital representation that was sent by the silicon load.

The localization offset (LO) is due to dynamic voltage deviations relative to the behavior of the PDN and the on-die decoupling local to the functional unit where the load changes. This error is a function of the variations that can happen locally where the instantaneous voltage may be different in locale depending upon where one measures it on the die. Usually, there are multiple high-performance components within a functional unit that switch at different times (and have different current capabilities) and can induce a slight change to the local voltage. Thus, power analysts affix an offset voltage to this local error, which then becomes part of the accumulation of the error as shown in Figure 3.28.

The final error, and often the one that gets the most attention, is the tester guardband. When the device is placed in a tester, a golden reference voltage is applied to it where the device undergoes a series of tests to determine what frequency it will operate at for a given *p-state* or set of *p-states*. The tester voltage, however, has a small error due to the test setup and the power supply that creates the voltage source. This tester guardband is part of the tester error in the test setup. Because the device is placed in an environment (test socket, etc.), the power path will be somewhat different than the actual use case which is the final product. Though the voltage sense point is usually at the same pins or bumps of the device, there can be differences between the tester and the product, which need to be accumulated in the tester guardband. This is the source of the error that is shown in the figure for the tester.

Sometimes to show how load line and voltage guardband can affect the system, it is appropriate to give an example. The following example illustrates some of the impacts that may be caused by changing the stackup or varying the load line.

Example 3.2   A power analyst is examining the effects on the power consumption in a system through changing the load-line slope in the highest *p-state* while increasing the guardband slightly. He feels that if he can trade off the guardband for the load line it will yield a net benefit in power for a particular steady-state workload on a server-based platform. Based on the (repetitive) current waveform for the load (Figure 3.29) and given the load line and voltage guardband data (initial and subsequent changes), determine the energy consumed and the differences between the two approaches. The maximum current is shown in the simple plot along with the drop to the minimum current. Assume the waveform repeats many times over 60 seconds. Use the fact that the active power ramps as the square of the voltage and that the leakage power ramps as the cube of the voltage. In this case, also use

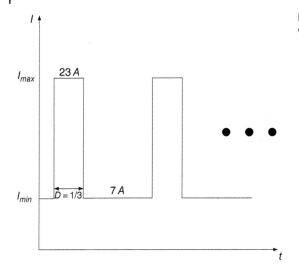

Figure 3.29 Current waveform for load.

the fact that the leakage power is approximately 22% of the overall power at I_{max} for the initial settings. Assume that our engineer has consulted a power integrity expert and has determined that the droop will be mitigated by the load line (Table 3.3).

Solution:
The current workload operates between two distinct points. First, at 23 A and then at 7 A. It is very periodic, so we shall assume that for this duration (60 seconds), we may determine the average power and thus the energy consumed in each state based on the load line. We first compute the energy consumed for the initial state. Since we have been given the voltage in the highest *p-state* as 1.2 V at the load (meaning this is the lowest voltage at I_{max}), we add the guardband and then compute the relative power in this state. We may determine the differences in both active and leakage power at the two ends of the spectrum given that we know one of the data points.

The power consumed at the high end of the spectrum initially was,

$$P_{init} = V_I \cdot I_I = (1.2 + .055) \cdot (23) = 28.86 \text{ W} \tag{3.62}$$

Table 3.3 Load line and voltage guardband comparisons.

Metric	Value
RLL_inital	5 mΩ
RLL_change	4.5 mΩ
VGB_inital	55 mV
VGB_change	65 mV

This is the total power that is consumed at I_{max}. We now need to compute the leakage power. This is a function of the cube of the voltage for this particular problem. Since we know the percentage, we may easily compute the contribution,

$$P_{lk\_init} = 0.22 \cdot P_{init} = 6.35 \text{ W} \tag{3.63}$$

Thus, the difference between these two values is the active power consumption at I_{max} and with the assigned voltage guardband at that particular point.

We know from Eq. (3.56) that the leakage power is a function of the cube of the voltage ($\alpha = 3$),

$$P_L = kV^3 \tag{3.64}$$

We can easily compute the constant k for this process so that we can compute the leakage for the design point where we choose a new guardband and load-line slope. Thus,

$$k = \frac{V^3}{P_L} \cong 3.12 \tag{3.65}$$

Before we compute the new values, let us compute the minimum power at the low end of the current for the system. We already know that the current (total) is 7 A in this range. This gives us the *change* in current to compute the *change* in voltage.

$$\Delta V = \Delta I \cdot R_{LL} = (23 - 7)(5 \text{ m}\Omega) = 80 \text{ mV} \tag{3.66}$$

Thus, the voltage at I_{min} is now the I_{max} voltage plus guardband, plus the change in voltage in Eq. (3.66) or $1.2 + 0.055 + 0.08 = 1.335$ V. The total power is simply the total current times this voltage or 9.345 W. The leakage power is based on Eq. (3.64) and is easily computed to be $\sim$7.64 W. The active power is the difference between these two ($\sim$1.7 W). Note that the active power is quite low within the low current portion of the load line, indicating that there is very little activity in this particular functional unit during this workload window. For a real processing unit (relatively speaking), this is quite plausible.

Now the question is, what is the power consumed with the new settings for voltage guardband and load line? For the I_{max} power with the new voltage guardband, we may compute the total power by increasing the voltage per the new specifications. There are two components to this. First, there is an active power increase, which varies as the square of the voltage. Second, there is a leakage power increase, which varies as the cube of the voltage. Using the previous data, we increase both by the ratios of the voltage

$$P_{A\_new} = P_{A\_init} \cdot \left(\frac{V_{init}}{V_{new}}\right)^2 \cong 22.87 \text{ W} \tag{3.67}$$

$$P_{lk\_new} = P_{lk\_init} \cdot \left(\frac{V_{init}}{V_{new}}\right)^3 \cong 6.50 \text{ W} \tag{3.68}$$

As expected, the total power during this period of activity increased (from 28.86 to 29.38 W). The next question is what is the change in current and voltage with the new settings for the load line and voltage guardband? Since the current is dependent upon the voltage, we cannot use one to compute the other to achieve where the new settings would be. However, we can determine where the two lines intersect; that is, we can determine where the power would be equivalent. Since the question is which one will result in the higher power, all we need to know is whether or not the two lines intersect before the 7 A point or after (e.g. lower than 7 A). This is because the shallower load line is starting at a higher point. Using the very simple slope-intercept method, we may easily find the equations for the two lines, where $y = V$ and $x = I$,

$$y_{init} = (-.005)x_{init} + 1.37, y_{new} = (-.0045)x_{new} + 1.369 \qquad (3.69)$$

The new intercept is at $x = -0.984$ A. Thus, we now know that the power consumed by the new settings will be higher than the old settings. To find out where the voltage ended up, let us assume that the minimum load-line point for the new settings was at $I = 7$ A, the same as the initial settings. Then the new voltage for the load line would be

$$y_{new} = v = (-.0045)(7) + 1.369 = 1.338 \text{ V} \qquad (3.70)$$

This is voltage at the high end of the load line. Though it is only 3 mV higher than the initial settings, it will result in a *higher* power consumption than where the engineer started from.

The energy difference between the two, using 7 A as the I_{min} limit, will be

$$E_\Delta = E_{init} - E_{new} = 951.1 - 962.2 = 11.1 \text{ J} \qquad (3.71)$$

Often these checks are helpful to determine if changing guardband or the load line helps or hurts the system.

In Section 3.3, we investigate noise from different sources in power integrity and how it can affect voltage delivered to the load, and therefore the load line and voltage guardband.

3.3   System Noise Considerations in Power Integrity

In the Section 3.2.1.2, concepts around the internals of the PDN and how they affected the system were examined. Additionally, the load line and voltage guardband were introduced as part of the interaction between the load and the operation of the voltage regulator, all the while taking into account the effects of the system droop and other elements. In this section, we analyze how noise affects the power delivery with respect to the previous discussions.

Noise that affects the power delivery path can encroach upon the load in different ways. The source of the dominant noise elements typically come from only a few electrical devices in the platform. This is because designers are required to decouple most noise sources from interacting with other electronics in the system to prevent a potential malfunction. Many circuits are highly susceptible to noise (such as clocks, wireless receivers, etc.) which makes good noise mitigation techniques and proper decoupling essential in the design of the product. Thus, it is imperative that system designers and power integrity engineers have a strong grasp of the causes and interactions of such noise-producing elements to prevent the disruption of normal operation.

When it comes to noise coupling into the power rail, because of the filtering in the PDN and the main power supply filter, most noise generators are readily mitigated. However, the two main sources of noise, which most PI engineers contend with, come from the load itself, while others come from the power regulator. Both the source and the load contain high switching currents that can increase noise in the power delivery path. As was seen in the previous section, a good portion of the load-line slope and voltage guardband is designed to address this noise from the load (the local offset error being one of them). Thus, if this noise can be mitigated to some extent, the power consumed by the loads in the system may also be reduced, resulting in a more efficient system.

However, one might ask, what about noise from other sources outside of the power supply and the load, how are those dealt with in the platform, and how do they affect the overall voltage budget? The regulator generates noise simply because it is switching large transistors (bridges) on and off in a periodic fashion (Chapter 2), generating large currents to deliver power to the load. These currents commutate into the inductor, and thus ripple is seen at the output of the main filter capacitor. However, as shown in Figure 3.30, the currents also make their way back to the source by circulating through an impedance in the ground path. If the impedances between other systems are in the right amplitude and frequency band, this noise can couple beyond the VR and its filter.

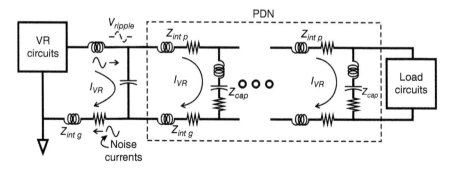

Figure 3.30 Noise current generation in power path.

There are two main components to the noise emanating from the VR that can affect the power delivery path: *high-frequency* edge-rate dependent noise, and *low-frequency* switching noise. Both types of noise are normally considered differential. *Differential mode* noise means that the currents generated propagate outward from the source and are differential in nature, meaning that they are equal and opposite on the transmission path as shown in Figure 3.31. *Common-mode* noise means that the currents which propagate down the power delivery path are not necessarily equal nor opposite and may propagate with the same phase or a different one down power and ground.

The differential currents which are generated from the switching nature of the buck converter are typically slow and manifest themselves as periodic ripple on both the voltage and return path. However, if the impedances along the voltage and ground paths are different from each other (which they often are), the currents may become out of phase and the signal may be attenuated higher down one of the paths relative to the other. In this case, if one were to measure the currents on the power and ground emanating from the periodic switching from the voltage regulator, they would appear as currents that may have started as differential in nature but ended up looking very much like common-mode to the casual observer when they arrive at the end of the transmission path.

As we examine this noise further, it is important to assess the source in terms of how it may affect the overall system components that were discussed in the previous sections. Specifically, how does the noise affect the guardband and tolerance band for the load line in the system? This may be answered when one considers how the noise propagates through the PDN as shown in Figure 3.30. If we assume a static current load for the moment, it is easy to see the effect the noise may have at the load independent of how the noise was generated. By adding the appropriate decoupling, the PDN can mitigate this noise to a very large extent, especially since the frequency content is relatively low (at or near the switching frequency of the VR). Thus, the *filtering* in the power delivery path typically reduces this noise to an acceptable level by attenuating

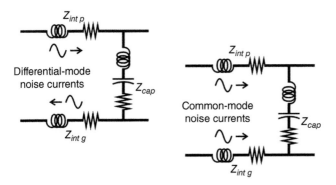

Figure 3.31 Differential and common-mode currents.

it through the PDN. This residual noise, what is left over, strictly as a function of the VR switching, typically results in a very small source of error at the load with respect to the ToB and is added to the voltage ripple portion as we saw earlier. Note that this is at the *load*, not at the source, which is where the accumulation of the error is intended to have its largest effect on the system margins.

The second component of noise from the regulator that is of interest here is the high-frequency noise. Though it was stated that this is due to the HF switching of the edge rate of the VR, it can also manifest itself from coupling from other VRs or other HF noise sources in the system through the ground system or through other electromagnetic coupling mechanisms. Fortunately, if the VR design was done properly, this noise should be mainly shunted local to the voltage regulator and not emanate out into the power delivery path. This is the reason why many VR and system designers add HF capacitance along with the bulk capacitance to the VR LC filter as shown in Figure 3.32.

However, sometimes the noise can propagate into the path, in which case the PDN must also address this noise. Once again, because of the frequency content, the noise may manifest itself as common mode rather than differential mode. However, the question is what category does this noise fall into? Is this in the voltage guardband margins or a form of ToB that we have seen earlier? In this case, the noise is usually bucketed into either a new voltage guardband area for generic HF power supply noise or lumped into the existing structure as shown in Figure 3.28. This noise is statistical in nature since it is usually considered part of the random system noise that can affect voltage guardband.

The other main source of noise is typically the load itself. With respect to the voltage guardband, most of this noise is taken into account. The main reason is because the near load PDN decoupling is intended to attenuate this noise and reduce it to a level where the load-line slope and guardband have incorporated them into the overall budget.

Normally the low-frequency ripple and droop effects are larger than the high-frequency noise that is generated near the load. However, there are instances where sometimes the large currents inside the functional blocks of

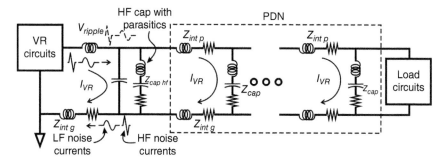

Figure 3.32 HF noise and cap added to VR filter.

a device can generate high voltages that exceed the low-frequency droop of the system. Basically, this means that the on-die voltage excursions for short durations have exceeded those in the lower frequency regime (see Chapters 9 and 10). When this occurs, usually a resonance has been excited within the local die PDN in a frequency band where the impedance is higher than desired. Because of this, the noise may start to encroach on the limits of operation of the system. The implications of course is that this noise can cause data corruption, performance degradation, or even full system disruption. But, this is not always the case. The reason is because not all voltage droop is capable of distorting the timing relationship between the clock and data [13]. Because localized die phenomena are shorter than PDN droop events and because the short duration voltage excursions normally affect both the clock and data circuits simultaneously, the logic is usually less susceptible to the short droop event than it is to the larger one.

The conservative approach to guardbanding against such noise in the power delivery path is to consider all the noise – voltage based on the power rail – as equal with respect to how it affects the set point; that is, if a voltage droop event occurs, no matter what the frequency dependency is, that noise has the potential (unless guardbanded against) to cause the timing to be affected. However, there is a *locality* and *temporal* aspect with respect to voltage relationship and the propagation of the signals throughout a given system. As the clock and data signals propagate across a functional block, the timing through a given set of logic circuits must be maintained. If the voltage sags as the signals propagate along their path, both the clock and data pulses may stretch. However, this is highly dependent upon the duration of the droop. If the event occurs for 1–3 ns, the effect may be minimal to none [13]. This is because the effect is very *localized* and the number of gates that are affected, relative to both the clock and data paths and time it takes to propagate through the logic, are minimized since it is only possible that the delays are affected only by a few gates if any, and normally, both the clock and data are affected almost equally. Conversely, if the voltage droop is sustained for a relatively long time, say greater than 10 ns up to hundreds of nanoseconds, depending upon how low the voltage droop lasts, the effects can be catastrophic. This is because the number of gates the signals will propagate through can cause both local and global delays, which can affect data integrity and performance. This is discussed in more detail in later chapters, in particular Chapter 10.

The Section 3.4 discusses the effects that noise from PDN can have on emissions in the system.

3.4 EMI and Power Integrity

EMI, is considered mainly a platform or system issue. Normally, power integrity engineers are not involved in the testing or mitigation of such noise in the system, mainly because the sources of such noise are the active components in the system (e.g. logic and power devices). However, this does not mean that PI engineers should ignore emissions, particularly when designing their PDN networks or when modeling the power delivery path. The reason is because the choice of the capacitors in the network can have a large effect on the noise generated in the platform. Though capacitors are chosen for decoupling purposes, they can also resonate at frequencies which can compromise emissions on the platform. This is particularly true for low-frequency capacitors which have been targeted for energy storage rather than for noise mitigation. This is fundamentally why systems engineers often add additional decoupling to specific rails after the power integrity engineer has done their analyses because often they are trying to attenuate additional noise in the system which can exacerbate either conducted or radiated emissions.

EMC standards should be understood by most engineers, even if their jobs are not directly related to certification. Computer platforms sold today are required to meet certain electromagnetic emissions requirements (e.g. FCC Class A or B), and thus the generation of noise at the chassis, board, and silicon is something that affects all designers. For the power integrity engineer, it is important to understand the design, layout, and interaction of the PDN and ensure that the power delivery maintains, or rather, does not disrupts the certification process for the platform.

Many small computer systems, such as tablets, are normally required to meet FCC Class B regulations for domestic environments. For larger systems such as servers, they must meet Class A requirements. Some products intended for nonresidential/domestic environments may also be used for residential – but, because these products may emit more emissions, the user may need to take care that the interference is low enough that it does not interrupt other devices. The requirements are given in *decibels* of noise as measured in a chamber or an *OATS* (Open Area Test Site). The unit of measure is either the *electric* or *magnetic* field. Thus, for the electric field, the relative emitted noise measured will be a ratio. This ratio is relative to a given noise floor; for the FCC standards, this value is $1\,\mu V$. Thus, if one $0.1\,mV$ of noise were measured from a radiating source, the amount of noise in decibels would be

$$\text{decibels of noise} = 20\,\log\left(\frac{1e-4\,V/m}{1e-6\,V/m}\right) = 40\,dB \qquad (3.72)$$

EMC engineers usually make their measurements in terms of decibels since most of the regulations are defined in these terms.

There are two types of emissions that one is concerned with when discussing EMC at the system level: *radiated* and *conducted*. Paul [30] discusses both types in detail, and the authors recommend that the reader review this reference for an excellent overview. Conducted emissions are simpler than radiated in that the measurement is located at the power cord and is usually done by placing a current probe around the cord to pick up the conducted noise at

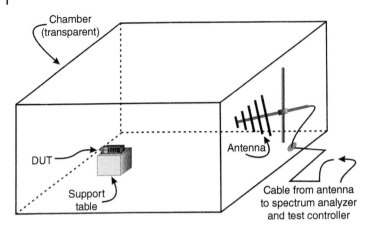

Figure 3.33 Illustration of conducted noise path in a chassis.

this point. Figure 3.33 shows how noise may move from the circuit board and chassis, out through the power cord. The noise may start from any active device where either *common-mode* or *differential mode* currents may be generated. Common-mode currents are currents that travel along a path in the same direction away from (typically) the source. This path may consist of a power/ground plane pair or simply two parallel conductors. Differential mode currents are currents that travel in equal and opposite directions along conductors. The main difference between them with respect to EMI is that common-mode currents generate more emissions and are often harder to limit in a system.

Both currents are generated from some type of *source* – the source may be a device on a printed circuit board that switches periodically or aperiodically. Most offenders for the source of noise are high-frequency clocks, high speed IOs, high-performance computing devices, power converters, and others. If the ground loops are not well contained for the device, switching noise may find its way across the printed circuit board and into the power supply at the back. In that case, if proper filtering was not employed, the noise may find its way onto the power cord and thus emit into space. For conducted emissions, however, the frequency limits are fairly low (under 30 MHz), and thus after this frequency the more critical emissions will be of the radiated type. In a number of cases, it may be required to add a filter (such as a common-mode choke) to reduce these emissions to an acceptable level. The filter absorbs the conducted energy and the frequencies of interest, which reduces the electromagnetic noise to an acceptable level.

Radiated emissions are more complex since there are many different ways in which noise may radiate out of the chassis and off the PCB. Moreover, the bandwidth for the regulations is higher than the conducted emission regulations. Due to the possibility of interfering with RF communications, today in many systems, EMI must be kept to a minimum in order to ensure the integrity of the radio waves (and therefore, data) being transmitted through space. The measurements for such noise is typically done in a chamber where the DUT (device under test) or source is located in one part of the chamber and the antenna, or receptor, is located a given distance away (depending upon the chamber this could be 1–3 m). The chamber is shielded from outside noise and is calibrated to a given noise floor as previously defined. The measurement equipment is typically a type of spectrum analyzer (SA), which measures the noise that is emitted through a given limited bandwidth window. Figure 3.34 shows a simple illustration of a typical measurement that may be made in a chamber to determine the emissions from a particular chassis/DUT. Depending upon whether the noise is common mode or differential mode, the noise limits within a given bandwidth will be different.

Regulations for FCC Class A and B at 50 MHz are given in Table 3.4 [38].

The fields radiating from a DUT are complex and the electric and magnetic fields that radiate are usually *spherical* in nature. Though the EMC engineer will measure the results directly from the closed chassis, measurements may also be made off the PCB itself in an open environment. Often, the EMC engineer can probe parts of the board to determine where the noise is coming from and

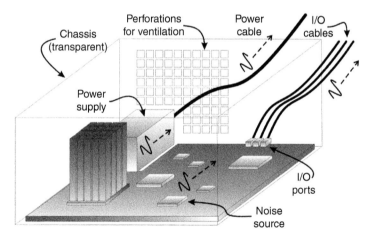

Figure 3.34 Example of radiated emissions measurement setup.

Table 3.4 FCC class limits class A and B at 50 MHz.

Regulation	Limit (µV/m)	Distance (m)
FCC Class A	90	10
FCC Class B	100	3

what the cause is. The radiated noise in the far field rolls of as $1/r$, where r is the radius of the field to the measurement point from the point of origin – in this case the DUT or board. Near-field amplitudes typically roll off at $1/r^2$ and $1/r^3$ for magnetic and electric fields respectively. The distinction between near and far field for a *dipole*, for example, comes when the wavelength of the field is 1/6 that of the radius, or

$$r = \frac{\lambda}{2\pi} \cong \frac{1}{6}\lambda \tag{3.73}$$

Additionally, a crude way of estimating the noise off a board is to represent the source as an *electric dipole*. Referring to Figure 3.35, the dipole resides on the z axis as shown.

The dipole may be used to represent a source carrying current on a PCB, such as a power plane with current. Since the interest here is in relatively high frequencies with respect to radiated emissions, the discussion will be limited to 200 MHz and above. Below this frequency, EMI engineers usually rely on good chassis design to shield the noise due to the larger wavelengths.

The electric and magnetic fields for an electric dipole, in the far field, are shown here for reference:

$$\vec{E}_{FF} = j\eta_0\beta_0 \frac{\hat{I}l}{4\pi} \sin\theta \, \frac{e^{j\beta_0 r}}{r} \vec{u}_\theta \tag{3.74}$$

$$\vec{H}_{FF} = j\beta_0 \frac{\hat{I}l}{4\pi} \sin\theta \, \frac{e^{j\beta_0 r}}{r} \vec{u}_y \tag{3.75}$$

The *magnetic* dipole, or *current loop* equations, may be found in [30]. Here, the terms in Eqs. (3.74) and (3.75) are as follows: η_0 is the intrinsic impedance of free space, the current vector I is the current at the source of the dipole in the z direction, and l is the length of the dipole. Note the vector directions of the fields in the far field – where the electric and magnetic fields are orthogonal to each other (as expected). The fields are essentially *plane waves* propagating in air at this point. To use these equations to approximate the noise from a PCB, the

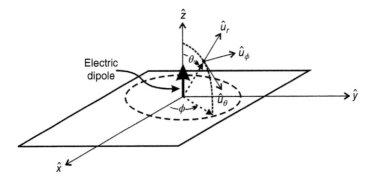

Figure 3.35 Electric dipole in spherical coordinates.

results may be simplified in Eq. (3.75). For example, if one has a power/ground plane pair on the surface of a board, pointing in the z direction, where the currents are strictly differential mode, the magnitude of the electric field will be

$$|\vec{E}_{FF}| \cong 1.316 \times 10^{-14} \frac{|\hat{I}||f^2 ls}{d} \tag{3.76}$$

where s is the separation of the planes (center to center) and d is the distance from this center to the measuring point (e.g. the antenna). This is the result as given by Paul [30]. As another example, if one had a PDN strip of 3 cm on the surface of a PCB with a differential mode current of 15 mA at 200 MHz, separated by a distance of 3 mm, in a 3 m chamber, one would expect the electric field strength to be,

$$|\vec{E}_{FF}| \cong 1.316 \times 10^{-14} \frac{(15e-3)(200e6)^2(3e-2)(3e-3)}{3m} \cong 240 \,\mu V \tag{3.77}$$

or ~47.5 dB. It should be noted that some assumptions must be made to ensure that this is a *reasonable* result – that the width of the planes must be small relative to the other dimensions (s and d). Also, that the currents at the frequency of interest are uniform – which, in most cases, is usually a reasonable assumption. In some cases, as discussed previously, the EMC engineer may make near-field measurements for the PI engineer to determine where the noise comes from. These measurements, of course, assume that the board design is finished somewhat and that the *final* design is now *after the fact*. This is sometimes at the request of the PI engineer to get an understanding of whether or not the PDN layout and design create excessive emissions. Though the PI engineer's PDN may not be the source of the noise that is generated in a system, it is crucial that the decisions made with respect to the power distribution layout and overall system power delivery do not limit the EMC engineer from making sure the system is certified. This means that not only must the PI engineer implement a satisfactory PDN for the system but he must also not create additional noise, or allow this noise to propagate unnecessarily across the planes and traces of the PCB.

There are a number of mechanisms in the design of the planes and the PDN, which may lead to the generation of possible emission problems. To discuss the details on each one would be beyond the scope of this book. However, a brief understanding and description of some of the more common mechanisms is warranted. Many of these problems may appear to be simply a matter of *common sense* for the designers involved but when layout and system constraints come to bear, often basic noise mitigation strategies can take a back seat to more seemingly pressing matters. Thus, it is a good idea to have a level of understanding of what can happen if the engineer is not vigilant. Some of the causes of noise-generating sources are listed in Table 3.5.

Table 3.5 Typical PDN-related issues causing potential EMI noise.

Noise-generating causes
Signal vias through a power bus
Power plane discontinuities
Ground or power plane inductive routes
Poor placement of decoupling capacitors or poor choice
Poor routing of power/ground connections through connector or socket with silicon
Poor choice of local and/or global system ground point relative to PDN ground

Signal vias through a power bus – It is often required to route signals through a quiet power bus due to routing constraints. However, there is an opportunity to couple noise into the plane – particularly if the signals are noisy, such as high-speed IO or clocks [39]. This is particularly true if the same signal serpentines through the plane from one layer to the next stitching its way from one side to the other.

Power plane discontinuities: This is similar to the signal via issues above but is more general. This is mainly around the power plane distribution having discontinuities in it – such as abrupt necking down in regions, capacitor bridging across planes, and large regions where the areas are void of capacitors along the PDN. The main problem here is that discontinuities can generate common-mode currents, which can generate significant EMI in a system.

Ground and/or power plane inductive routes: Another problem is where one needs to route the power plane through multiple regions, like a signal trace. This results in potentially large inductances that can resonate with signal noise if noise couples into them. As usual, it is best to try to minimize this layout problem to help mitigate system-level EMI.

Poor placement of decoupling capacitors or poor choice: Depending upon the frequency of noise in the system, poor placement of capacitors (e.g. large inductive connections to the plane or placement relatively far away from the power/ground planes) can accentuate noise issues as well. This is more common than one may realize – since often the layout engineer and PI engineer may not be fully engaged at the proper time (or sometimes not at all). Moreover, it is easy to put down the wrong type of capacitors for the portion of the distribution network when the schematic is given – unless specific instructions are also given to the layout engineer to do so.

Poor routing of power/ground connections through connector or socket with silicon: This problem occurs mostly when the package designer either does not have enough power/ground connections in the package available or the placement of those connections is not optimized due to other constraints.

The result is usually discontinuities in the routing, resulting in noise coupling into the planes and thus generation of platform-level EMI noise.

Poor choice of local and/or global system ground point relative to PDN ground: Usually there are two schools of thought with respect to grounding a PCB to a chassis: local grounding at a single point and global grounding throughout. For the PI engineer, it is critical that the PDN grounds are local to the silicon and power converter to keep ground loops contained close to their sources. When the board is grounded – whether it is local or global – the PI engineer should make sure that their PDN ground reference has a low impedance *local* loop and does not have a significantly large impedance associated with it – at any particular point – back to the main chassis to board ground to help mitigate the cause of significant EMI noise.

When considering EMC, for the power integrity engineer, it is always good to maintain proper practices at the beginning of the design phase to ensure that noise is mitigated at the beginning, rather than at the end. Thus, establishing a rapport with the EMC engineer during the development phase can help mitigate possible EMI issues later on.

Section 3.5 briefly discusses noise mitigation with respect to the PDN for power integrity.

3.5   Brief Discussion on Noise Mitigation for Power Integrity

This section is more of a qualitative discussion on noise mitigation than quantitative since we discuss this subject in more detail later on in the text. This is mainly with respect to understanding the effects of adding capacitors to the PDN to aid in noise mitigation in the platform.

In general, adding smaller high-frequency decoupling to the PDN does not affect the effectiveness of the PDN which is normally one of the concerns PI engineers have when they need to reanalyze their networks after decoupling has been added. However, there are a few legitimate concerns one should have when adding capacitance.

The first concern is replacing existing capacitors with other capacitors. Normally, this has the largest impact. In general, it is always good to rerun simulations after the PDN has been changed, even if the changes are minor. But in the case of replacing an existing capacitor (or more than one) on a platform, it is imperative to redo, at least in part, the modeling for this.

The second issue is moving capacitive structures around to accommodate noise mitigation for the platform. The main problem here is with dropping down interconnects for the capacitors into existing planes or strips. Though subtle, if a capacitor is moved to a side of a plane where the connection has relatively higher impedance than it was previously, this could cause issues with resonances with other structures; in particular, the impact to localized loads can have a significant impact here. Thus, it is good practice to work very closely with the systems engineer to make sure the change to the location of the device has not significantly affected its impedance relative to the previous layout.

Finally, the choice of capacitor is always an important consideration. Some devices not only resonate electrically but can mechanically oscillate due to both placement and the type of electrical excitation that can cause the device to move. Normally MLCCs of good quality and small size do not have these issues, but it is important to ensure that if a new device is chosen that its dielectric quality over the life of the device and due to temperature is not changed significantly from that of the original device. This is likely one of the most common problems encountered on high volume products. Often the reason has to do with a vendor change and/or cost reduction. Moreover, this can happen very late in the development process when the power integrity engineer has moved on to other projects. Unfortunately, if an issue arises, it probably occurs after the change has been made and the problem has been uncovered either in production testing or (even worse) due to a field failure. There is little one can do to fix this. One thing that can help is to ensure the PI engineer carefully documents the limits to which the PDN decoupling can operate. This includes not only the values but also their tolerances over the life of operation. Normally, this can be done in their final report with a copy sent to the systems engineer.

3.6 Summary

In this chapter, we covered many of the system elements which make up the components in the PDN. The physical elements are simply resistive, capacitive, and inductive in nature when it comes to the interconnection in the system. We analyzed some of these components, in particular the resistance between two points in a plane where we found that the current density is not uniform in many cases. Understanding what the resistance is can help PI engineers decide whether or not the placement between devices will have a negative effect on their PDN design when they develop their SPICE models.

We also examined how to analyze the inductance by using partial inductance techniques. These are analytical methods which yield simple results that may be used to check against extracted models later on.

Another important learning in this chapter was the operation and breakdown of MLCC capacitors. Here we discussed in detail on how capacitors were constructed and operated. We learned that capacitors have polarization to them and vary dramatically over temperature and time, which is why it is critical when analyzing a device in a system that the engineer understands both SoL and EoL operational limits to the device in question.

Another important aspect to how the PDN is linked to the system was how it related to the load with respect to system guardband and load line. The load line is the slope of the system voltage relative to the static current delivered by the power supply. In many advanced platforms and power rails, the voltage regulator has a load line programmed into it to help save power and improve performance. The load line has an intrinsic error associated with it called the tolerance band. The ToB can change over the current range depending upon its design and the errors in the circuits in the VR. The load line can also relate directly to the *p-state* in the silicon device. The *p-state* is defined as the voltage/frequency pair that the functional unit is operating at when active. The second part to the system margins is the guardband of the system. This is a combination of either statistical or fixed errors that accumulate to a voltage offset in the system when power is supplied to the device. The system guardband is put in place to ensure the data integrity of the device that is being powered.

Finally, system noise was discussed briefly in terms of both general noise and EMI. Normally emission is not in the purview of the power integrity engineer. However, the design of the PDN can have a direct impact on noise in the system which is why it is important for the PI engineer to understand some of the system impacts that their analysis and design can have on the platform.

Problems

3.1 Determine the resistance of a trace 5 cm long, 2 cm wide, and 0.2 mm thick. Assume the current is uniform throughout and the temperature is 40 °C. If the current increases and causes the temperature to rise to 80 °C, what is the new resistance?

3.2 If the signal propagating through the trace in the previous problem is 200 MHz, what is the skin depth and the subsequent resistance (AC) of the trace at 40 °C?

3.3 Do both of the integrations in Eqs. (3.20) and (3.21). Write out explicitly the steps in the integration(s).

3.4 Ignoring the constants of integration in Eq. (3.22), plot $I(x,5)$ from 0 to 20. Assume $J_0 = 3$. Is the shape of the current what you expected? Compare with the current density plot in Figure 3.4.

3.5 Plot the impedance of an MLCC capacitor for $R = 5$ mΩ, $L = 100$ pH, and $C = 2.2$ µF. Where is the resonance at for this capacitor? Based on the information in the chapter, is this a HF decoupling device or a low-frequency decoupling device? Why?

3.6 How much as a percentage has a Z5U capacitor degraded after 15 K hours of usage? After 150 K hours? How much time would it take (approximately) for the device to degrade to 60% of its capacitance from SoL?

3.7 Determine the load line for a VID setting in a functional unit of 1.2 V with a functional voltage setting of 1.04 V and a current of 5 A?

3.8 Determine the leakage in a device when $\alpha = 2.4$, with the constant of 1.7 and a voltage of 1.1 V. What is the approximate active power if the dynamic capacitance is 10 nF, the activity factor is 11%, and the frequency of operation is 700 MHz? What is the ratio of leakage to active power?

3.9 How many decibels would you read if you measured 300 µV in a chamber from a radiating source in a platform?

Bibliography

1 Paul, C.R. (1982). *Introduction to Electromagnetic Fields*. McGraw-Hill.
2 Balanis, C.A. (1989). *Advanced Engineering Electromagnetics*. Wiley.
3 Ruehli, A.E. (1972). Inductance calculations in a complex integrated circuit environment. *IBM Journal of Research and Development* 16: 470–481.
4 Paul, C.R. (2010). Partial inductance. In: *IEEE EMC Society Newsletter*, Summer 2010, Issue No. 226, 34–42. IEEE, Inc.
5 Pan, M.J. and Randall, C. (2010). A brief introduction to ceramic capacitors. *DEIS* 26: 44–50.
6 Campbell, C.K. et al. (1992). Aspects of modeling high voltage ferroelectric nonlinear ceramic capacitors. *IEEE Transactions on Components, Hybrids, and Manufacturing Technology* 15: 245–251.
7 Leverich, J. et al. (2009). Power management of datacenter workloads using per-core power gating. *IEEE Computer Architecture Letters* 8: 48–51.
8 Gill, A. et al. (2015). Investigation of short channel effects in Bulk MOSFET and SOI FinFET at 20 nm node technology. *IEEE (INDICON)*. IEEE.
9 Zhang, M.T. (2001). Powering intel pentium 4 generation processors. *EPEP*. IEEE.
10 Koertzen, H.W. (2003). Impact of die sensing on CPU power delivery. *APEC*. IEEE.
11 Ted DiBene, J. II, (2013). *Fundamentals of Power Integrity*. Wiley.
12 Oboril, F. et al. (2014). Aging aware design of microprocessor instruction pipelines. *IEEE Transactions on Computer-Aided Design of Integrated Circuits and Systems* 33: 704–716.
13 Waizman, A. and Chung, C.-Y. (2000). Extended adaptive voltage positioning (EAVP). *EPEP*. IEEE.

14 Gupta, V. and Rincon-Mora, G.A. (2002). Predicting and designing for the impact of process variations and mismatch on the trim range and yield of bandgap references. In: *MWSCAS* (21–23 March 2005), 503–508. San Jose, CA: IEEE, Inc.

15 Eirea, G. and Sanders, S.R. (2006). High precision load current sensing using on-line calibration of trace resistance. *PESC*. IEEE.

16 Waizman, A. and Chung, C.-Y. (2001). Resonant free power network desing using extended adaptive voltage positioning (EAVP) methodology. *IEEE Transactions on Advanced Packaging* 24: 236–244.

17 Jeong, K., Kahng, A.B., and Samadi, K. (2009). Impact on guardband reduction on design outcomes: a quantitative approach. *IEEE Transactions on Semiconductor Manufacturing* 22: 552–565.

18 Cui, W. et al. (2003). DC power-bus noise isolation with power-plane segmentation. *IEEE Transactions Electromagnetic Compatibility* 2: 889–903.

19 Cui, W. et al. (2001). DC power-bus isolation with power islands. *EMC Symposium*. IEEE.

20 Hu, H., Hubing, T., and Van Doren, T. (2001). Estimation of printed circuit board bus noise at resonance using a simple transmission line model. *EMC Symposium*. IEEE.

21 Mao, J. et al. (2002). Estimating DC power bus noise. *EMC Symposium*. IEEE.

22 Hubing, T. et al. (1999). Power-bus noise reduction using power islands in printed circuit board designs. *EMC symposium*. IEEE.

23 Liu, Z.H. et al. (2005). Study of power-bus noise isolation using SPICE compatible method. *EMC Symposium*. IEEE.

24 Hueting, J.E. et al. (2004). Gate-drain charge analysis for switching in power trench MOSFETs. *IEEE Transactions on Electron Devices* 51: 1323–1330.

25 Lopez, T. and Elferich, R. (2008). Accurate performance predictions of power MOSFETs in high switching frequency synchronous buck converters for VRM. *PESC*. IEEE.

26 Matoglu, E. et al. (2005). Voltage regulator module noise analysis for high-volume server applications. *EPEP*. IEEE.

27 Feng, G. et al. (2006). Analysis of noise coupling result from overlapping power areas within power delivery networks. *EMC Symposium*. IEEE.

28 Knighten, J.L. et al. (2006). PDN design strategies III. Planes and materials – are they important factors in power bus design? *EMC Society*. IEEE.

29 Paul, C.R. (1992). *Introduction to Electromagnetic Compatibility*. Wiley.

30 Okoshi, T., Uehara, Y., and Takeuchi, T. (1976). The segmentation method – an approach to the analysis of microwave planar circuits. *IEEE MTT Transactions* 24: 662–668.

31 Wang, C. et al. (2005). An efficient approach for power delivery network design with closed-form expressions for parasitic interconnect inductances. *IEEE Transactions on Advanced Packaging* 29 (2): 320–334.

32 Swaminathan, M. et al. (2010). Designing and modeling for power integrity. *IEEE Transactions on Electromagnetic Compatibility* 52 (2): 288–310.

33 Van Valkenburg, M.E. (1974). *Network Analysis*, 3e. Prentice Hall.

34 Paul, C.R. (2011). Physical dimensions vs. electrical dimensions, and modeling for power integrity. *IEEE EMC Society Newsletter* (230): 43–38.

35 Shi, H., Sha, F., Drewniak, J. et al. (1997). An experimental procedure for characterizing interconnects to a DC power bus on a multilayer printed circuit board. *IEEE Transactions on Electromagnetic Compatibility* 39 (4): 279–285.

36 Hubing, T.H., Drewniak, J., Van Doren, T., and Hockanson, D. (1995). Power bus decoupling on multilayer printed circuit boards. *IEEE Transactions on Electromagnetic Compatibility* 37 (2): 155–166.

37 Ott, H.W. (2009). *Electromagnetic Compatibility Engineering*. Wiley.

38 Cui, W. et al. (2000). EMI resulting from signal via transitions through the DC power bus. *IEEE International Symposium on Electromagnetic Compatibility*. IEEE.

39 Waizman, A. et al. (2004). Integrated power supply frequency domain impedance meter (IFDIM). *EPEP*. IEEE.

40 Since, H.J.H. et al. (2007). Study of high speed current excitation reverse engineering methodology using measured voltage and PDN impedance profile from a running microprocessor. *EPEP*. IEEE.

41 Thirugnanam, R. et al. (2007). On channel modeling for impulse-based communications over a microprocessor's power distribution network. *ISPLC Symposium*. IEEE.

4

Electromagnetic Concepts for Power Integrity

Though some equations for the passive components related to elements found in power integrity applications were introduced in the previous Chapter 3, the foundation for these equations were not expounded upon. In this chapter, some of the fundamental concepts are introduced along with some important equations for analyzing power integrity problems.

To understand electrically the structures for power integrity applications, it is important to comprehend the assumptions that are be made with respect to the frequency of operation. In nearly all the cases, even with today's high-frequency systems operation, the power integrity structures may be analyzed using static concepts. That is, electrostatic and magneto-static concepts should apply across the board. All of the methods have their bases in Maxwell's equations. Thus, the fundamentals will start with a review of these.

The static approximation is an important one and is discussed in the context of transmission line theory in the next chapter as well as a very brief discussion below. A short divergence into the basis behind this assumption now follows.

The path from the source to the load is truly a transmission-like path for the power signals to traverse. This has been illustrated already throughout this text, but a variant of this network is shown in Figure 4.1.

If a signal is sent from one side of the network to the other, it will take a certain amount of time to propagate to the other side. This time is governed by the components that exist between the two points. Given the characteristics of the transmission path (its impedance, etc.), we can estimate this time of flight from the phase velocity, assuming a *lossless* transmission line for now. With the impedance for the network shown in the figure, assuming that the resistive elements are small, we will see, without proof at this time that the impedance of the path will be

$$Z_0 = \sqrt{\frac{L}{C}} \tag{4.1}$$

Thus, the impedance is only dependent upon the inductance and capacitance of the structure for a lossless system. It should be noted that for a power

Power Integrity for Electrical and Computer Engineers, First Edition. J. Ted Dibene II and David Hockanson.
© 2020 John Wiley & Sons, Inc. Published 2020 by John Wiley & Sons, Inc.

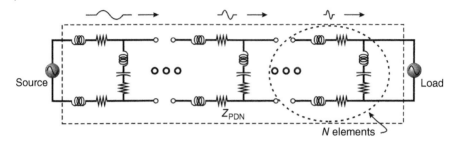

Figure 4.1 Distribution from the source to the load.

distribution network (PDN), this is a very crude assumption. To be literal, the components that make up the transmission path vary in value depending upon where they are physically located and for what purpose they solve. However, for the moment, we will assume that each section is broken down into very small elements as in Figure 4.1 and the variation in the impedance is considered small. Thus, we will assume the values for each element (L, R, and C) are essentially the same in all sections. We also assume there are n elements that make up the distribution from one side to the other.

The velocity of propagation is related to the characteristic impedance of the line by the following relation for a lossless transmission line:

$$v_0 = \frac{1}{\sqrt{LC}} = \frac{1}{C\sqrt{\frac{L}{C}}} = \frac{1}{Z_0 C} \tag{4.2}$$

The wavelength is related to the velocity and thus the frequency of the signal or

$$\lambda = \frac{v_0}{f} \tag{4.3}$$

We have seen in the previous chapter the type of noise that may be generated from voltage regulator power signals. In general, frequency-dependent effects on transmission paths do not occur until the wavelength of the signal is at least ¼λ or larger; that is, the frequency of the signal has to be high enough such that at least a quarter wavelength of the signal may *stretch* across the transmission path. The quarter wavelength assumption is a good general rule and applies to standing wave phenomena on transmission lines. We make use of this fact for the PDN to gain an insight into the static approximations that we will assume are necessary in this chapter to simplify our equations and analyses.

To check the static assumption, we can estimate the electrical path length for a common power delivery network in the following example.

Example 4.1 Estimate the electrical path length for a power delivery network from the power supply to the load. The actual physical length from source to load is 18 mm. Assume the losses are small and that the transmission characteristics are $\sim 0.7c$, where c is the speed of light. Assume the highest frequency component in the transmission path is no greater than 500 MHz.

Solution:
We know the frequency of the signal and the velocity based on the transmission path. The wavelength in meters is then

$$\lambda = \frac{v_0}{f} = \frac{0.7\,c}{f} = \frac{0.7 \times (3 \times 10^8)}{500 \times 10^6} = 0.42 \text{ m} \equiv 420 \text{ mm} \tag{4.4}$$

A quarter wavelength is then $420/4 = 105 \text{ mm} \gg 18 \text{ mm}$, the physical length of the transmission path. Thus, the static approximation would be a valid one in this case.

It should be noted that the 18 mm path length is reasonable for a good many power delivery applications (though some may be longer or shorter depending upon the product). Also, though losses can have an effect on the velocity, PDN series resistive losses are normally quite small (relative to other system losses such as the VR and load). Depending upon the change in impedance (with or without losses), if the difference is small, it usually means that the lossless approximation to evaluate the wavelength is a reasonable one.

There are many texts today that introduce electromagnetic concepts, and at the end of this chapter, we give a reasonable list for the reader to look over. However, most texts are not intended to focus on power integrity and give the reader a much more general view of the fundamentals. Though this is the best way to learn EM concepts, the approach is intended for readers who are mainly interested in EM-based physics in general. In this text, these concepts will be more focused around what is important for students interested in power integrity. Thus, the equations here will be centered around how to analyze power distribution components on the circuit board, package, and silicon.

Herein, it will be assumed that the reader is familiar with basic EM theory, so this chapter will leave out a number of the details that are usually accompanied in general texts. The reader will be steered toward the references in the back should they need a refresher in this area. We will start with a quick review of coordinate systems and Maxwell's equations and then move into a review of static approximations (mainly fundamental equations) that we will use throughout the chapter.

4.1 Coordinate Systems

Different problems may be better solved in different coordinate systems depending upon how the problem is posed and the relationship with geometry of the structure. Structures with circular dimensions that extend in parallel for relatively long lengths may be better set up in cylindrical coordinates rather than Cartesian. Understanding which coordinate system to use for a particular problem is one of the first steps an engineer undertakes when starting the solution process.

There are numerous texts on electromagnetics and other disciplines, which discuss coordinate systems in detail [1, 4, 5]. For a thorough review of this area, the reader is referred to the references at the end of the chapter. In this section, we will touch upon the three most common coordinate systems: *Cartesian*, *Cylindrical*, and *Spherical*. All of these systems are orthogonal. In addition, they represent three dimensions in real space. To describe a point in space requires only three metrics. This is also true when describing a vector in space. Though there are non-orthogonal coordinate systems that are intended to help describe systems for specialized problems, we will only be concerned with the three here in this text.

The most commonly used coordinate system is *Cartesian*. It is the one that virtually all students become familiar with when they first encounter basic algebra in their early school years. For problems that we will encounter throughout this text and in power integrity relating to electromagnetics, it is important to be able to solve vector problems. The reader should be familiar with vector math, which will be used throughout this text. As mentioned earlier, we need only three components to describe a vector in three dimensions as shown in Eq. (4.5).

$$\mathbf{A} = A_x + A_y + A_z \tag{4.5}$$

The vector $\mathbf{A}$ here is comprised of three vectors, each pointing in the direction of one of the axis. Normally, it is best to describe a vector in any of these spaces as a weighted coefficient multiplied by a unit vector. This is shown in Eq. (4.6).

$$\mathbf{A} = A_x \hat{u}_x + A_y \hat{u}_y + A_z \hat{u}_z \tag{4.6}$$

where $\hat{u}_i$ is a unit vector in the direction of ith axis in this case. A unit vector has a magnitude of 1. For example, the magnitude of the unit vector in the x direction may be found by taking the magnitude of the x directed component and dividing it by its magnitude.

$$\hat{u}_x = \frac{A_x}{|A_x|} \tag{4.7}$$

A unit vector in the direction of the vector $\mathbf{A}$ may be found in a similar manner.

$$\hat{u}_A = \frac{\mathbf{A}}{|\mathbf{A}|} \tag{4.8}$$

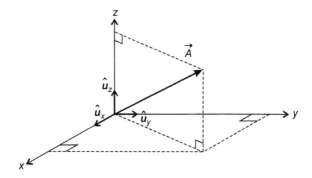

Figure 4.2 Illustration of a vector A in Cartesian coordinates.

The magnitude of the vector may be found by taking the components in Eq. (4.8) and then the square root.

$$|\mathbf{A}| = \sqrt{A_x^2 + A_y^2 + A_z^2} \tag{4.9}$$

We can choose to represent a specific vector **A** from the origin to a point in space in Cartesian coordinates. This is shown in Figure 4.2. This vector may be represented in any of the three coordinate systems without changing its characteristics.

In vector analysis, it is important to be able to add and multiply vectors. This is particularly true when it comes to vector calculus with respect to electromagnetic problems. As stated previously, there are many excellent texts that emphasize how to manipulate vectors mathematically, and a number of them are given as references at the end of this chapter. We also provide some basic techniques in the appendix for the reader. The next example illustrates some introductory math for manipulating vectors in Cartesian coordinates. In this example, we introduce the dot and cross product of two vectors as well.

Example 4.2 Given the two vectors **A** and **B** below, do the following: add them together, find the magnitude of the result, and then take the dot product and the cross products of the two vectors.

$$\mathbf{A} = 6\hat{u}_x + 2\hat{u}_y + 9\hat{u}_z \tag{4.10}$$

$$\mathbf{B} = 4\hat{u}_x + 3\hat{u}_y + 5\hat{u}_z \tag{4.11}$$

Solution:
First, we add the two vectors. Because they are already broken down into their unit vector representations with coefficients, the addition is straightforward.

$$\mathbf{A} + \mathbf{B} = \mathbf{C} = (6+4)\hat{u}_x + (2+3)\hat{u}_y + (9+5)\hat{u}_z = 10\hat{u}_x + 5\hat{u}_y + 14\hat{u}_z \tag{4.12}$$

$$|\mathbf{C}| = \sqrt{10^2 + 5^2 + 14^2} = \sqrt{267} \cong 17.9 \tag{4.13}$$

The first method we use when multiplying each vector is to take the dot product. The dot product of two vectors **A** and **B** may be found using the following formula:

$$\mathbf{A} \cdot \mathbf{B} = |A||B|\cos\theta \tag{4.14}$$

The result of the dot product of two vectors is scalar. We may also determine the dot product by multiplying the like coefficients together.

$$\mathbf{A} \cdot \mathbf{B} = (A_x\hat{u}_x + A_y\hat{u}_y + A_z\hat{u}_z) \cdot (B_x\hat{u}_x + B_y\hat{u}_y + B_z\hat{u}_z)$$
$$= (A_x \cdot B_x + A_y \cdot B_y + A_z \cdot B_z) \tag{4.15}$$

Doing this for the two vectors in above, we get

$$\mathbf{A} \cdot \mathbf{B} = (6 \cdot 4 + 2 \cdot 3 + 9 \cdot 5) = 75 \tag{4.16}$$

The cross product is the second method for multiplying two vectors by each other. The result of the cross product of two vectors is another vector. The definition for the cross product is given in Eq. (4.17).

$$\mathbf{A} \times \mathbf{B} = |A||B|\sin\theta\,\hat{u}_n \tag{4.17}$$

We may find the cross product of two vectors also through the determinant as shown in Eq. (4.18).

$$\mathbf{A} \times \mathbf{B} = \begin{vmatrix} \hat{u}_x & \hat{u}_y & \hat{u}_z \\ A_x & A_y & A_z \\ B_x & B_y & B_z \end{vmatrix}$$
$$= \hat{u}_x(A_yB_z - A_zB_y) - \hat{u}_y(A_xB_z - A_zB_x) + \hat{u}_z(A_xB_y - A_yB_x) \tag{4.18}$$

Thus, using Eq. (4.18), the cross product of the two vectors in question is

$$\mathbf{A} \times \mathbf{B} = \hat{u}_x(2 \cdot 5 - 9 \cdot 3) - \hat{u}_y(6 \cdot 5 - 9 \cdot 4) + \hat{u}_z(6 \cdot 3 - 2 \cdot 4)$$
$$= -17\hat{u}_x + 6\hat{u}_y + 10\hat{u}_z \tag{4.19}$$

The direction of the resulting vector is found by the right-hand rule as well where one points their fingers in the direction of the first vector, curling them toward the second, then using their thumb to point in the direction of the new vector. This new vector will also be orthogonal to the plane where the two original vectors resided.

4.1.1 The Cylindrical Coordinate System

As stated previously, there are some problems that are better posed within the framework of a coordinate system that aligns better to the geometry. The cylindrical coordinate system works well with geometries such as wires and objects

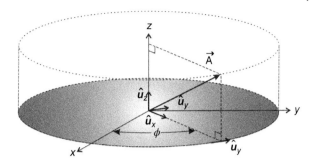

Figure 4.3 Cylindrical coordinate system.

where the fields revolve around the objects in a circular fashion. A vector in three dimensions such as those described in Section 4.1.1 may be represented in cylindrical coordinates as well.

For example, we can choose to represent the vector **A** in Cartesian coordinates and in cylindrical coordinates using the following equation:

$$\mathbf{A} = A_r \hat{u}_x + A_\phi \hat{u}_y + A_z \hat{u}_z \tag{4.20}$$

This vector is shown in Figure 4.3.

We may convert a vector in one coordinate system into another through simple coordinate translation. For example, let us say we want to convert vector **A** in Cartesian coordinates into cylindrical coordinates. This means we want to determine

$$A_\rho, A_\phi, A_z \text{ in terms of } A_x, A_y, A_z \tag{4.21}$$

The translation in the variables is shown in the following equations.

$$x = \rho \cos \phi, \quad y = r \sin \phi, \quad z = z, \rho = \sqrt{x^2 + y^2}, \quad \tan \phi = \frac{y}{x} \tag{4.22}$$

We could derive these pictorially (see Appendix C) as well. The translation from Cartesian to cylindrical is then

$$A_\rho = A_x \cos \phi + A_y \sin \phi, \quad A_\phi = -A_x \sin \phi + A_y \cos \phi, \quad A_z = A_z \tag{4.23}$$

One of the more important mathematical manipulations that we will use in solving electromagnetic problems is through vector calculus. We can represent a differential element in space in these coordinate systems just like we have represented the vectors. In Cartesian coordinates, a differential vector is similar to a regular vector except that the vector $d\mathbf{l}$ is made up of differential elements.

$$d\mathbf{l} = dx\,\hat{u}_x + dy\,\hat{u}_y + dz\,\hat{u}_z \tag{4.24}$$

In cylindrical coordinates, the ϕ term is an angle so that we need to represent the physical length.

$$d\mathbf{l} = dr\,\hat{u}_x + r\,d\phi\,\hat{u}_y + dz\,\hat{u}_z \tag{4.25}$$

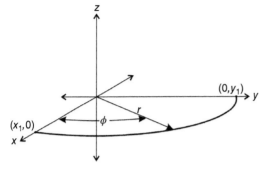

Figure 4.4 Arc in $z = 0$ plane.

Once again, the only difference between the representations of the vectors is the coordinate system. The next problem illustrates where posing a problem in one coordinate system makes the solution easier to find than in another.

Example 4.3 Find the length of the arc in Figure 4.4 using both Cartesian and cylindrical coordinate systems.

Solution:
Let us take the easier problem first. We know the answer by inspection is

$$l = \frac{\pi r}{2} \tag{4.26}$$

We may find this by using cylindrical coordinates as well such that the arc is in the direction of the coordinate ϕ, or

$$d\mathbf{l} = r \, d\phi \, \hat{u}_\phi \tag{4.27}$$

We wish to integrate from point x_1 to y_1 or from the angle 0 to $\pi/2$.

$$l = \int_0^{\frac{\pi}{2}} r \, d\phi = \frac{\pi r}{2} \tag{4.28}$$

Now let us perform the same integration but in Cartesian coordinates. We have a differential element in both x and y now. We need to find the differential length in terms of one variable to make the integration simpler. A picture here helps. If we examine the arc in just the x–y plane, we can get a better idea of what the differential element is.

From the picture (Figure 4.5), we know the differential element is

$$dl^2 = dx^2 + dy^2 \tag{4.29}$$

or,

$$dl = dx \sqrt{1 + \frac{dy^2}{dx^2}} \tag{4.30}$$

Figure 4.5 Differential elements in x–y plane.

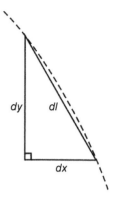

The equation for the arc comes from the equation of a circle

$$y^2 = r^2 - x^2 \tag{4.31}$$

Taking the derivative and substituting into Eq. (4.30), we get the integral

$$l = \int_1^0 \frac{r}{\sqrt{r^2 - x^2}} dx \tag{4.32}$$

In Section 4.6, we verify that this may be solved through substitution, which eventually ends up back at the integral in Eq. (4.28).

4.1.2   The Spherical Coordinate Systems

The last coordinate system that we will consider will be spherical. Though the other two will probably be used more extensively than this one, it is still important to understand this as part of the fundamentals. To translate from Cartesian to spherical, let us first examine Figure 4.6.

We see that there is now another angle involved in the development of this coordinate system. To translate from Cartesian, we have to relate the old variables to the new variables. Thus, we wish to convert a vector in terms of the new coordinate system or

$$A_r, A_\theta, A_\phi \text{ in terms of } A_x, A_y, A_z \tag{4.33}$$

By doing this, we arrive at the new translation

$$x = r \sin\theta \cos\phi, \quad y = r \sin\theta \sin\phi, \quad z = r \cos\theta,$$

$$r = \sqrt{x^2 + y^2 + z^2}, \quad \theta = \tan^{-1}\frac{\sqrt{x^2 + y^2}}{z}, \quad \phi = \tan^{-1}\frac{y}{x} \tag{4.34}$$

A differential length is also similar to the cylindrical coordinate system and may be represented as

$$d\mathbf{l} = dr\,\hat{\boldsymbol{u}}_r + r\,d\theta\,\hat{\boldsymbol{u}}_\theta + r\,\sin\theta\,d\phi\,\hat{\boldsymbol{u}}_\phi \tag{4.35}$$

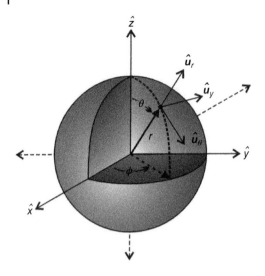

Figure 4.6 Spherical coordinate system.

In many cases, we are asked to find the differential surface of a region relative to a flux field or other metric. In this instance, a differential surface may be represented as

$$ds_r = r^2 \sin\theta \, d\phi \, d\theta \, \hat{u}_r \tag{4.36}$$

$$ds_\phi = r \, d\theta \, dr \, \hat{u}_\phi \tag{4.37}$$

and

$$ds_\theta = r \sin\theta \, dr \, d\phi \, \hat{u}_\theta \tag{4.38}$$

The next example illustrates how to use the differential area with vectors.

Example 4.4 Find the area of 1/8 of a section of the sphere from Figure 4.6.

Solution:
We should be able to determine the area first by inspection. Let us take the easier problem first. The area of a sphere is determined to be

$$S_1 = 4\pi r^2 \tag{4.39}$$

Thus, we know the answer by inspection that 1/8 the area is then

$$S_{1/8} = \frac{\pi r^2}{2} \tag{4.40}$$

We now use calculus to find this area. The vector that we are most interested in is the one pointing in the direction of r. This is orthogonal to the surface of interest. Thus, the differential vector that we want is from Eq. (4.36). We now need to integrate over this surface. We set up the integral that we are interested in

$$S_{1/8} = \int_0^{\pi/2} \int_0^{\pi/2} r^2 \sin\theta \, d\phi \, d\theta \tag{4.41}$$

Integrating this equation, we get

$$\int_0^{\pi/2} \int_0^{\pi/2} \sin\theta \, d\phi \, d\theta = r^2 \frac{\pi}{2} \int_0^{\pi/2} \sin\theta \, d\theta = \frac{\pi r^2}{2} \tag{4.42}$$

This is the answer we predicted by inspection.

4.2 EM Concepts – Maxwell's Equations

As in all cases, electromagnetic concepts begin with Maxwell's equations. Though the general representations are *time* and *spatially* dependent, we will quickly narrow down the focus to the static or DC approximations for these equations.

The four main equations that we are interested in are the following:

$$\nabla \times \vec{\mathbf{H}} = \frac{\partial \vec{\mathbf{D}}}{\partial t} + \vec{\mathbf{J}} \tag{4.43}$$

$$\nabla \times \vec{\mathbf{E}} = -\frac{\partial \vec{\mathbf{B}}}{\partial t} \tag{4.44}$$

$$\nabla \cdot \vec{\mathbf{B}} = 0 \tag{4.45}$$

$$\nabla \cdot \vec{\mathbf{D}} = \rho \tag{4.46}$$

The notation used here will be to use **bold** letters where the variables are vectors, with or without the vector notation as shown.[1] When the values are *scalars*, we will use italics and typically will not bold the variables per convention with many other texts.

Each variable in Maxwell's equations is assumed to be *time* and *space* dependent as noted above; that is, in general, a variable is governed by both its position and the time relative to its particular functional relationship as in the electric flux density vector as an example.

$$\vec{\mathbf{D}} = \vec{\mathbf{D}}(r, t) \tag{4.47}$$

In this case, r represents the position relative to a specific coordinate system. The spatial dependent variable r may also be a vector. As we shall see, the equations simplify greatly if there is no time dependency.

---

1 Note that we have been using this notation already in the earlier chapters.

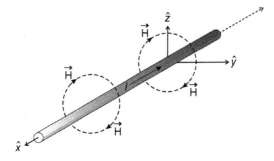

Figure 4.7 Illustration of Ampere's law for a current-carrying wire.

The first equation that we will examine is Ampere's law, which relates the magnetic field around a loop to the current going through that loop. Figure 4.7 illustrates the concept. The equation shown here is in *differential* form.

In Eq. (4.43), **H** is the magnetic field, **D** is the electric flux, and **J** is the enclosed conduction current density. The electric flux term with **D** was added later by Maxwell to complete the more general relationship between the electric and magnetic fields. Essentially, the equation shows that if a current flows through a wire, it will generate a magnetic field outside of that wire in the direction that is orthogonal to the flow of that current. This direction is defined by the right-hand rule and is represented by the curl of the vector **H**.

Here again, we see that all of the variables have both spatial and time dependence for the general case. For the specific case where the fields are considered *static*, we may expand the two sides in Cartesian coordinates to see the effects.

$$\nabla \times \mathbf{H} = \begin{vmatrix} \hat{u}_x & \hat{u}_y & \hat{u}_z \\ \dfrac{\partial}{\partial x} & \dfrac{\partial}{\partial y} & \dfrac{\partial}{\partial z} \\ H_x & H_y & H_z \end{vmatrix} = \hat{u}_x \left(\frac{\partial H_z}{\partial y} - \frac{\partial H_y}{\partial z} \right)$$

$$- \hat{u}_y \left(\frac{\partial H_x}{\partial z} - \frac{\partial H_z}{\partial x} \right) + \hat{u}_z \left(\frac{\partial H_y}{\partial x} - \frac{\partial H_x}{\partial y} \right) \tag{4.48}$$

Equation (4.48) is the expanded version of the left-hand side of Eq. (4.43). The curl of a vector in three dimensions may be found by taking the determinant of the three spatial components. This is a simple way of determining the curl relationship without having to remember the exact formula. Here, we have taken the determinant in Cartesian coordinates. Expanding the right-hand side now, we get

$$\frac{\partial \mathbf{D}}{\partial t} + \mathbf{J} = \frac{\partial}{\partial t} (D_x \hat{u}_x + D_y \hat{u}_y + D_z \hat{u}_z) + J_x \hat{u}_x + J_y \hat{u}_y + J_z \hat{u}_z \tag{4.49}$$

Since there is no time dependence, the electric flux terms in **D** go to zero. Also, the conduction current is only flowing in the *negative x* direction per the diagram. This means the only part of the vector which is relevant is the one pointing in this direction. There are no other components in the static approximation. Thus, Ampere's law for the *quasi-static* case simplifies to the following:

$$-\widehat{u}_x \left(\frac{\partial H_z}{\partial y} - \frac{\partial H_y}{\partial z} \right) = -J_x \widehat{u}_x \tag{4.50}$$

or, eliminating the vector notation (since it is implied)

$$\left(\frac{\partial H_z}{\partial y} - \frac{\partial H_y}{\partial z} \right) = J_x \tag{4.51}$$

We may integrate both sides of Eq. (4.51) to get the total current enclosed by the area outside of the wire.

$$\int_S \left(\frac{\partial H_z}{\partial y} - \frac{\partial H_y}{\partial z} \right) dS = \int_S J_x dS \tag{4.52}$$

The integral on the left-hand side may be turned into a contour integral using Stokes' theorem.

$$\oint_C \mathbf{H} \cdot dr = \int_S \mathbf{J} \cdot dS \tag{4.53}$$

where r is in the direction of the path that circumscribes the wire; in this case, the angle is ϕ and S is the surface area surrounding the wire. To illustrate the point, we can put limits on these integrals and compute them as *definite* integrals over a region. Notice that the coordinate system was changed from Cartesian to cylindrical for simplicity. Assuming no fringing fields and that the length of the wire is very long while the radius is very small, we can apply the appropriate limits. The integration over a wire that stretches between length $-a$ and a would be

$$\oint_0^{2\pi} H_\phi r \, d\phi = \int_{-a}^{a} \int_0^{2\pi} J_x \, dx \, r \, d\phi = I \tag{4.54}$$

The term on the right-hand side of Eq. (4.54) is simply the total current enclosed. Given the constraints on the wire and that, the current is *static*, we may easily see that the magnetic field will also be constant around a contour.[2] Performing the integration on the left-hand side and solving for the magnetic field yields

$$H_\phi = \frac{I}{2\pi r} \tag{4.55}$$

This simply says that the field truly does vary as the radial distance from the wire but is constant for a *fixed* radius. From the previous chapter, given this current source, it was possible to compute the total flux from this current. If

---

2 Note that this is NOT true for the field as it radiates outward from the wire. Because the wire is circular, this necessitates that the magnetic field around a given uniform circle through the angle ϕ will yield constant field strength.

the problem is bounded with a given return path that has symmetry relative to the position of the wire, then it is possible to compute the inductance as well which was done in the previous chapter. This shows a simple, but direct, application to Ampere's law.

The next equation to examine is based on Faraday's law. As in Ampere's law, this equation is once again fundamental and can be used to compute the parasitics of various common structures with the goal of using them in circuits to determine the overall behavior of the network.

Essentially, Faraday's law states that the induced electromotive force (EMF; voltage) in any closed circuit is equal to the negative time rate of change of the magnetic flux enclosed by that circuit. Or mathematically

$$\varepsilon = -\frac{d\Phi_B}{dt}, \quad \text{volts} \tag{4.56}$$

Note that previously from Eq. (4.44), the electric field has dimensions of volts/meter, which means that after taking the curl, the left-hand side in that equation has dimensions of volts/square meter, which is essentially a *density* function. If both sides are integrated with respect to its spatial dependence, the result is Eq. (4.56) which is now in volts.

As an illustration of Faraday's law, we refer to Figure 4.8. Suppose we wish to compute the inductance of a solenoid. We may use the static field assumption in this example of Faraday's law to analyze this problem.

In Figure 4.8, it is assumed that there are N turns in the solenoid. If there exists a time-changing magnetic field induced in the solenoid, say through moving a steel bar in and out of the center of the solenoid, a voltage will be created across the inductor. Thus, the EMF of the system is related to a change in the flux in the center of the coil and the number of turns in the solenoid by the following equation:

$$\varepsilon = -N\frac{d\Phi_B}{dt} \tag{4.57}$$

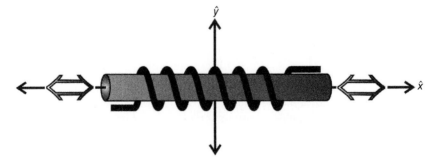

Figure 4.8 Example with solenoid.

In turn, the change in flux, as discussed in Chapter 2, is related to the magnetic flux density and the magnetic field.

$$\Delta \Phi = \int_A \Delta \mathbf{B} \cdot d\mathbf{A} \tag{4.58}$$

In this case, depending upon the movement (direction) the steel bar is going (into the coil or out), the magnetic flux flows in either the y or $-y$ direction. The area of integration is over the cross-section of the inner part of the solenoid. Thus,

$$\Delta \Phi = \Delta BA \tag{4.59}$$

Thus, the total flux will be

$$\Phi = BA = LI \tag{4.60}$$

This was shown in Chapter 2 when we analyzed a typical inductor structure. The question now is what is the magnetic flux density *inside* of the coil? If the current flows in the wire, from Ampere's law, we may compute the magnetic field by integrating around a wire and do this for N times for the number of turns, or

$$\oint_C H \cdot d\mathbf{l} \rightarrow H = \frac{NI}{l}$$

where l is the length of the solenoid as shown in Figure 4.8. Substituting in for the magnetic field in Eq. (4.60) now, we get

$$LI = BA = \mu_0 \frac{NI}{l} A = \mu_0 \frac{NI}{l} \pi r^2 \rightarrow L = \mu_0 \frac{N}{l} \pi r^2 \tag{4.61}$$

Note that the self-inductance here only depends on the number of turns (N), the length l, and the radius inside of the solenoid. Also, note that we left off the direction of the magnetic field. However, it may be easily deduced by using the right-hand rule by wrapping one's hand around one of the wires and looking at the direction of the magnetic field. In this case, it would be directed along the x axis.

Equation (4.45) is Gauss' law for magnetism. It shows that there are no magnetic monopoles or magnetic-free charges. By taking the divergence of the magnetic flux, we find that there is no net flux coming from a source. We know by analyzing the current loops in magnetic materials that there are only magnetic *dipoles*. This is in contrast to Gauss' law for the electric flux where there is a source of charge and a net electric flux flows outward from its source.

If a sphere of charge exists in space as in Figure 4.9, the charge density is a function of charge per unit volume of the sphere. We can expand Eq. (4.46) in terms of spherical coordinates here to get the electric flux components.

$$\nabla \cdot \vec{\mathbf{D}} = \frac{1}{r^2} \frac{\partial (r^2 D_r)}{\partial r} + \frac{1}{r \sin \theta} \frac{\partial}{\partial \theta} (D_\theta \sin \theta) + \frac{1}{r \sin \theta} \frac{\partial D_\varphi}{\partial \varphi} = \rho \tag{4.62}$$

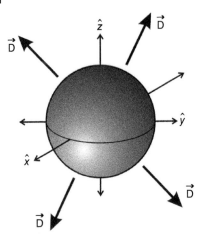

Figure 4.9 Sphere in space with electric flux emanating radially.

In this case, the electric field will radiate outward from the sphere, and thus the only electric field component will be in the radial r direction.

$$D_r = D(r) = \epsilon_0 E(r) = \frac{Q}{4\pi r^2} \tag{4.63}$$

The voltage is the integral over the electric field from the surface of the sphere to a point in space.

$$V(r) = -\int_a^r E(r)dr = \frac{-Q}{4\pi\epsilon_0} \int_a^r \frac{dr}{r^2} = \frac{Q}{4\pi\epsilon_0} \left(\frac{1}{r} - \frac{1}{a}\right) \tag{4.64}$$

If we were to place another hollow sphere concentrically around the one with the charge, we could find the voltage between them. Therefore, the voltage between the two spheres will be

$$V(a) - V(b) = \frac{Q}{4\pi\epsilon_0} \left(\frac{1}{a} - \frac{1}{b}\right) \tag{4.65}$$

We know that the charge is related to the capacitance from the previous chapter. Thus, the capacitance is simply the ratio of the charge to the voltage, which yields the following result:

$$C = \frac{Q}{V} = 4\pi\epsilon_0 \left(\frac{ab}{b-a}\right) \tag{4.66}$$

where a and b are the inner and outer radii respectively. Section 4.2.1 discusses an important formula in magneto-statics. This is the Biot–Savart Law and its equation.

4.2.1 The Biot–Savart Law

One of the key equations that is useful in a number of ways for solving inductance problems is based on the Biot–Savart law which allows one to compute the fields from a current source at a distance. The assumption is that the current is essentially constant through the element or wire that contains it, and that a field will be generated at some random point from the current source itself.

The equation is usually expressed in integral form.

$$\mathbf{B}(\mathbf{r}) = \frac{\mu_0}{4\pi} \int_C \frac{I \times (\mathbf{r} - \mathbf{r}')}{|\mathbf{r} - \mathbf{r}'|^3} dl \tag{4.67}$$

At first glance, this equation looks daunting given all of the variables. However, a simple dimensional analysis on both sides reveals quite a bit about the equation. If we do a simple check of the dimensions, we find that they match (as we would expect). The term on the left-hand side is in Webers/square meter, which is Henries-Amperes/square meter. This tells us that it is the same flux density that we have seen in the previous chapters. If we then expand the right-hand side of Eq. (4.67), we should see the same dimensions as on the left-hand side. Starting with the left-hand side and working our way across, we get

$$\text{left-hand side} \rightarrow \frac{H \cdot A}{m^2} \quad \text{right-hand side} \rightarrow \left(\frac{H}{m} \cdot \frac{A \cdot (m)}{m^3} \right) \cdot m = \frac{H \cdot A}{m^2} \tag{4.68}$$

This shows that dimensionally both sides are equivalent. It is often good to understand the Biot–Savart equation by examining each term individually and then understanding the whole later.

A variant of the Biot–Savart equation is:

$$\mathbf{B}(\mathbf{r}) = \frac{\mu_0 I}{4\pi} \int_C \frac{d\mathbf{l} \times \hat{\boldsymbol{u}}_R}{R^2} \tag{4.69}$$

In this case, we have pulled out the current under the integral since it is now constant. The only confusion with this equation is that $\hat{\boldsymbol{u}}_R$ is now a unit vector in the direction of R; thus, it has been normalized so that it is *unit-less*. The denominator is now just the distance from the source to the point of where the field is (Figure 4.10). The left-hand side is understandable except for possibly the variable $\mathbf{r}$. The vector $\mathbf{r}$ represents the position of the magnetic flux due to the source current at a distance $\mathbf{r}'$ from the wire. This is an important distinction in the cause-effect relationship between how the source generates the field and where each portion of that source current has a corresponding field. The differential form of the Biot–Savart Law is where one usually starts when using it to compute the magnetic flux density or possibly the inductance of a structure analytically. This is shown in Eq. (4.70):

$$d\mathbf{B}(\mathbf{r}) = \frac{\mu_0 I}{4\pi} \frac{d\mathbf{l} \times \hat{\boldsymbol{u}}_R}{R^2} \tag{4.70}$$

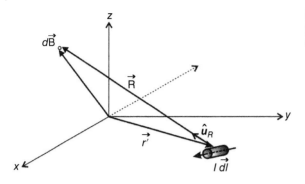

Figure 4.10 Biot–Savart physical representation for differential current source.

The usefulness of the Biot–Savart equation may be found through an actual example as shown here.

Example 4.5   Determine the magnetic flux density for a very long wire in space as in Figure 4.11 using the Biot–Savart Law.

Solution:
This is a classic problem that illustrates the usefulness of the Biot–Savart equation. First, we orient the wire on the z axis for simplicity. Second, we choose a point in space as shown in Figure 4.11. The unit vector $\hat{u}_R$ is in the direction of where we want to know the field strength at point P. Note that this is **NOT** the unit vector in the direction of the coordinate r (cylindrical coordinates). The first vector we need to find is $d\mathbf{l}$. This is the differential length of the wire in the direction of the z axis. We determine the vector $d\mathbf{l}$ as,

$$d\mathbf{l} = dz\,\hat{u}_z \tag{4.71}$$

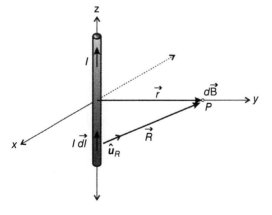

Figure 4.11 Long wire in space.

where $\hat{u}_z$ is a unit vector in the same direction as the axis z. Note that the length is a differential length here. The vector for $\hat{u}_R$ is in the direction of both z and r.

$$\hat{u}_R = \frac{z\hat{u}_z + r\hat{u}_r}{|\mathbf{R}|} \tag{4.72}$$

The magnitude of the vector $\mathbf{R}$ in the direction of field to point P is,

$$|\mathbf{R}| = \sqrt{z^2 + r^2} \tag{4.73}$$

Therefore, the total differential flux field using Eq. (4.70) is

$$d\mathbf{B} = \frac{\mu_0 I}{4\pi} \frac{d\mathbf{l} \times \hat{u}_R}{R^2} = \frac{\mu_0 I}{4\pi} \frac{dz\,\hat{u}_z \times (z\hat{u}_z + r\hat{u}_r)}{(z^2 + r^2)^{\frac{3}{2}}} \tag{4.74}$$

We note that the cross product of two vectors in the same direction is zero. The second term though shows an orthogonal direction to the third axis in cylindrical coordinates or

$$d\mathbf{B} = \frac{\mu_0 I}{4\pi} \frac{d\mathbf{l} \times \hat{u}_R}{R^2} = \frac{\mu_0 I}{4\pi} \frac{r\,dz\,\hat{u}_y}{(z^2 + r^2)^{\frac{3}{2}}} \tag{4.75}$$

This was found by observing that

$$\hat{u}_z \times \hat{u}_r = \hat{u}_y \tag{4.76}$$

We now integrate Eq. (4.75) over the length of the z axis where the length is assumed to be infinite.

$$\int d\mathbf{B} = \frac{\mu_0 I r}{4\pi} \int_{-\infty}^{\infty} \frac{dz}{(z^2 + r^2)^{\frac{3}{2}}} \hat{u}_y \tag{4.77}$$

Note that r is constant over the integration. To integrate the function above, we make a trigonometric substitution where

$$z = r\tan(u) \tag{4.78}$$

This comes from noticing that

$$1 = \cos^2 u + \sin^2 u \rightarrow r^2 + r^2\tan^2 u = r^2\sec^2 u \tag{4.79}$$

then,

$$dz = r\sec^2 u\,du \tag{4.80}$$

Substituting Eq. (4.80) into Eq. (4.77) gives

$$\frac{\mu_0 I r}{4\pi} \int_{-\infty}^{\infty} \frac{dz}{(z^2 + r^2)^{\frac{3}{2}}} = \frac{\mu_0 I r}{4\pi} \int_{-\infty}^{\infty} \frac{r\sec^2 u\,du}{(r^2\sec^2 u)^{\frac{3}{2}}} = \frac{\mu_0 I}{4\pi r} \int_{-\infty}^{\infty} \cos u\,du \tag{4.81}$$

Integration is now straightforward which results in the following solution:

$$\mathbf{B} = \frac{\mu_0 I r}{4\pi} \sin u \Big|_{-\infty}^{\infty} \hat{u}_y = \frac{\mu_0 I r}{4\pi} \sin u \Big|_{-\infty}^{\infty} \hat{u}_y \tag{4.82}$$

We now need to replace the sine function with the variables in terms of z and r. This can be found from a simple identity. From Eq. (4.78) and using the identity,

$$\sin^2 u = \frac{z^2}{r^2}\cos^2 u = \frac{z^2}{r^2}(1 - \sin^2 u) \rightarrow \sin u = \frac{z}{\sqrt{z^2 + r^2}} \tag{4.83}$$

The solution now falls out of Eq. (4.82).

$$\mathbf{B} = \left.\frac{\mu_0 I r}{4\pi}\sin u\right|_{-\infty}^{\infty}\hat{\mathbf{u}}_y = \left.\frac{\mu_0 I r}{4\pi r^2}\frac{z}{\sqrt{z^2 + r^2}}\right|_{-\infty}^{\infty} = \frac{\mu_0 I}{2\pi r}\hat{\mathbf{u}}_y \tag{4.84}$$

Section 4.2.2 discusses a very powerful addition to using the Biot–Savart equation and that is the magnetic vector potential.

4.2.2 The Magnetic Vector Potential

The magnetic vector potential normally does not have a physical representation but is very powerful mathematically and analytically for many electromagnetic problems. The magnetic vector potential may be related to the magnetic flux through the following simple equation:

$$\mathbf{B} = \nabla \times \mathbf{A} \tag{4.85}$$

We can present without proof the relationship between the magnetic vector potential and the current density through the next equation.

$$\mathbf{A} = \int_V \frac{\mu_0 \mathbf{J}}{4\pi R}dv \tag{4.86}$$

where $\mathbf{J}$ is the current density in Amperes/square meter and the integration is over the volume. We can verify once again that the dimensions are equivalent on both sides by doing a dimensional analysis. Using Eq. (4.85), we know that the magnetic vector potential has one less spatial relation in the denominator than $\mathbf{B}$ does. Starting on the left-hand side and moving to the right-hand side, we get

$$\text{left-hand side} \rightarrow \frac{H \cdot A}{m} \quad \text{right-hand side} \rightarrow \left(\frac{H}{m} \cdot \frac{A}{m^3}\right) \cdot m^3 = \frac{H \cdot A}{m} \tag{4.87}$$

This obviously checks out. We can now use this equation to determine the magnetic flux for different geometries as we did in Section 4.2.1 using the Biot–Savart Law. We analyze a similar problem to the one in the previous example here.

Figure 4.12 Short wire in space.

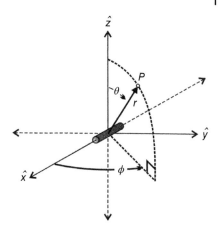

Example 4.6 Determine the magnetic flux density for a relatively short wire (a in length) in space carrying a current as in Figure 4.12 using the magnetic vector potential.

Solution:
We orient the wire along the x axis here for simplicity in the mathematical analysis. The current is flowing in the x direction as per Figure 4.12. We use Eq. (4.86) above to set up the problem.

$$\mathbf{A} = \int_V \frac{\mu_0 \mathbf{J}}{4\pi R} dv \tag{4.88}$$

The first thing we note is that we can integrate out the current density term since the current is already given. Thus, we are left with a line integral since the integral of the current density is over an area

$$\mathbf{A} = \frac{\mu_0 I}{4\pi} \int_{-\frac{a}{2}}^{\frac{a}{2}} \frac{dl}{R} \hat{\boldsymbol{u}}_x \tag{4.89}$$

We then make the assumption that the length a is much smaller than the radius r. Thus, integrating, we get

$$A_x = \frac{\mu_0 I a}{4\pi r} \tag{4.90}$$

From Eq. (4.48), earlier in the chapter, we may compute the magnetic flux density from the curl relation.

$$\mathbf{B} = \frac{\partial}{\partial z} A_x \hat{\boldsymbol{u}}_y - \frac{\partial}{\partial y} A_x \hat{\boldsymbol{u}}_z \tag{4.91}$$

To differentiate here, we write r in terms of Cartesian coordinates

$$r = \sqrt{x^2 + y^2 + z^2} \tag{4.92}$$

Substituting back into Eq. (4.90) and taking the derivatives in Eq. (4.91), we get

$$\frac{\partial}{\partial z}A_x = \frac{\mu_0 I a}{4\pi}\frac{-z}{(x^2+y^2+z^2)^{\frac{3}{2}}} \tag{4.93}$$

and,

$$\frac{\partial}{\partial y}A_x = \frac{\mu_0 I a}{4\pi}\frac{-y}{(x^2+y^2+z^2)^{\frac{3}{2}}} \tag{4.94}$$

Substituting back into Eq. (4.91) now yields

$$\mathbf{B} = \frac{\mu_0 I a}{4\pi}\frac{-z}{(x^2+y^2+z^2)^{\frac{3}{2}}}\widehat{\boldsymbol{u}}_y + \frac{\mu_0 I a}{4\pi}\frac{y}{(x^2+y^2+z^2)^{\frac{3}{2}}}\widehat{\boldsymbol{u}}_z \tag{4.95}$$

We could leave the solution in terms of Cartesian coordinates if we wanted to. To make it more elegant though, we use spherical coordinates based on the angles in Figure 4.12. Thus, the final solution is now

$$\mathbf{B} = \frac{\mu_0 I a}{4\pi r^2}(-\cos\theta\widehat{\boldsymbol{u}}_y + \sin\theta\sin\phi\widehat{\boldsymbol{u}}_z) \tag{4.96}$$

Performing a coordinate conversion yields

$$\mathbf{B} = \frac{-\mu_0 I a}{4\pi r^2 \cos\phi}(1+\tan\phi)\widehat{\boldsymbol{u}}_\theta \tag{4.97}$$

The next few sections make use of these concepts where we compute some very simple closed-form relations to be used for analytical purposes.

4.3 Some Useful Closed-Form Equations

There are a number of capacitive, inductive, and resistive structures that are commonly found when analyzing problems in power integrity interconnects. This means that if the geometry can be approximated by a reasonable closed-form equation, it makes for analyzing the problem later, using numerical methods or for just getting a first-principles approximation, an easier proposition for the PI engineer. Sometimes a developer simply needs a first pass analysis, but due to time constraints he cannot perform the electrical extractions right away and is required to get an answer fairly quickly. This is where a number of these formulas can increase one's productivity.

The issue with any closed-form solution is that in most cases, it is only an approximation to the exact solution, or stated more appropriately, an approximation to a more accurate one. Closed-form solutions are never intended to replace electrical parameter extractions, such as those that come from field solvers (as discussed in Chapter 7). However, they normally can provide very good insights into complex problems without having to resort to numerical computations. Moreover, they are very good checks against modeling errors with such programs and help to keep the analyses bounded.

Though many physical structures around the power distribution network are complicated, usually there is a reasonable equivalent form to the structure that may be approximated with an equation that yields insight to the circuit behavior of that structure.

This is not to say that these equations are by any means exact solutions to the problem at hand as we just discussed. On the contrary, many of these closed-form equations are approximations themselves of an even *simpler* problem, which can often yield a result that may end up being useless to an engineer who is trying to gain some level of accuracy. This is when it is important to understand the limitations of the equation as it relates to how it may apply to the problem at hand. When dealing with such problems, which involve static fields in this case, there are two important questions that should be asked. First, is the equation a reasonable estimate, if not an exact, solution to the similar canonical problem that is being analyzed? And second, are the fields well approximated for the geometry, materials, environment in which the actual problem is posed for?

To answer the first question, it is necessary to examine the equation that one intends to apply and understand its limitations and how it varies from the actual problem that is being solved. If the equation has a number of assumptions and approximations, these criteria need to be well understood before it is to be applied to the problem. Also, the engineer must comprehend how far from the canonical problem the actual one is to ensure its limitations are acceptable (in addition to the expectation on the accuracy of the result).

All of these formulae relate in some fashion back to Maxwell's equations and are circuit approximations from the fields that are generated from some real or fictitious source. As an example, an extracted external inductance computation comes about from assuming a current is flowing through some physical conductor, and thus we know there are fields generated outside of this structure, which result in estimating it as an inductive circuit component (and possibly other components). One problem that often occurs when using simple equations is that the actual problem may have conductors that are much closer than the closed-form solution was intended for. If this is the case, the fields may concentrate such that the estimate may not be a good approximation to the actual solution.

If that occurs, it is often necessary to break the problem down further or use another formula or formulae. The concept of partial inductances was introduced in the previous chapter to help in this understanding. This is a method whereby a larger more complex structure is broken down into smaller ones with the goal of simplifying (hopefully) the results to allow for an easier analysis. Though this is not always successful, it is often the first step in the analysis process.

In this section, we analyze and derive some of the more common problems with respect to the power integrity PDN path. The limitations of such derivations are also discussed along with the assumptions that are intended to get to the results we are looking for. In the following section, we apply a few of these equations to some real-world examples and generate results, giving the engineer a starting point for their analytical thought process.

4.3.1 Simple Plane-Pair Inductance

One of the simpler, but important elements to the interconnect of a PDN is the inductance between two components, such as two capacitors on a printed circuit board. Because this path is typically either the path on the motherboard or the package, the connections are often through wide traces or power and ground planes. If the width to separation ratio is large enough (typically 5× or more), then a reasonable approximation may be made to represent the inductance of these planes for use in a circuit analysis later on.

The simplest approximation assumes that the current is uniform throughout the upper and lower planes. Interestingly, given that many interconnect paths today are made of relatively thin traces, this approximation is normally good enough to represent the inductive path. We will examine a more general case later, for the resistance, where the current is not uniform but emanates outward, say, from a group of vias to another group of vias.

Figure 4.13 illustrates the problem. We start by assuming that one plane is located in the xy plane ($z = 0$ axis) and that the other is completely parallel and is in the $z = s$ axis. It may also be assumed that the current flows uniformly in both the top and bottom planes and that the fringing fields are considered negligible compared to the main fields in this particular analysis.

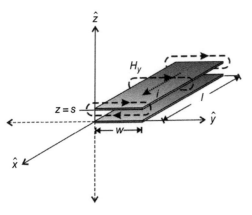

Figure 4.13 Plane pair with uniform current density for inductance computation.

We may certainly use some of the tools from the previous sections, but for this problem, it is possible to estimate the magnetic fields by observing a few important concepts. First, we may use Ampere's law to compute the magnetic field between the two planes. If the current is flowing in the positive x direction as shown in Figure 4.13, using the right-hand rule, we can show that the magnetic field would then be in the positive y direction between the planes and would circle around the upper plane heading in the negative y direction. We should be able to interpret this directly by looking at a modified version of Eq. (4.48), which has been reduced to only include the x-directed current

$$\left(\frac{\partial H_z}{\partial y} - \frac{\partial H_y}{\partial z}\right)_x = \left(0 - \frac{\partial H_y}{\partial z}\right)_x = J_x \tag{4.98}$$

Note that the first term goes to zero because H_z is negligible by definition. To solve this problem, we need to determine the magnetic flux density first, which may be found by solving for the magnetic field. To get this we can integrate Eq. (4.98) on both sides.

$$\iint \frac{dH_y}{dz} dz\, dx = \oint H_y\, dx = \iint J_x dz\, dx = I \tag{4.99}$$

Note that we used Stokes' theorem to turn the integral on the left-hand side into a contour integral. The integral on the right-hand side is simply the total current enclosed. Since the metal plates are assumed to be much wider than their separation, the fields on the top side are also negligible. Thus, we can integrate the left-hand side by making the approximation in the integral and then solving for the y directed magnetic field.

$$I = \oint H_y\, dx \cong \int_0^w H_y\, dx = H_y w \rightarrow H_y = \frac{I}{w} \tag{4.100}$$

We know that the total magnetic flux may be determined by integrating over the region between the planes here.

$$\psi = \iint B_y \cdot dA = \iint \mu H_y\, dz\, dx = \int_0^l \int_0^s \mu \frac{I}{w} dz\, dx = \frac{\mu l s I}{w} \tag{4.101}$$

Since we know that the flux is related to the inductance, we can now solve for it.

$$\psi = LI = \frac{\mu l s I}{w} \rightarrow L = \frac{\mu l s}{w} \tag{4.102}$$

Again, this is only an approximation to the exact result for many problems.

Let us examine some of the assumptions in this formulation of the inductance between two planes. First, we assumed that the length and the width of the planes were larger than the separation. In this case, we used an estimate that is 5× larger. This is because once the geometries become smaller than 5×, we begin to see larger fringing fields which will affect the result. One way to observe this is by comparing our inductance with another canonical problem, that of two wires in space. As we shrink the geometries down in one direction and increase the separation, the problem starts to look very much like our two-wire problem. We now examine this problem in Section 4.3.2.

4.3.2 Inductance of Two Wires in Space

In a few earlier sections in this chapter, we examined the magnetic flux density at a given point in order to gain insights into using equations such as the Biot–Savart law or the magnetic vector potential. We can now examine what the inductance would be between two wires in space using some of these concepts. Figure 4.14 shows two wires.

We assume to begin with that the wire diameter is small enough to be neglected and that its length is much greater than its separation or diameter. Also, we assume that the currents are differential in nature; that is, there is an equal and opposite current flowing in one verses the other shown in Figure 4.14.

From Section 4.2, we can determine the total current from Ampere's law where,

$$\oint H_\phi r \, d\phi \cong \int_0^{2\pi} H_\phi r \, d\phi = H_\phi 2\pi r = I \rightarrow H_\phi = \frac{I}{2\pi r} \qquad (4.103)$$

Note that we are using cylindrical coordinates to set up the problem here.

It is now a simple matter to find the total flux and thus the inductance through integration over the area between the wires. Before we set up the integration, we should note that the flux from each wire goes out into space ad-infinitum. However, because we are integrating to find the *total* flux, we note by the right-hand rule that the flux on both the right-hand side of the right wire and the left-hand side of the left wire cancels as shown in Figure 4.15. Thus, we are left with only the flux between them. Interestingly, this flux adds rather than subtracts, which is why we get twice the flux in the middle.

Figure 4.14 Two wires in space.

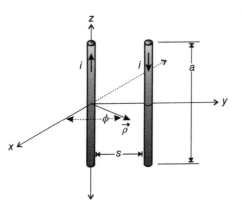

Figure 4.15 Flux in different regions.

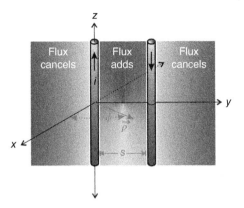

The integral now that we set up is similar to the one from Section 4.3.1

$$\psi = \iint B_\phi \cdot dA = \iint \mu H_\phi \, dz \, dr$$

$$= 2 \int_{r_w}^{s} \int_0^a \mu \frac{I}{2\pi r} \, dz \, dr = \frac{\mu a I}{\pi} \ln\left(\frac{s}{r_w}\right) \tag{4.104}$$

Thus, the inductance is now simply

$$\psi = LI \rightarrow L = \frac{\mu a}{\pi} \ln\left(\frac{s}{r_w}\right) \tag{4.105}$$

Often the inductance is expressed in terms of a per unit length

$$\frac{L}{m} = \frac{\mu}{\pi} \ln\left(\frac{s}{r_w}\right) \text{(H/m)} \tag{4.106}$$

As a note, we can see that the integration from the wire diameter to the separation between the two wires neglected the wire diameter itself since we assumed that the separation was much larger than the wire radius. Also, we can apply a similar rule here for accuracy sake when applying this formula. If the separation is 5× greater or more than the wire radius, then the formula should be reasonably accurate. Moreover, the length of the wire should be approximately 5× larger than the separation. These rules of thumb become handy when performing estimations in the future.

4.3.3 Resistance Between Two Vias in a Plane

A valuable formula which estimates the resistance between two circular regions in a plane is the approximation to the resistance between two vias. In many cases, the basic calculation of the resistance in a plane where the current is uniform suffices for an estimate. However, when the plane is much larger than the source and/or sink, it is often necessary to get a more accurate estimate.

The derivation is slightly more complex than the previous ones in that we need to determine first the current density function to compute the resistance (our final objective). We may place two circular areas to represent the source and sink and locate them on the x axis. If we do so, eventually, the computation may be easier to be solved in Cartesian coordinates.

The first step in the problem is to identify where we want the computation to go. To get the resistance, we need to compute the current and the voltage. From Faraday's law we know that

$$V_A - V_B = -\int \mathbf{E} \cdot d\mathbf{l} = -\int \frac{\mathbf{J}}{\sigma} \cdot d\mathbf{l} \rightarrow R = \int \frac{\mathbf{J}}{I\sigma} \cdot d\mathbf{l} \qquad (4.107)$$

where **J** is the current density. Thus, our objective is to find **J**. We may start with the picture shown in Figure 4.16.

We assume that the z axis is out of the paper toward us. The current flows from left to right in the figure from one via toward the other. The thickness of the plane is t and the current is uniform in the z direction.

The current density at a point in cylindrical coordinates is

$$\mathbf{J}_r = \frac{I}{2\pi tr}\hat{u}_r \qquad (4.108)$$

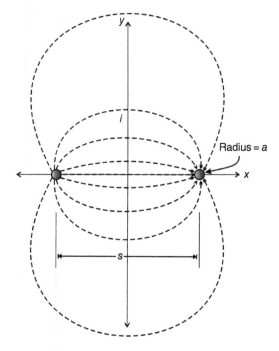

Figure 4.16 Current density between two vias in a plane.

We can verify that this is indeed the function by integrating it over the entire area to get the total current I. Converting to Cartesian coordinates yields

$$\mathbf{J}_{xy} = \frac{I}{2\pi t \sqrt{x^2 + y^2}} \left(\frac{x}{\sqrt{x^2 + y^2}} \hat{\mathbf{u}}_x + \frac{y}{\sqrt{x^2 + y^2}} \hat{\mathbf{u}}_y \right) \tag{4.109}$$

This simplifies to

$$\mathbf{J}_{xy1} = \frac{-I}{2\pi t \left(\left(x + \frac{s}{2} \right)^2 + y^2 \right)} \left(\left(x + \frac{s}{2} \right) \hat{\mathbf{u}}_x + y \hat{\mathbf{u}}_y \right), \quad -\frac{s}{2} < x < 0 \tag{4.110}$$

We should note that we used the coordinate translations from Section 4.1.1 to derive this. Note that this function is monotonically *decreasing* as it moves within the range defined. This means that this function is actually negative; for example, the current density is a negative function in this area. Alternatively, the current density rate is *decreasing*. As the data in x traverses from the x axis (0) toward the other via (positive x), the current density is *increasing*. Thus, this function is *positive*. This is the second function

$$\mathbf{J}_{xy2} = \frac{I}{2\pi t \left(\left(x - \frac{s}{2} \right)^2 + y^2 \right)} \left(\left(x - \frac{s}{2} \right) \hat{\mathbf{u}}_x + y \hat{\mathbf{u}}_y \right), \quad 0 < x < \frac{s}{2} \tag{4.111}$$

The resistance is now found through Eq. (4.107). Here, we make the simplification that we integrate along the $y = 0$ axis.

$$R = \int \frac{\mathbf{J}}{I\sigma} \cdot d\mathbf{l} = \frac{1}{2\pi\sigma t} \int_{-\frac{s}{2}+a}^{\frac{s}{2}-a} \left[\frac{1}{\left(x + \frac{s}{2} \right)} - \frac{1}{\left(x - \frac{s}{2} \right)} \right] dx \tag{4.112}$$

where a is the radius of both of the vias. After integration, we get

$$\frac{1}{2\pi\sigma t} \ln \frac{\left(x + \frac{s}{2} \right)}{\left(x - \frac{s}{2} \right)} \Bigg|_{-\frac{s}{2}+a}^{\frac{s}{2}-a} = \frac{1}{2\pi\sigma t} \left[\ln \left(\frac{s-a}{-a} \right) - \ln \left(\frac{a}{-s+a} \right) \right]$$

$$= \frac{1}{\pi\sigma t} \ln \left(\frac{s-a}{a} \right) \tag{4.113}$$

We will apply this formula shortly to a few simple problems at the end of the chapter. We will also do a quick check on each of these formulae to see if they make sense.

4.3.4 Inductance of Small Wire or Trace Above Plane Using Image Theory

Another important formula is the inductance of a wire above a plane. In actual, this problem is no different than the two-wire problem we just solved earlier in Section 4.3.2. The reason is that we can use image theory to compute the inductance, assuming the distance from the plane is large enough [1]. There is an exact formula that was derived in [2] that is slightly more complex.

We first orient the wire above a plane as shown in Figure 4.17. The wire is once again coming out of the z axis toward us.

We assume it is very long relative to the diameter ($l = a$) and the distance from the plane which is located on the $y = 0$ axis ($s/2$). If we place a theoretical wire symmetrically below the plane as shown in the figure, we can imagine that the field lines will be perfectly symmetric. This is now the same problem that we solved for (essentially) in Section 4.3.2. We know that the inductance was found to be

$$L = \frac{\mu a}{\pi} \ln\left(\frac{s}{r_w}\right) \tag{4.114}$$

But since we are only examining half the space (the region above the plane), we know that the inductance must be

$$L = \frac{\mu a}{2\pi} \ln\left(\frac{s}{r_w}\right) \tag{4.115}$$

Notice that in this case we integrated essentially from the radius to $s/2$ rather than to entire separation distance s.

Section 4.4 gives some examples on how to use these formulasand their applicability to certain problems.

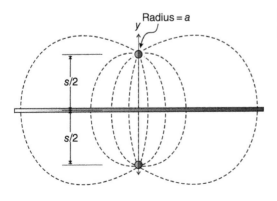

Radius = a

Figure 4.17 Wire above ground plane.

4.4   Examples of Using Equations

We end this chapter by analyzing a few key structures that are common in power integrity problems for interconnections between components (mainly capacitors) in a PCB or package. As a reminder, these problems make use of the previous formulae with the assumption that they are approximations only.

The first thing we need to discuss here are the approximation limits. In each of the aforementioned equations, we made some important assumptions with regard to the dimensions. The reason is that in all of these cases, if these basic geometrical assumptions are not met, we cannot assure the accuracy of the results. Our expectation with most of these closed-form expressions is to achieve an accuracy of 20% or better. Though this seems like a large error, in fact, it is a reasonable expectation given the complexity of many of the problems that they are applied to as we shall see. The basic problem is that the formulae break down quickly when the limits are violated. This is because the field patterns can no longer be approximated, and both fringing fields and charge concentrations within the conductors begin to dominate.

Thus, as we examine each problem independently, we will first check to see if the conditions are met.

4.4.1   Power Trace Above a Plane Between Capacitors

Figure 4.18 illustrates a classic problem that is seen on a PCB as examined from a top view. We place the structure within a Cartesian coordinate system for simplicity. In this case, there are capacitors placed at both ends of the interconnect of interest.

Our goal is to estimate (or approximate) the inductance between the two planes. One of the planes is larger than the other, which means that the problem is not exactly the same as what our formula is intended to predict. However, it may be close enough to get a reasonable estimate. First, we determine the dimensions and then check to see if they meet our overall limits. The dimensions for the traces are shown in Table 4.1.

The width is the dimension of the smaller plane (top one in this case). The separation is the distance between the two planes (dielectric thickness). To get

Table 4.1 Dimensions for problem.

Dimensions	Metric
l (cm)	7.8
w (cm)	1.8
s (mm)	0.315

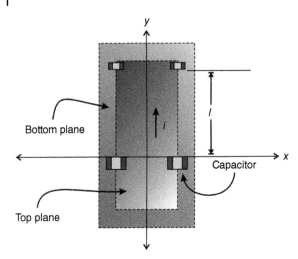

Figure 4.18 Capacitors on either side of a plane structure.

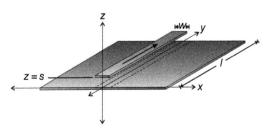

Figure 4.19 Trace above a plane.

the entire view, we place the two planes within all of the axes, as shown in Figure 4.19.

The length is taken to the edges of the interconnect of the vias into the planes for the capacitors themselves; that is, we want to estimate where the current sink will be from one end to the other. This determines the length of the trace in the table. We do not want to use the center of vias since this is not where the current is flowing. The inductance is determined from the edge of one end where the vias egress downward to the other end.

The length and width should be approximately 5× the separation between the two planes to ensure the approximation is useful. A quick check of the dimensions clearly shows this is the case (note the separation s is in millimeters). The other question we need to ask is why we do not use the width of the bottom plane? This is because the field patterns will be focused between the top trace and the bottom trace. Since the field patterns are essentially between them (H and E), and the separation is small, we should not expect to see a large amount of current on the bottom trace outside of the shadow of the top trace or plane.

We now check the limit of the function to see if it makes sense. If the separation s goes to zero, we would expect that the inductance would also go to zero.

$$\lim_{s \to 0} \frac{\mu l s}{w} = 0 \tag{4.116}$$

Also, if the length extends to infinity, we would expect the inductance to go there as well, which it does by inspection. The final one is the width. What if the width goes to infinity? Why does the inductance go to zero? We can examine the magnetic field for a moment to see why. First, the magnetic field is estimated from Eq. (4.100).

$$I = \oint H_y \, dx \cong \int_0^w H_y \, dx = H_y w \to H_y = \frac{I}{w} \tag{4.117}$$

Since the current enclosed is a constant, as we integrate over the width, the field between the planes gets smaller and smaller. Thus, in the limit, the magnetic field goes to zero and thus the inductance.

We can now estimate the inductance from our simple formula

$$L \cong \frac{\mu l s}{w} = \frac{(4\pi e - 7)(7.8e - 2)(3.15e - 4)}{(1.8e - 2)} = 1.7 \text{ nH} \tag{4.118}$$

We can check this when we get to Chapter 7 where we solve a number of these problems numerically. Static problems such as these may be analyzed in two dimensions due to their linearity along a given axis. Therefore a two-dimensional field solver or a simple piece of code may be written to determine the solution.

4.4.2 Inductance of a Trace Over a Plane

We now use the formula from Section 4.3.4. The illustration below shows a classic problem that this formula may be used for. We have a trace over a plane where the length and width of the plane are large compared to the trace width and separation or height off of the plane. Normally, this problem is solved using a field solver as will be discussed in Chapter 7. However, in this case, we wish to estimate the inductance. Our goal is to get within 20% of the actual value. We will use a closed-form expression to estimate the correct value here. There are a number of these equations that estimate the inductance of a trace over a plane. Some of these are discussed in Ref. [15]. However, for illustration purposes, we use the aforementioned basic formula.

If the wire width is small enough, we may approximate it using a circular wire. The dimensions for the structure are given in Table 4.2.

In this case the wire width is only $\frac{1}{2}$ the separation. We will see in a moment if our computation is good enough for estimating the inductance (Figure 4.20).

Table 4.2 Dimensions for wire above plane.

Dimensions	Metric
l_{tp} (cm)	7.8
w_t (mm)	0.157
s (mm)	0.315
w_p (cm)	3.4

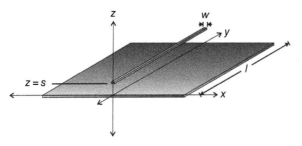

Figure 4.20 Wire above ground plane.

Let us compute the inductance using Eq. (4.115). We approximate the radius with ½ the width of the inductor for now. The inductance is then approximated as

$$L \cong \frac{\mu l}{2\pi} \ln\left(\frac{s}{\frac{w}{2}}\right) = (2e - 7)\ln\left(\frac{2 \times 0.315}{0.157}\right) = 21.7 \,\text{nH} \tag{4.119}$$

We may compare this with an approximate expression for a land from Ref. [8], which yields

$$L \cong \frac{\mu l}{2\pi} \left[\ln\left(\frac{s}{\frac{w}{2}}\right) + \frac{1}{2}\right] = 29.5 \,\text{nH} \tag{4.120}$$

This is a 26% error. This is larger than desired for an estimate (usually). It is interesting to note that Eq. (4.120) is very similar to the one in Section 4.3.2. The equation assumes that the length is much larger than the width, which is the case. Thus, we would expect this equation to be somewhat accurate. However, the one is Section 4.3.4 does not agree well with this one. If we increase the separation by say 7× that of the width, we would get a 16% error, which is within our target accuracy. In this case, we can see that the dimensions actually matter. Moreover, sometimes it is more prudent to use another formula rather than one that is not necessarily developed for the problem at hand.

4.5 Summary

In this chapter, we reviewed a number of important concepts for electromagnetic structures as a basis for the upcoming analyses for the power integrity problems. This introduction was limited and discussed only the concepts that were applicable to the overall subject at hand. The reader was reminded throughout to review the references at the end of this chapter, and other subject matter, for a more thorough study.

The chapter began with a review of vectors to give a basis for the later analyses for the vector calculus equations that were to be introduced. The discussion then moved onto the three most common coordinate systems: *Cartesian, Cylindrical*, and *Spherical*. Many problems will be encountered in the study of power integrity with just the first two.

The main concepts of this chapter were around static fields; here, the focus was on electrostatic and magneto-static fundamentals, which are crucial to the understanding of extracting circuit elements from first principal analyses. These basics are a subset of Maxwell's equations which were introduced in brief. Within these two areas, a number of important relations were discussed, including *Ampere's law* for static fields, *Faraday's law*, and *Gauss's law*.

The next subject is an overview of the Biot–Savart law and the magnetic vector potential. These equations are useful both conceptually and practically for solving many simple geometrical structures for estimation of key parameters such as inductance and resistance.

We then moved into introducing some simple but useful closed-form relations for analyzing structures using first principles. The four structures are a simple-plane pair, inductance of two parallel wires, resistance between vias in a plane, and finally, the inductance of a small wire above a plane. Though these formulae are simplified representations of actual circuits on a PCB or package, they are nonetheless useful for many approximations. Also discussed were a few of the limitations in the boundaries of these formulae.

Finally, we discussed two simple examples of how to use the formulas, specifically, a power trace above a plane and the inductance of a small trace above a plane. Both examples were intended to illustrate how the approximations may be reasonable in some cases versus others where the result may not be accurate enough. More examples of using these formulae are given in the "Problems" section at the end of the chapter.

Problems

4.1 Convert the following vector from Cartesian coordinates to cylindrical and spherical.

$$\vec{F} = 5\hat{u}_x + 23\hat{u}_y + 13\hat{u}_z \tag{4.121}$$

Table 4.3 Dimensions for Problem 4.8.

Dimensions	Metric
l (cm)	3
w (cm)	1.2
s (mm)	0.315

4.2 Multiply the following two vectors together using both the cross product and the dot product.

$$\vec{A} = 2\hat{u}_x - 4\hat{u}_y + 11\hat{u}_z \tag{4.122}$$

$$\vec{B} = -\hat{u}_x + 7\hat{u}_y + 4\hat{u}_z \tag{4.123}$$

Plot the result on a graph using the same coordinate positions as in Figure 4.2

4.3 Add the two vectors together from the previous problem. Plot the result. Now convert them to cylindrical coordinates and plot them in a graph as in Figure 4.3.

4.4 Convert the following vector from spherical coordinates to Cartesian. Plot the result.

$$\vec{E} = 7\hat{u}_r + 2\hat{u}_\theta + 4\hat{u}_z \tag{4.124}$$

4.5 Shorten the wire in Example 4.5 to a distance a, where the length is much smaller than the observed field strength. Compute the flux density vector **B** for the short wire.

4.6 Compute the same flux density as in the previous problem except use the magnetic vector potential. Use Eq. (4.86).

4.7 Determine the flux density of a long wire now as in Example 4.5 using the magnetic vector potential.

4.8 Compute the inductance of a simple plane-pair using equation for the dimensions in Table 4.3.

4.9 Determine the inductance between two wires given the dimensions in Table 4.4.

Table 4.4 Dimensions for Problem 4.9.

Dimensions	Metric
a (cm)	11
r_w (mm)	0.2
s (cm)	4

Table 4.5 Dimensions for Problem 4.10.

Dimensions	Metric
s (cm)	7
r_a (mm)	0.2
t (mm)	0.3

Table 4.6 Dimensions for Problem 4.11.

Dimensions	Metric
l_{tp} (cm)	7.8
w_t (mm)	0.157
s (mm)	0.04
w_p (cm)	1.7

4.10 Determine the resistance between two vias given the dimensions in Table 4.5.

4.11 Determine the inductance of a wire above a plane given the dimensions in Table 4.6. Use both Eqs. (4.115) and (4.120). What is the percent difference in accuracy?

Bibliography

1 Paul, C.R. (1992). *Introduction to Electromagnetic Compatibility*. Wiley.
2 Paul, C.R. and Nasar, S.A. (1987). *Introduction to Electromagnetic Fields*, 2e. Wiley.
3 Marshall, S.V., DuBroff, R.E., and Skitek, G.G. (1996). *Electromagnetic Concepts and Applications*, 4e. Prentice Hall.

4 Hayt, W.H. (1981). *Engineering Electromagnetics*, 4e. McGraw-Hill.

5 Ramo, S., Whinnery, J.R., and Van Duzer, T. (1994). *Fields and Waves in Communication Electronics*, 3e. Wiley.

6 Balanis, C.A. (1989). *Advanced Engineering Electromagnetics*. Wiley.

7 Sadiku, N.O. (1992). *Numerical Techniques in Electromagnetics*. CRC Press.

8 Grover, F.W. (1952). *Inductance Calculations*. Dover.

9 Chew, W.-C., Jin, J.-M., Lu, C.-C. et al. (1997). Fast solution methods in electromagnetics. *IEEE Transactions on Antennas and Propagation* 45 (3): 533–543.

10 Goddard, K.F., Roy, A.A., and Sykulski, J.K. (2005). Inductance and resistance calculations for isolated conductors. *IEE Proceedings – Science, Measurement and Technology* 152 (1): 7–14.

11 Kim, J., Fan, J., Ruehli, A. et al. (2011). Inductance calculations for plane-pair area fills with vias in a power distribution network using a cavity model and partial inductances. *IEEE Transactions on Microwave Theory and Techniques* 59 (8): 1909–1924.

12 Ruehli, A.E. (1972). Inductance calculations in complex integrated circuit environment. *IBM Journal of Research and Development* 16 (5): 470–481.

13 Hernandez-Sosa, G. and Sanchez, A. (2012). Analytical calculation of the equivalent inductance for signal vias in parallel planes with arbitrary P/G via distribution, in complex integrated circuit environment. In: *ICCDCS* (14–17 March 2012). Playa del Carmen: IEEE, Inc.

14 Zhou, F., Ruehli, A.E., and Fan, J. (2010). Efficient mid-frequency plane inductance calculation. In: *POWERCON* (25–30 July 2010). Fort Lauderdale, FL: IEEE, Inc.

15 Leferink, F.B.J. (1995). Inductance calculations; methods and equations. In: *IEEE International Symposium on Electromagnetic Compatibility, Symposium Record* (14–18 August 2010), 16–22. Atlanta, GA: IEEE, Inc.

16 Leferink, F.B.J. and Van Doom, M.J.C.M. (1993). Inductance of printed circuit board ground planes. In: *IEEE International Symposium on Electromagnetic Compatibility, Symposium Record* (9–13 August 1993), 327–329. Dallas, TX: IEEE, Inc.

17 Leferink, F.B.J. (1996). Inductance calculations; experimental investigations. *IEEE International Symposium on Electromagnetic Compatibility, Symposium Record* (19–23 August 1996). IEEE, Inc., Santa Clara, CA. pp. 235–240.

18 Hoer, C. and Love, C. (1965). Exact inductance equations for rectangular conductors with applications to more complicated geometries. *Journal of Research of the National Bureau of Standards – C. Engineering and Instrumentation* 69C (2): 127–137.

Part II

Tools for Power Integrity Analysis

5

Transmission Line Theory and Application

In the previous chapter, we noted at the beginning that transmission line behavior played a role in the understanding of power integrity problems. The physical layout of a power distribution design typically includes structures that are reasonably continuous or long with respect to frequency (e.g. power rails, power planes). This was evident in the first example in Chapter 4. In many cases, transmission line analysis can provide a fast and accurate representation for evaluating the efficacy of a design. Furthermore, understanding the behavior of transmission lines can yield insight in simulation and experimental validations. Thus, an abbreviated introduction into transmission line theory follows.

5.1 Telegrapher's Equations

The development of the typical transmission line equations, or telegrapher's equations, as they are often referred to, begins with a few assumptions. Figure 5.1 shows an arbitrary transmission line geometry from which development can take place. Two conductors, each of arbitrary cross-section, are illustrated. The conductors are long relative to their separation and thickness, and are parallel along the entire length without a change in cross-section. The primary assumption for employing the transmission line analysis is that the electric and magnetic fields associated with the geometry are orthogonal to each other as well as the direction of propagation. The orthogonality in this relationship is called the transverse electromagnetic (TEM) mode and is dominant in power integrity applications [1]. The source and load may couple into higher-order modes, but are typically inconsequential for power integrity.

The second assumption is that the cross-section is an equipotential surface, and consequently the current is uniform through the conductor cross-section. Lastly, the separation between the two conductors must be electrically small compared to the shortest wavelength of interest. The combination of the

Power Integrity for Electrical and Computer Engineers, First Edition. J. Ted Dibene II and David Hockanson.
© 2020 John Wiley & Sons, Inc. Published 2020 by John Wiley & Sons, Inc.

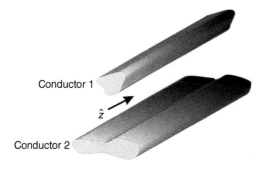

Figure 5.1 Two long, parallel conductors of arbitrary cross-section comprising a general transmission line.

Conductor 1

$\hat{z}$

Conductor 2

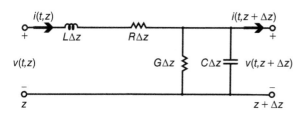

$i(t,z)$ $L\Delta z$ $R\Delta z$ $i(t,z + \Delta z)$

$v(t,z)$ $G\Delta z$ $C\Delta z$ $v(t,z + \Delta z)$

z $z + \Delta z$

Figure 5.2 Schematic representation of a segment of a transmission line with per-unit length parameter.

described assumptions allows for considering voltages and currents, instead of a full-wave analysis.

The constancy of the geometry along the length (the z direction in Figure 5.1) provides the opportunity to distribute the general parameters of resistance, inductance, capacitance, and conductance in sections along the path that are small enough to ensure that a significant phase change does not occur along a modeled segment.

Figure 5.2 shows a schematic drawing of a reduced section of a transmission line. The transmission line is reduced to an electrically small segment of length Δz. The characteristic parameters are given in per-unit length. The resistance per-unit-length models the resistive losses in the conductors. The conductance per-unit-length models the dielectric losses of the media between and around the conductors. The per-unit-length capacitance and inductance model the stored electric and magnetic fields, respectively. The voltage and current in the transmission line segment are functions of time and space. Consequently, the voltage and current equations can be developed to define values over the entire length of the transmission line.

Using basic circuit analysis, the following equations for voltage and current may be derived:

$$\frac{v(t, z + \Delta z) - v(t, z)}{\Delta z} = -Ri(t, z) - L\frac{\partial i(t, z)}{\partial t} \tag{5.1}$$

$$\frac{i(t, z + \Delta z) - i(t, z)}{\Delta z} = -Gv(t, z + \Delta z) - C\frac{\partial v(t, z + \Delta z)}{\partial t} \tag{5.2}$$

where the variables and parameters are described in Figure 5.2. Equations (5.1) and (5.2) detail the approximate voltage and current along a two-conductor transmission line. However, the exact voltage and current at any point on the transmission line can be found by taking the limit as $\Delta z \to 0$. The equations are then reduced to

$$\frac{\partial v(t, z)}{\partial z} = -Ri(t, z) - L \frac{\partial i(t, z)}{\partial t} \tag{5.3}$$

$$\frac{\partial i(t, z)}{\partial z} = -Gv(t, z) - C \frac{\partial v(t, z)}{\partial t} \tag{5.4}$$

These describe the voltage and current at every point on the transmission line, and are the general telegrapher's equations in the time domain. The equations appear similar to one-dimensional differential forms of Faraday's and Ampere's laws.

The equations can be further manipulated: essentially decoupled. Differentiating Eqs. (5.3) and (5.4) with respect to the length z yields

$$\frac{\partial^2 v(t, z)}{\partial z^2} = -R \frac{\partial i(t, z)}{\partial z} - L \frac{\partial^2 i(t, z)}{\partial z \partial t} \tag{5.5}$$

$$\frac{\partial^2 i(t, z)}{\partial z^2} = -G \frac{\partial v(t, z)}{\partial z} - C \frac{\partial^2 v(t, z)}{\partial z \partial t} \tag{5.6}$$

Substituting Eqs. (5.3) and (5.4) appropriately into Eqs. (5.5) and (5.6) results in the following travelling wave equations:

$$\frac{\partial^2 v(t, z)}{\partial z^2} = LC \frac{\partial^2 v(t, z)}{\partial t^2} + (RC + LG) \frac{\partial v(t, z)}{\partial t} + RGv(t, z) \tag{5.7}$$

$$\frac{\partial^2 i(t, z)}{\partial z^2} = LC \frac{\partial^2 i(t, z)}{\partial t^2} + (RC + LG) \frac{\partial i(t, z)}{\partial t} + RGi(t, z) \tag{5.8}$$

Equations (5.3), (5.4), (5.7), and (5.8) comprise the necessary tools to analyze transmission line voltages and currents. It should be pointed out that Eqs. (5.7) and (5.8) are very similar in appearance to the wave equations for electromagnetic fields [2]. Further details can be achieved by continuing in the time domain or moving to the frequency domain which we will do in Section 5.1.1.

5.1.1 Damped Transmission Line Approximation

One of the applications to power integrity that is valuable to the engineer is the damped transmission line approximation. This is where we assume that as the signal propagates down the transmission line, the series resistance increases as a function of distance as compared to being the same value for every subsegment.

We start with Figure 5.3 to illustrate the concept. In most cases, the conductance portion of the transmission line for power integrity applications

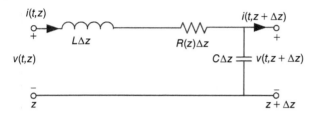

Figure 5.3 Damped transmission line representation.

is considered very large, and thus, it is removed from the transmission line equations and is considered an open circuit.

We may reanalyze the transmission line equations based on the changes to Figure 5.2.

$$\frac{\partial v(t,z)}{\partial z} = -R(z)i(t,z) - L\frac{\partial i(t,z)}{\partial t} \tag{5.9}$$

and,

$$\frac{\partial i(t,z)}{\partial z} = -C\frac{\partial v(t,z)}{\partial t} \tag{5.10}$$

As before, we may decouple the previous equations by differentiating both with respect to z to get,

$$\frac{\partial^2 v(t,z)}{\partial z^2} = LC\frac{\partial^2 v(t,z)}{\partial t^2} - \frac{\partial R}{\partial z}i(t,z) + RC\frac{\partial v(t,z)}{\partial t} \tag{5.11}$$

and,

$$\frac{\partial^2 i(t,z)}{\partial z^2} = RC\frac{\partial i(t,z)}{\partial t} + LC\frac{\partial^2 i(t,z)}{\partial t^2} \tag{5.12}$$

Equation (5.11) is not completely decoupled. However, we can use Eq. (5.9) to decouple this. Substituting in now, we get,

$$\frac{\partial^2 v(t,z)}{\partial z^2} = LC\frac{\partial^2 v(t,z)}{\partial t^2} - \frac{\partial R}{\partial z}\frac{1}{R}\left(\frac{\partial v(t,z)}{\partial z} + LC\frac{\partial v(t,z)}{\partial t}\right) + RC\frac{\partial v(t,z)}{\partial t} \tag{5.13}$$

where the functional relationship for R is implied. Equations (5.12) and (5.13) make up the two equations that may define the damped transmission line approximation.

From a practical perspective, one can imagine that as the length of the transmission line increases, the series resistance also increases to the point where the transmission line becomes an open circuit in the limit. This is essentially the effect the damped transmission line can have on the physical structure.

The importance of this relation may be seen when analyzing certain power integrity problems where the ideal solution resembles that of a damped transmission line for certain applications. For example, for very fast loads where it is required that the local impedance from the capacitance be very low for dampening purposes, we see that near the load, the impedance should be small. However, as one makes their way back toward the source, the impedance may be allowed to increase. This is similar to having the series components increase back toward the voltage source, dampening the lower frequency components or the signals that are further away from the load.

We will revisit the damped transmission line approximation in a practical sense in Chapter 8. Section 5.2 discusses the solution to the voltage and current equations for a standard transmission line using frequency-domain analysis.

5.2   Frequency-Domain Analysis Fundamentals

Steady-state analysis for transmission lines is more easily performed in the frequency domain. A common approach for solving partial-differential equations is to assume that the solution is the product of multiple functions each with one independent variable or

$$v(t, z) = T(t)V(z) \tag{5.14}$$

This technique is called *separation of variables* and is discussed in detail in Chapter 7. A functional solution to $T(z)$ in Eqs. (5.7) and (5.8) is $T(t) = e^{\mp j\omega t}$, where the electrical engineering standard of $j = \sqrt{-1}$ is used. The development continues with the further electrical engineering standard that the time solution only includes the positive argument, so $T(t) = e^{j\omega t}$ (note that any scalar multiplier can be combined with the z functional). Substituting the product forms of $v(t,z)$ and $i(t,z)$ into Eqs. (5.3) and (5.4) yields

$$e^{j\omega t}\frac{\partial V(z)}{\partial z} = -e^{j\omega t}RI(z) - j\omega e^{j\omega t}L\,I(z) \tag{5.15}$$

$$e^{j\omega t}\frac{\partial I(z)}{\partial z} = -e^{j\omega t}GV(z) - j\omega e^{j\omega t}C\,V(z) \tag{5.16}$$

Factoring out $e^{j\omega t}$ results in the general telegrapher's equations for the frequency domain:

$$\frac{\partial V(z)}{\partial z} = -I(z)(R + j\omega L) \tag{5.17}$$

$$\frac{\partial I(z)}{\partial z} = -V(z)(G + j\omega C) \tag{5.18}$$

Similarly, the wave equations of (5.7) and (5.8) can be represented in the frequency domain as

$$\frac{\partial^2 V(z)}{\partial z^2} = (R + j\omega L)(G + j\omega C)V(z) \tag{5.19}$$

$$\frac{\partial^2 I(z)}{\partial z^2} = (R + j\omega L)(G + j\omega C)I(z) \tag{5.20}$$

The general solution to Eq. (5.19) is

$$V(z) = V^- e^{\sqrt{(R+j\omega L)(G+j\omega C)}z} + V^+ e^{-\sqrt{(R+j\omega L)(G+j\omega C)}z} \tag{5.21}$$

Equation (5.21) details waves propagating in the $-z$ and z directions. In the electrical engineering fashion, the temporal argument was chosen to be $j\omega t$, and time advancement is assumed to increase positively. Consequently, the exponential argument $-\sqrt{(R + j\omega L)(G + j\omega C)}z$ represents a $+z$ directed wave, while the $-z$ directed wave is captured with the $+\sqrt{(R + j\omega L)(G + j\omega C)}z$. The values for the scalar multipliers V^- and V^+ require source and load equations to be determined.

A similar generic solution can be given for Eq. (5.20); however, considering basic circuit constructs may achieve better insight into the behavior of currents and voltages on transmission lines. Substituting Eq. (5.21) into Eq. (5.17) yields the generic current solution:

$$I(z) = -V^- \sqrt{\frac{G + j\omega C}{R + j\omega L}} e^{\sqrt{(R+j\omega L)(G+j\omega C)}z} + V^+ \sqrt{\frac{G + j\omega C}{R + j\omega L}} e^{-\sqrt{(R+j\omega L)(G+j\omega C)}z} \tag{5.22}$$

Equation (5.22) was generated from the sign conventions identified in Figure 5.2. The negative sign assigned to the scalar V^- indicates that the $-z$ directed current is indeed in the reverse polarization.

Considering only the positively directed voltage and current yields the equations

$$V^+(z) = V^+ e^{-\sqrt{(R+j\omega L)(G+j\omega C)}z} \tag{5.23}$$

$$I^+(z) = V^+ \sqrt{\frac{G + j\omega C}{R + j\omega L}} e^{-\sqrt{(R+j\omega L)(G+j\omega C)}z} \tag{5.24}$$

For a given z, the ratio of the $+z$ directed voltage and current is given by

$$\frac{V^+(z)}{I^+(z)} = \sqrt{\frac{R + j\omega L}{G + j\omega C}} \equiv Z_o \tag{5.25}$$

where Z_o is defined as the characteristic impedance of the transmission line and identifies the relationship between the directional components of the travelling waves. The same ratio results from analyzing the $-z$ directed voltage and current after accounting for the change in the polarity of the current. The next example illustrates a use for the previous equations.

Example 5.1   *Characteristic Impedance of a Coaxial Transmission Line*
An infinitely long coaxial transmission line is shown in Figure 5.4. The conductors are perfect electric conductors (PECs). The dielectric between the conductors has a permittivity and permeability given by ε and μ, respectively. Find the characteristic impedance.

Solution:
To establish the convention, assume that the center conductor is the positive voltage and the outer conductor is the reference conductor. The PECs indicate that the current is distributed on the inner surface of the outer conductor and the outer surface of the inner conductor. Furthermore, the PECs and the lossless nature of the dielectric result in $R, G = 0$.

By symmetry, the current on the inner and outer conductors is uniformly distributed around the surface. The magnetic flux density vector $\vec{B}$ would also be uniformly distributed and polarized in the $\hat{\Phi}$ direction. Employing Ampere's law in between the two conductors on paths of constant distance ρ results in

$$\frac{1}{\mu} \int_0^{2\pi} (B_\phi \hat{\phi}) \cdot (\rho \, d\phi \hat{\phi}) = I \tag{5.26}$$

$$\vec{B}(\rho) = \frac{I\mu}{2\pi\rho} \hat{\phi} \tag{5.27}$$

The magnetic flux circulates around the center conductor. The total flux Ψ_m can be calculated as

$$\Psi_m = \int_a^b \frac{I\mu}{2\pi\rho} \hat{\phi} \cdot d\rho \, \hat{\phi} \, \Delta z = \frac{I\mu}{2\pi} \ln \frac{b}{a} \, \Delta z \tag{5.28}$$

Figure 5.4 A depiction of a coaxial transmission line for use in Example 5.1. The dimensions are given pictorially. The conductors are perfect electric conductors (PECs), and the dielectric in between is lossless with permittivity and permeability values of ε and μ.

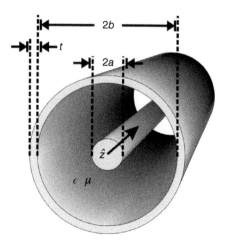

Finally, the inductance per unit length can be calculated.

$$L_{coaxial} = \frac{\Psi_m}{I \Delta z} = \frac{\mu}{2\pi} \ln \frac{b}{a} \text{ (H/m)} \tag{5.29}$$

In a similar fashion, the capacitance per unit length can be calculated. A uniformly distributed charge Q may be assumed around the surface of the center conductor, and an equal but opposite charge is likewise on the inner surface of the outer PEC. Consequently, the electric field varies only with respect to ρ, and is polarized in the $\hat{\rho}$ direction. Employing Gauss's law yields

$$\oint\!\!\!\oint_s \varepsilon \overrightarrow{E}(\rho) \cdot \overrightarrow{ds} = \int_0^{2\pi} \varepsilon E(\rho) \rho \, d\phi \Delta z = Q \tag{5.30}$$

The electric field vector is parallel to the surfaces that form the cross-section of the coaxial cable, and therefore the dot product is zero. The uniformity along the length of the wire results in the integration described in Eq. (5.30). Solving the integral results in

$$\overrightarrow{E}(\rho) = \frac{Q}{2\pi\varepsilon\rho \, \Delta z} \hat{\rho} \tag{5.31}$$

Integrating along a path from the interior conductor to the outside conductor gives the voltage between the two:

$$V = \int_a^b \frac{Q}{2\pi\varepsilon\rho \, \Delta z} \hat{\rho} \cdot d\rho \hat{\rho} = \frac{Q \ln\left(\frac{b}{a}\right)}{2\pi\varepsilon \, \Delta z} \tag{5.32}$$

Using this information in the defining equation for capacitance per unit length yields

$$C \equiv \frac{Q}{V \Delta z} = \frac{2\pi\varepsilon}{\ln\left(\frac{b}{a}\right)} \text{ (F/m)} \tag{5.33}$$

Having detailed the development of the per-unit-length inductance and capacitance in this lossless case, the characteristic impedance Z_o is then given by

$$Z_o = \sqrt{\frac{\frac{\mu}{2\pi} \ln \frac{b}{a}}{\frac{2\pi\varepsilon}{\ln \frac{b}{a}}}} = \sqrt{\frac{\mu}{\varepsilon}} \ln\left(\frac{b}{a}\right) \text{ (}\Omega\text{)} \tag{5.34}$$

Reviewing the exponent of Eqs. (5.21)–(5.24), the exponent is given as a complex variable that is not expressed in a form that readily yields separate real and imaginary parts.

$$\gamma \equiv \alpha + j\beta = \sqrt{(R + j\omega L)(G + j\omega C)} \tag{5.35}$$

Equation (5.35) can be rearranged, and complex math can be applied to yield

$$\gamma = j\omega\sqrt{LC}\,\sqrt{1-\left(\frac{R}{\omega L}\right)^2}\,\sqrt{1-\left(\frac{G}{\omega C}\right)^2}\,e^{-j\frac{1}{2}\tan^{-1}\left(\frac{R}{\omega L}\right)}e^{-j\frac{1}{2}\tan^{-1}\left(\frac{G}{\omega C}\right)} \quad (5.36)$$

Applying Euler's formula and half-angle trigonometric identities provides an expression that separates the real and imaginary parts from the previous equation.

$$Re(\gamma) = \frac{\omega\sqrt{LC}}{2}\left[\frac{\sqrt{\sqrt{\left(\frac{R}{\omega L}\right)^2+1}-1}\sqrt{\sqrt{\left(\frac{G}{\omega C}\right)^2+1}+1}}{\sqrt{\sqrt{\left(\frac{R}{\omega L}\right)^2+1}+1}\sqrt{\sqrt{\left(\frac{G}{\omega C}\right)^2+1}-1}}\right]$$

$$(5.37)$$

$$Im(\gamma) = \frac{\omega\sqrt{LC}}{2}\left[\sqrt{\sqrt{\left(\frac{R}{\omega L}\right)^2+1}+1}\sqrt{\sqrt{\left(\frac{G}{\omega C}\right)^2+1}+1} - \sqrt{\sqrt{\left(\frac{R}{\omega L}\right)^2+1}-1}\sqrt{\sqrt{\left(\frac{G}{\omega C}\right)^2+1}-1}\right]$$

$$(5.38)$$

The real part of the exponent γ constitutes the loss of the signal on the transmission line. In the general form, the loss is a function of frequency. For power integrity, we have seen that the loss is a significant concern as it is not only the costs of supplying power that is wasted in a product but also the costs for lengthening the operating time for the growing number of battery-operated systems and the efficiency requirements on nonmobile products.

The imaginary part of the exponent γ captures the wave nature of the signals in the transmission line. Of particular interest in characterizing the signal is the wavelength in the transmission line and the phase velocity v_p, which can be determined in conjunction with the wavelength given that $\lambda f = v_p$. Looking at a span of $\Delta z = \lambda$, and observing that this span requires a total phase change of 2π yields

$$\frac{\omega\sqrt{LC}}{2}\left[\sqrt{\sqrt{\left(\frac{R}{\omega L}\right)^2+1}+1}\sqrt{\sqrt{\left(\frac{G}{\omega C}\right)^2+1}+1} - \sqrt{\sqrt{\left(\frac{R}{\omega L}\right)^2+1}-1}\sqrt{\sqrt{\left(\frac{G}{\omega C}\right)^2+1}-1}\right]\lambda = 2\pi$$

$$(5.39)$$

Substituting $\omega = 2\pi f$ to the left-hand side gives

$$
\frac{\sqrt{LC}}{2}\left[\frac{\sqrt{\sqrt{\left(\frac{R}{2\pi fL}\right)^2+1}+1}\sqrt{\sqrt{\left(\frac{G}{2\pi fC}\right)^2+1}+1}}{\sqrt{\sqrt{\left(\frac{R}{2\pi fL}\right)^2+1}-1}\sqrt{\sqrt{\left(\frac{G}{2\pi fC}\right)^2+1}-1}}\right] 2\pi f \lambda
$$

(5.40)

Finally, substituting in $\lambda f = v_p$, and rearranging results in the phase velocity as

$$
v_p = \frac{2}{\sqrt{LC}}\left[\frac{\sqrt{\sqrt{\left(\frac{R}{2\pi fL}\right)^2+1}+1}\sqrt{\sqrt{\left(\frac{G}{2\pi fC}\right)^2+1}+1}}{-\sqrt{\sqrt{\left(\frac{R}{2\pi fL}\right)^2+1}-1}\sqrt{\sqrt{\left(\frac{G}{2\pi fC}\right)^2+1}-1}}\right]^{-1}
$$

(5.41)

Example 5.2 *Phase Velocity of a Coaxial Transmission Line* Determine the phase velocity for the transmission line geometry described in Example 5.1.

Solution:
The lossless nature of the transmission line simplifies the phase-velocity equation substantially:

$$
v_p = \frac{1}{\sqrt{LC}}
$$

(5.42)

Substituting in the previously determined expressions for per-unit capacitance and inductance yields

$$
v_p = \frac{1}{\sqrt{LC}} = \frac{1}{\sqrt{\frac{\mu}{2\pi}\ln\frac{b}{a}}\sqrt{\frac{2\pi\varepsilon}{\ln\frac{b}{a}}}} = \frac{1}{\sqrt{\mu\varepsilon}}
$$

(5.43)

Of particular note is that the phase velocity is only dependent on the material properties, and the result is the same as a radiating wave.

Lossless line models are useful and sufficiently accurate for electrically short transmission lines. However, in practice, the phase velocity is a function of frequency as shown in Eq. (5.41). Consequently, a generic waveform consisting of multiple frequency components will be distorted as it travels along the transmission line. As an example, a square wave launched at one end of the transmission line will not be a square wave down the line because the phase fronts of the various frequency components travel at different speeds. This type of distortion is generally called *dispersion*.

An interesting result falls out of the solution for the exponent γ if the transmission line is specifically built such that

$$\frac{R}{L} = \frac{G}{C} = \xi \tag{5.44}$$

Substituting into Eq. (5.41) yields

$$v_p = \frac{2}{\sqrt{LC}} \left[\frac{\sqrt{\sqrt{\left(\frac{\xi}{2\pi f}\right)^2 + 1} + 1} \sqrt{\sqrt{\left(\frac{\xi}{2\pi f}\right)^2 + 1} + 1}}{-\sqrt{\sqrt{\left(\frac{\xi}{2\pi f}\right)^2 + 1} - 1} \sqrt{\sqrt{\left(\frac{\xi}{2\pi f}\right)^2 + 1} - 1}} \right]^{-1} \tag{5.45}$$

Simplifying Eq. (5.45) results in

$$v_p = \frac{2}{\sqrt{LC}} \left[\sqrt{\sqrt{\left(\frac{\xi}{2\pi f}\right)^2 + 1} + 1} - \sqrt{\sqrt{\left(\frac{\xi}{2\pi f}\right)^2 + 1} + 1} \right]^{-1} = \frac{1}{\sqrt{LC}} \tag{5.46}$$

Designing a transmission line with the parameters of Eq. (5.46) results in a distortion-less transmission line.

The signal decay portion of the exponent α can also be reduced by substituting Eq. (5.46) into the real part in Eq. (5.37):

$$\alpha = \frac{\omega\sqrt{LC}}{2} \left[\frac{\sqrt{\sqrt{\left(\frac{\xi}{\omega}\right)^2 + 1} - 1} \sqrt{\sqrt{\left(\frac{\xi}{\omega}\right)^2 + 1} + 1}}{+\sqrt{\sqrt{\left(\frac{\xi}{\omega}\right)^2 + 1} + 1} \sqrt{\sqrt{\left(\frac{\xi}{\omega}\right)^2 + 1} - 1}} \right] \tag{5.47}$$

Simplifying Eq. (5.47) yields

$$\alpha = \omega\sqrt{LC} \sqrt{\sqrt{\left(\frac{\xi}{\omega}\right)^2 + 1} - 1} \sqrt{\sqrt{\left(\frac{\xi}{\omega}\right)^2 + 1} + 1}$$

$$= \omega\sqrt{LC} \sqrt{\left| \sqrt{\left(\frac{\xi}{\omega}\right)^2 + 1} - 1 \right| \left| \sqrt{\left(\frac{\xi}{\omega}\right)^2 + 1} + 1 \right|} \tag{5.48}$$

and finally

$$\alpha = \omega\sqrt{LC}\sqrt{\left(\frac{\xi}{\omega}\right)^2 + 1 - 1} = \omega\sqrt{LC}\,\frac{\xi}{\omega} = \sqrt{LC}\,\xi = \sqrt{LC\xi^2} = \sqrt{RG}$$

(5.49)

Great care must necessarily be taken in order to design distortion-less transmission line while minimizing the signal decay from loss. Furthermore, R and G are material properties that may themselves be frequency dependent (e.g. FR-4), which adds more difficulty in designing the distortion-less line if the required bandwidth is large.

Fortunately, in power integrity, dispersion is far less impactful in comparison to signal integrity. Of greater concern is a smooth transfer of power and minimizing the power lost to the resistive part of the path. Continuing the observations obtained from the frequency domain brings the discussion to analyzing and calculating input impedance.

Figure 5.5 shows the basic schematic for a loaded transmission line. Using previously established equations and nomenclature, the voltage and current relations at the load are given by

$$\frac{V(0)}{I(0)} = Z_L = \frac{V^- + V^+}{-\frac{V^-}{Z_o} + \frac{V^+}{Z_o}} = Z_o\frac{\frac{V^+}{V^-} + 1}{\frac{V^+}{V^-} - 1} \implies \frac{V^+}{V^-} = \frac{Z_L + Z_o}{Z_L - Z_o}$$

(5.50)

Given the relationship between the positive negative propagating waves established in Eq. (5.50), the input impedance at a distance l from the load can be calculated as

$$Z_{in} = \frac{V(-l)}{I(-l)} = \frac{V^+e^{\gamma l} + V^-e^{-\gamma l}}{\frac{V^+}{Z_o}e^{\gamma l} - \frac{V^-}{Z_o}e^{-\gamma l}}$$

$$= Z_o\frac{\frac{V^+}{V^-}e^{\gamma l} + e^{-\gamma l}}{\frac{V^+}{V^-}e^{\gamma l} - e^{-\gamma l}} = Z_o\frac{\frac{Z_L+Z_o}{Z_L-Z_o}e^{\gamma l} + e^{-\gamma l}}{\frac{Z_L+Z_o}{Z_L-Z_o}e^{\gamma l} - e^{-\gamma l}}$$

$$= Z_o\frac{(Z_L + Z_o)e^{\gamma l} + (Z_L - Z_o)e^{-\gamma l}}{(Z_L + Z_o)e^{\gamma l} - (Z_L - Z_o)e^{-\gamma l}}$$

(5.51)

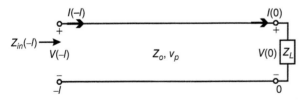

Figure 5.5 Schematic representation with a load for analyzing transmission line input impedance with a load impedance of Z_L.

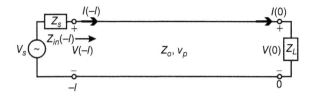

Figure 5.6 Schematic representation with a load for analyzing transmission line input impedance with a load impedance of Z_L, an equivalent source V_s, and a source impedance Z_s.

or

$$Z_{in} = Z_o \frac{Z_L(e^{\gamma l} + e^{-\gamma l}) + Z_o(e^{\gamma l} - e^{-\gamma l})}{Z_L(e^{\gamma l} - e^{-\gamma l}) + Z_o(e^{\gamma l} + e^{-\gamma l})} = Z_o \frac{Z_L + Z_o \frac{(e^{\gamma l} - e^{-\gamma l})}{(e^{\gamma l} + e^{-\gamma l})}}{Z_L \frac{(e^{\gamma l} - e^{-\gamma l})}{(e^{\gamma l} + e^{-\gamma l})} + Z_o}$$

$$= Z_o \frac{Z_L + Z_o \tanh(\gamma l)}{Z_L \tanh(\gamma l) + Z_o} \tag{5.52}$$

Notice that if the line is matched to the load ($Z_L = Z_o$), then the input impedance is simply Z_o. This *matching* has implications that will be more apparent later.

One more component is required to completely solve the differential equations: V^+ and V^- are still unresolved. Adding the source model completes the equations. Figure 5.6 shows the schematic to be used for the final analysis.

Applying basic circuit theory provides the necessary equation.

$$V(-l) = V_s \frac{Z_{in}}{Z_s + Z_{in}} \tag{5.53}$$

Substituting in the general form constructed for $V(-l)$ and the result of Eq. (5.50) yields

$$V^- e^{-\gamma l} + V^+ e^{\gamma l} = V^- \left(e^{-\gamma l} + \frac{V^+}{V^-} e^{\gamma l} \right) = V^- \left(e^{-\gamma l} + \frac{Z_L + Z_o}{Z_L - Z_o} e^{\gamma l} \right)$$

$$= V_s \frac{Z_{in}}{Z_s + Z_{in}} \tag{5.54}$$

Rearranging the result in Eq. (5.54) gives

$$V^- \left(\frac{(Z_L - Z_o)e^{-\gamma l} + (Z_L + Z_o)e^{\gamma l}}{Z_L - Z_o} \right) = V^- \left(\frac{Z_L(e^{\gamma l} + e^{-\gamma l}) + Z_o(e^{\gamma l} - e^{-\gamma l})}{Z_L - Z_o} \right)$$

$$= 2V^- \left(\frac{Z_L \cosh(\gamma l) + Z_o \sinh(\gamma l)}{Z_L - Z_o} \right)$$

$$= 2 \cosh(\gamma l)V^- \left(\frac{Z_L + Z_o \tanh(\gamma l)}{Z_L - Z_o} \right) = V_s \frac{Z_{in}}{Z_s + Z_{in}} \tag{5.55}$$

Equation (5.52) is then substituted into Eq. (5.55) to yield

$$2\cosh(\gamma l)V^{-}\left(\frac{Z_L + Z_o\tanh(\gamma l)}{Z_L - Z_o}\right)$$

$$= V_s\frac{Z_o\frac{Z_L + Z_o\tanh(\gamma l)}{Z_L\tanh(\gamma l) + Z_o}}{Z_s + Z_o\frac{Z_L + Z_o\tanh(\gamma l)}{Z_L\tanh(\gamma l) + Z_o}}$$

$$= V_s\frac{Z_o(Z_L + Z_o\tanh(\gamma l))}{Z_s(Z_L\tanh(\gamma l) + Z_o) + Z_o(Z_L + Z_o\tanh(\gamma l))} \tag{5.56}$$

Equation (5.56) can be rearranged and simplified to solve for V^{-}. Back substituting into Eq. (5.51) provides the solution for V^{+}. The solutions are given by

$$V^{-} = \frac{V_s}{2\cosh(\gamma l)}\frac{Z_o(Z_L - Z_o)}{Z_s(Z_L\tanh(\gamma l) + Z_o) + Z_o(Z_L + Z_o\tanh(\gamma l))} \tag{5.57}$$

$$V^{+} = \frac{V_s}{2\cosh(\gamma l)}\frac{Z_o(Z_L + Z_o)}{Z_s(Z_L\tanh(\gamma l) + Z_o) + Z_o(Z_L + Z_o\tanh(\gamma l))} \tag{5.58}$$

As discussed in relation to Eq. (5.52), matching the load to the line has an interesting impact. The voltage (and current) propagating in the $-z$ direction is zero, as observed in Eq. (5.57).To summarize, the voltage and current can be calculated anywhere on the transmission line with the final assembled equations:

$$V(z) = \frac{V_sZ_o}{2\cosh(\gamma l)}\frac{(Z_L - Z_o)e^{\gamma z} + (Z_L + Z_o)e^{-\gamma z}}{Z_s(Z_L\tanh(\gamma l) + Z_o) + Z_o(Z_L + Z_o\tanh(\gamma l))} \tag{5.59}$$

$$I(z) = \frac{V_s}{2\cosh(\gamma l)}\frac{-(Z_L - Z_o)e^{\gamma z} + (Z_L + Z_o)e^{-\gamma z}}{Z_s(Z_L\tanh(\gamma l) + Z_o) + Z_o(Z_L + Z_o\tanh(\gamma l))} \tag{5.60}$$

where

$$\gamma = \frac{\omega\sqrt{LC}}{2}\left[\sqrt{\sqrt{\left(\frac{R}{\omega L}\right)^2 + 1} - 1}\sqrt{\sqrt{\left(\frac{G}{\omega C}\right)^2 + 1} + 1}\right.$$
$$\left. + \sqrt{\sqrt{\left(\frac{R}{\omega L}\right)^2 + 1} + 1}\sqrt{\sqrt{\left(\frac{G}{\omega C}\right)^2 + 1} - 1}\right]$$
$$+ j\,\frac{\omega\sqrt{LC}}{2}\left[\sqrt{\sqrt{\left(\frac{R}{\omega L}\right)^2 + 1} + 1}\sqrt{\sqrt{\left(\frac{G}{\omega C}\right)^2 + 1} + 1}\right.$$
$$\left. - \sqrt{\sqrt{\left(\frac{R}{\omega L}\right)^2 + 1} - 1}\sqrt{\sqrt{\left(\frac{G}{\omega C}\right)^2 + 1} - 1}\right] \tag{5.61}$$

and

$$Z_o = \sqrt{\frac{R + j\omega L}{G + j\omega C}} \tag{5.62}$$

An example of how to use the previous equations follows.

Example 5.3 *Transmission Line Power Application* A new stadium design has expectations that all power delivery is housed in one location to ease service costs. A $10\,\Omega$ load is located $1000\,$m from the source that requires $10\,$V root mean square (RMS). The designer of the voltage regulator has provided a simple, but sufficiently accurate model whose Thevenin equivalent is an ideal source and a $0.1\,\Omega$ source resistance. Furthermore, the power is being delivered at $1000\,$Hz. A power integrity engineer is proposing a special transmission line with $R = 2.5\,\mu\Omega/$cm, $G = 25\,$nS/cm, $L = 10\,$nH/cm, and $C = 100\,$pF/cm. What must the source voltage be? How much power is delivered to the load? How much power is absorbed by the source? How much power is lost in the line?

Solution:
Figure 5.6 can be used to aid in the evaluation. The characteristic impedance of the line is given by

$$Z_o = \sqrt{\frac{R + j\omega L}{G + j\omega C}} = 10\,\Omega \tag{5.63}$$

The proposed transmission line is clearly distortion-less ($R/L = G/C$). The line is furthermore matched to the load ($Z_o = Z_L$), so the voltage at the source can be reduced to

$$
\begin{aligned}
|V(0)| &= \left| \frac{V_s Z_o}{2\cosh(\gamma l)} \frac{2Z_o}{Z_s(Z_o\tanh(\gamma l) + Z_o) + Z_o(Z_o + Z_o\tanh(\gamma l))} \right| \\
&= \left| \frac{V_s Z_o}{\cosh(\gamma l)} \frac{1}{(Z_s + Z_o)(\tanh(\gamma l) + 1)} \right| \\
&= \left| \frac{V_s Z_o}{(Z_s + Z_o)} \frac{1}{(\sinh(\gamma l) + \cosh(\gamma l))} \right| = \left| \frac{V_s Z_o}{(Z_s + Z_o)} e^{-\gamma l} \right| \\
&= \left| \frac{V_s Z_o}{(Z_s + Z_o)} e^{-\alpha l} e^{-j\beta l} \right| = \frac{|V_s| Z_o}{(Z_s + Z_o)} e^{-\alpha l} \to |V(0)| \frac{(Z_s + Z_o)}{Z_o} e^{\alpha l} \\
&= |V_s| \approx 10.4\, V_{RMS}
\end{aligned} \tag{5.64}
$$

Subsequently

$$|I_s| = \frac{V_s}{(Z_s + Z_o)} \approx \frac{11.2}{(0.1 + 10)} = 1.03\, A_{RMS} \tag{5.65}$$

The power numbers can now be calculated. The impedances of interest are all real, so the power delivered by the source can be calculated by

$$P_{del} = V_s I_s \approx 10.6 \text{ W} \tag{5.66}$$

Power absorbed by the source is given by

$$P_s = I_s^2 R_s \approx 0.1 \text{ W} \tag{5.67}$$

The load power can be similarly calculated as

$$P_L = \frac{V(0)^2}{R_L} = 10 \text{ W} \tag{5.68}$$

Finally, the power lost in the transmission line can be determined:

$$P_{TX} = P_{del} - P_s - P_L = 0.5 \text{ W} \tag{5.69}$$

Example 5.4 *Transmission Line Power Calculation II*   Another power integrity engineer observed that almost 5% of the total power delivered was lost in the transmission line. The second engineer looked at the transmission line and realized that another vendor had a line that reduced the value of R could be reduced to $R = 1\ \mu\Omega/\text{cm}$ and the shunt conductance to $G = 10\ \text{nS/cm}$. The inductance increased to $50\ \text{nH/cm}$, and the capacitance was reduced to $5\ \text{pF/cm}$. The engineer was sure that the reduction in the loss parameters would result in less wasted power and that the changes in the reactive elements would have minimal impact. Evaluate the problem again with the changes.

Solution:
To begin, the characteristic impedance can be calculated as

$$Z_o = \sqrt{\frac{R + j\omega L}{G + j\omega C}} = 97.6 e^{j0.152}\ \Omega \tag{5.70}$$

In order to supply the voltage to the load, the source voltage must be

$$10 = \frac{V_s Z_o}{2 \cosh(\gamma l)\, Z_s (Z_L \tanh(\gamma l) + Z_o) + Z_o(Z_L + Z_o \tanh(\gamma l))} \frac{2 Z_L}{}$$

$$\Rightarrow V_s \approx 32.5 e^{j1.27}\ V_{RMS} \tag{5.71}$$

The phase associated with the source voltage is referenced to the zero phase assumed in the load voltage. Similarly, the input impedance to the line is

$$Z_{in} = Z_o \frac{Z_L + Z_o \tanh(\gamma l)}{Z_L \tanh(\gamma l) + Z_o} \approx 33.7 e^{j1.23}\ \Omega \tag{5.72}$$

Knowing the voltages and impedances, the source current can then be calculated:

$$I_s = \frac{V_s}{(Z_s + Z_{in})} = 0.96\, A_{RMS} \tag{5.73}$$

As in the previous example, power numbers can now be calculated and are given by

$$P_{del} = V_s I_s \approx 10.6\ \text{W} \tag{5.74}$$

$$P_s = I_s^2 R_s \approx 0.1\ \text{W} \tag{5.75}$$

$$P_L = \frac{V(0)^2}{R_L} = 10\ \text{W} \tag{5.76}$$

and finally,

$$P_{TX} = P_{del} - P_s - P_L \approx 0.5\ \text{W} \tag{5.77}$$

Although the resistance of the line was reduced by more than 50%, the change in the characteristic impedance changes the current on the line. A reflected wave is now present, and the current has maxima and minima on the line. Ultimately, the reduction in the resistance and conductance had no impact on the total power.

The previous two examples involved delivering power at a specific frequency. Such an approach is more generally used in the microwave industry and may not be generally applicable to the power integrity engineer in the digital world. However, for completeness, there is another interesting concept apparent in the frequency-domain transmission line theory: the quarter-wave transformer.

Suppose power is delivered at a specific frequency over a long distance with respect to frequency. From Eq. (5.52), the input impedance of the line is related to the length and frequency through $\tanh(\gamma l)$. Generically expanding γ into its complex components $(\alpha + j\beta)$ results in

$$\tanh(\gamma l) = \frac{\tanh(\alpha l) + j \tan(\beta l)}{1 + j \tanh(\alpha l) \tan(\beta l)} \tag{5.78}$$

If the length of a section of a transmission line is held at a quarter wavelength, the $\tan(\beta l) = 0$, and the equation for input impedance is simplified to

$$Z_{in} = Z_o \frac{Z_L + Z_o \tanh(\alpha l)}{Z_L \tanh(\alpha l) + Z_o} \tag{5.79}$$

For a lossless line, the input impedance equals the load impedance regardless of the line impedance.

5.3 Power Planes, Grids, and Transmission Lines

In other chapters, the physical constraints and their influence on the frequencies of interest are discussed. Transmission line theory is applicable in various sections of the power delivery path depending on the application. Electrically long cables, bus bars, planes on a PCB, and even traces/wires on a large ceramic package may be areas where transmission line theory can streamline analysis.

Planes introduce another level of complexity, however, as compared to the coaxial cable discussed in Example 5.1. Planar structures are generally large in two dimensions, so a one-dimensional analysis, which was previously demonstrated, may be inapplicable. The common approach is to make distributed models in two dimensions and use a circuit simulator to provide the analysis. In Figure 5.7, a section of a voltage plane is gridded for modeling in a distributed fashion. The reference plane would be gridded in the same fashion, so the nodes are perpendicular to the planes, given that the separation between planes is small with respect to the frequencies of interest.

The nodes are located as shown in Figure 5.7; however, the values for the parameters are calculated as if the nodes were in the middle of the face, essentially having a uniform distribution across the face of the cell. The sides of the cells should be equal to ease the setup ($\Delta l = \Delta w$). The result of the model is essentially an averaging of the fields associated with the cell, not unlike numerical modeling with finite difference time domain (see Chapter 7).

Realistic losses in the materials limit the risk of underestimating high peaks, and a finer discretization may be used if there are concerns or a desire for greater accuracy.

A schematic representation of the plane in Figure 5.7 is shown in Figure 5.8. All parameters are included in the voltage plane in the figure, as is evident by the reference or "ground" symbol. The same approach was used in the previous

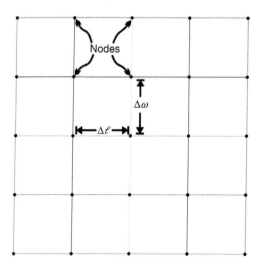

Figure 5.7 Depiction of a voltage-plane gridded for modeling. The grids are made in a square pattern, and the dark circles represent the nodes for the circuit model.

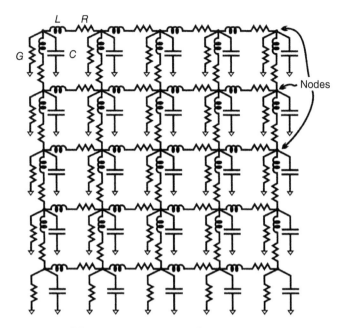

Figure 5.8 Schematic representation of the plane pair depicted in Figure 5.7.

one-dimensional approach. However, for power integrity, it is typically necessary to distribute the resistance and inductance on the ground plane as well with the conductance and capacitance (where necessary) terminating on the appropriate node of the ground plane. The inductance is simply halved from the original calculation and placed on the voltage branches and ground branches. The resistance should be calculated for each plane, which then captures possible differences in the foil thicknesses of the planes, and appropriately locates the resulting voltage drops.

Assuming that the edges of the plane depiction of Figure 5.6 represent the physical edge of the plane, one may observe that the fringing effects of the magnetic and electric flux, as well as changes in the resistance, are not accounted for in the model.

The errors associated with the edge approximation are typically small for the relatively thin dielectrics (compared with planar dimensions); however, engineers may wish to calculate specific values for edge parameters to improve accuracy.

The discretization of the planes as described earlier lends itself to easily locating source and load circuitry. Furthermore, the analysis can be used to determine maxima and minima in the voltage along the layers under various load conditions. Decoupling strategies can then be more effectively located for providing power and appropriately placed in the model for analysis. Furthermore, the distributed model can be extended to include multiple plane-pairs, and include interactions between the pairs and analyze optimal placement of decoupling capacitors.

Unfortunately, the lumped nondistributed approach to analysis of planar structures is considered impractical, although it would provide greater accuracy. The sources and loads are very localized on the planes. Consequently, the related voltages and currents propagate radially, adding a distance variable to the characteristic impedance and further complicating the transmission line equations. Such development is considered outside the scope of this work, as full-wave modeling is usually the tool of choice for that level of analysis.

Nonetheless, as we shall see in later chapters, analyzing the power distribution network (PDN) with lumped elements can yield many insights when examining such structures and are important toward understanding how the electrical behavior of placing capacitors and other lumped element circuits into a plane can either help or hinder the performance of the interconnect, or PDN. The reason is that when there is essentially only one source and load for the PDN, which is often the case, creating lumped element circuits allows one to quickly achieve results that can lead to the final solution more efficiently.

5.4 Summary

In this chapter, we covered transmission line basics. Although the PDN is not considered a transmission line, it has attributes that lend itself to transmission line analysis. In Section 5.1, we discussed telegrapher's equations. These are the basis for all transmission line analysis, and they start with a circuit-level description of the interconnect which consists of R, L, G, and C elements. We also discussed the damped transmission line approximation which assumes that the R element changes as a function of distance as the signals traverse the transmission line.

There was then a discussion on frequency-domain analysis with transmission lines where we assumed the signal was harmonic and then found a solution to the transmission line equations in this format. Finally, we discussed briefly power planes and grids and how these are used in analyzing PDNs for power integrity problems.

Problems

5.1 Determine the transmission line equations, neglecting the conductance term (G) assuming $L = L(z)$ and $C = C(z)$ in addition to the length dependency on R.

5.2 Determine the inductance, capacitance, and impedance of a coaxial cable using the values in Table 5.1.

Table 5.1 Values for coaxial calculation.

Metric	Value
a (mm)	0.1
b (mm)	1
Length (m)	1

How does the impedance compare with that in the example? Is this expected? Why?

5.3 What is the phase velocity of the cable in the previous problem?

5.4 If the resistance is equivalent to the inductance and the conductance to the capacitance (per unit length), determine the real and imaginary portions of the cable in Problem 5.2.

5.5 Redo Example 5.3 but reduce R by 25% while increasing the length by 2. What is the result? Was this expected if all the other inputs remained the same?

5.6 Redo Example 5.4 but now increase the resistance by 25% while decreasing the length by 2. What is the result? Was this also expected if all the other inputs remained the same?

5.7 Create a grid where $R = 1\,\Omega$, $G = 1\,\mu S$, $L = 10\,nH$, and $C = 0.1\,\mu F$. Build a model where you have a 5×5 grid. Use a 1 V input source at 100 MHz on the far left corner. Plot the voltages at each corner except the source point. What is the attenuation of the signal at each corner?

5.8 Redo Problem 5.7 by increasing G by 10× and reducing C by 3×. Was this what you expected?

Bibliography

1 Paul, C.R. and Nasar, S.A. (1987). *Introduction to Electromagnetic Fields*, 2e. Wiley.

2 Matick, R.E. (1969). *Transmission Lines for Digital and Communication Networks*. IEEE Press.

3 Paul, C.R. (1992). *Introduction to Electromagnetic Compatibility*. Wiley.

4 Ramo, S., Whinnery, J.R., and Van Duzer, T. (1994). *Fields and Waves in Communication Electronics*, 3e. Wiley.

5 Paul, C.R. (1994). *Analysis of Multiconductor Transmission Lines*. Wiley.

6 Zhu, Y. and Cangellaris, A.C. (2006). *Multigrid Finite Element Methods for Electromagnetic Field Solving*. Wiley. ISBN: 978-0-471-74110-7.

7 Sadiku, N.O. (1992). *Numerical Techniques in Electromagnetics*. CRC Press.

8 Balanis, C.A. (1989). *Advanced Engineering Electromagnetics*. Wiley.

9 Marshall, S.V., DuBroff, R.E., and Skitek, G.G. (1996). *Electromagnetic Concepts and Applications*, 4e. Prentice Hall.

10 Hayt, W.H. (1981). *Engineering Electromagnetics*, 4e. McGraw-Hill.

6

Signal Analysis Review

The paths and components of the power distribution network (PDN), associated with delivering power, are frequently complicated, and the source and load behaviors further exacerbate the challenges of a straightforward time-domain analysis. Many techniques exist for transforming the system equations and the various sources involved. In this text, it is assumed that the reader has knowledge of the common transform techniques discussed herein. A rigorous mathematical description is not offered here, but it can be found in numerous texts [1, 5]. The following review includes a practical reminder of the most common transforms and analytical tools used in power integrity.

6.1 Linear, Time-Invariant Systems

The foundation for signal analysis starts with setting the conditions under which transforms can be made. Although some important components in a power integrity design may not be linear or time invariant under all conditions, such approximations can be made with minimal impact on accuracy. As an example, a diode can be analyzed assuming that it is operating in the linear region, and the desired response can be checked in post analysis using Simulation Program with Integrated Circuit Emphasis (SPICE).

Starting with linearity, let $H[\cdot]$ be a system response. Furthermore, let $x(t)$ and $y(t)$ be possible stimuli and A be a scalar multiplier. The responses of $x(t)$ and $y(t)$ individually would be then

$$H[x(t)] = f(t) \tag{6.1}$$

and

$$H[y(t)] = g(t) \tag{6.2}$$

For linearity, a scalar multiplier must result in a scalar multiple of the response:

$$H[Ax(t)] = Af(t) \tag{6.3}$$

Power Integrity for Electrical and Computer Engineers, First Edition. J. Ted Dibene II and David Hockanson.
© 2020 John Wiley & Sons, Inc. Published 2020 by John Wiley & Sons, Inc.

and

$$H[x(t) + y(t)] = f(t) + g(t) \tag{6.4}$$

In general, circuit equations are linear, or are modeled in a linear fashion (such as a biased transistor), because multiplication, differentiation, and integration are linear operators.

A time-invariant system does not have restrictions on when it begins. Mathematically, the response to $x(t)$ is the same as $x(t - \Delta t)$, but delayed by Δt:

$$H[x(t - \Delta t)] = f(t - \Delta t) \tag{6.5}$$

As with linearity, circuit equations are time invariant, or are modified such that they can be modeled as such in an effort to use common analytical methods. A system that is generally not time invariant, for example, is road traffic, *especially* in metropolitan areas. Leaving for work at six in the morning may take a half hour; however, leaving at eight in the morning may take over an hour!

Given this brief reminder of linear time-invariant systems, analytical tools may be constructed. However, an important mathematical concept is necessary first, and a refresher on these basics is given in the following section.

6.2 The Dirac Delta Function

The Dirac delta function is not really a *function* in the strictest mathematical sense, but it proves to be a very useful tool in many analyses. This brief refresher on the delta function not just serves as a reminder but also ensures that the reader has the same nomenclature and conventions assumed in the material herein.

To begin, let a function $f(t)$ be given as:

$$f(t) = \frac{\epsilon - |t|}{\epsilon^2}, \quad \text{for } |t| \leq \epsilon, \text{ and } 0 \text{ otherwise} \tag{6.6}$$

The area under the curve described in Eq. (6.6) is one for $\epsilon \geq 0^+$. Now let the Dirac delta function $\delta(t)$ be the result of the limit as $\epsilon \to 0$. The area is still one, but technically the function goes to infinity at $t = 0$. The area is associated with the Dirac delta function and is called the *weight*.

An important utility of the Dirac delta function, hereafter simply called an *impulse function*, occurs under integration when in a product construction:

$$\int_{-\infty}^{+\infty} g(t)\delta(t - t_o)dt = g(t_o) \tag{6.7}$$

The result of Eq. (6.7) is used frequently in mathematics for a number of disciplines, including circuit and field analysis, as will be reviewed in the following section.

6.3 Convolution

The study of Power Integrity, like most signal analysis, typically requires an investigation of the system for a variety of stimuli. Rather than working through differential equations for each stimulus, a method for analyzing multiple responses is preferred. In the time domain, analysis is typically conducted by *convolution*.

To begin, let $H[\cdot]$ once again be a system response. A stimulus $x(t)$ results in $f(t)$:

$$f(t) = H[x(t)] \tag{6.8}$$

The stimulus $x(t)$ can be rewritten using Eq. (6.7), with a change of variables resulting in:

$$f(t) = H\left[\int_{-\infty}^{+\infty} x(\tau)\delta(t-\tau)d\tau\right] \tag{6.9}$$

Assuming that the operator $H[\cdot]$ can be moved inside of the integral operator,

$$f(t) = \int_{-\infty}^{+\infty} H[x(\tau)\delta(t-\tau)d\tau] \tag{6.10}$$

Note that the operator $H[\cdot]$ works on functions of t; consequently, $x(\tau)$ may be considered a scalar multiplier. The equation can then be rewritten as:

$$f(t) = \int_{-\infty}^{+\infty} x(\tau)H[\delta(t-\tau)]d\tau \tag{6.11}$$

The system response to the impulse function is called the *impulse response*. Given the linearity and time invariance of the operator, the impulse response is given by:

$$h(t) = H[\delta(t)] \tag{6.12}$$

By analyzing the impulse response, that is the result of a stimulus that is an impulse function, any other stimulus can be effectively analyzed by

$$f(t) = \int_{-\infty}^{+\infty} x(\tau)h(t-\tau)d\tau \tag{6.13}$$

Analyzing responses for various stimuli through Eq. (6.13) is called *convolution*.

Example 6.1   *Impulse Response of an RL Circuit*   An RL circuit is driven by an arbitrary voltage source. Find the impulse response for the current in the circuit.

Solution:
The current in the circuit is defined by the following equation:

$$v_{source}(t) = Ri(t) + L\frac{d}{dt}i(t) \quad \text{with } i(0-) = 0 \tag{6.14}$$

To begin, make the arbitrary source an impulse function:

$$\delta(t) = Rh(t) + L\frac{d}{dt}h(t) \tag{6.15}$$

Away from $t = 0$, Eq. (6.15) is homogeneous.

$$0 = Rh(t) + L\frac{d}{dt}h(t), \quad t \neq 0 \tag{6.16}$$

A solution to Eq. (6.16) is

$$h(t) = Ae^{-\frac{R}{L}t}u(t) \tag{6.17}$$

where $u(t)$ is the Heaviside step function, which is 1 for $t \geq 0$ and 0 for $t < 0$ and A is a scalar multiplier. To determine the value of A, we integrate Eq. (6.15):

$$\int_{0^-}^{\infty} \delta(t)dt = \int_{0^-}^{\infty} \left[Rh(t) + L\frac{d}{dt}h(t) \right] dt \tag{6.18}$$

The lower limit of the integral has been reduced to 0^- due to the zero functionality below the origin. The integral is then reduced to

$$1 = RA \int_{0^-}^{\infty} e^{-\frac{R}{L}t}dt + Lh(t \to \infty) - Lh(0^-) \tag{6.19}$$

Employing the initial condition from the setup of Eq. (6.14) and analyzing the remaining equation leaves the final solution for $h(t)$:

$$h(t) = \frac{1}{L}e^{-\frac{R}{L}t}u(t) \tag{6.20}$$

Equation (6.20) satisfies the initial conditions of the problem and the homogeneous solution away from $t = 0$. Considering the result intuitively, at $t = 0$, one observes that the function instantaneously changes from zero to $1/L$. The derivative therefore goes to infinity at that point, which supports the expectation of an impulse function as a source in Eq. (6.14). These cursory observations lead one to expect that the equation is accurate and can be used in calculating the current for multiple stimuli. A more rigorous mathematical verification may be employed if desired by the engineer, but is not included here for brevity.

Example 6.2 *Current in RL Circuit Calculations Using Convolution*
Find the currents in Example 6.1 for three excitations:

$$(1) \ V_0 u(t), \quad (2) \ V_0 u(t) \cdot \begin{cases} \frac{t}{\alpha} & t \leq \alpha, \alpha > 0 \\ 1 & \text{otherwise} \end{cases}, \quad \text{and} \quad (3) \ V_0 \sin(\omega t)$$

Solution:
The problems may be solved through straightforward integration using convolution.

(1) The first source is a step function with an amplitude of V_0. To apply convolution, we begin by substituting in for the impulse function in the argument:

$$h(t - \tau) = \frac{1}{L} e^{-\frac{R}{L}(t-\tau)} u(t - \tau) \tag{6.21}$$

Employing convolution as described in Eq. (6.13) yields:

$$i(t) = \frac{V_0}{L} \int_{-\infty}^{+\infty} u(\tau) e^{-\frac{R}{L}(t-\tau)} u(t - \tau) d\tau \tag{6.22}$$

Modifying the limits of integration according to the boundaries of the step functions leaves:

$$i(t) = \frac{V_0}{L} \int_0^t e^{-\frac{R}{L}(t-\tau)} d\tau, \quad t \geq 0 \tag{6.23}$$

Solving the integral gives the result for the current:

$$i(t) = \frac{V_0}{R} \left(1 - e^{-\frac{R}{L}t} \right) u(t) \tag{6.24}$$

(2) Analyzing the ramp of the second stimulus yields

$$i(t) = \frac{V_0}{L} \int_{-\infty}^{+\infty} u(\tau) e^{-\frac{R}{L}(t-\tau)} u(t - \tau) \begin{cases} \frac{\tau}{\alpha} & \tau \leq \alpha, \alpha > 0 \\ 1 & \text{otherwise} \end{cases} d\tau \tag{6.25}$$

The integral given in Eq. (6.25) must be done for two intervals: $t < \alpha$ and $t \geq \alpha$. Employing step functions to ensure the boundaries, the final solution is given as

$$\begin{aligned} i(t) = \frac{V_0}{R} \Big[\frac{1}{\alpha} &\left(t + \frac{L}{R} \left(e^{-\frac{R}{L}t} - 1 \right) \right) u(t) u(\alpha - t) \\ &+ \left(1 + \frac{L}{R\alpha} e^{-\frac{R}{L}t} \left(1 - e^{\frac{R}{L}\alpha} \right) \right) u(t - \alpha) \Big] \end{aligned} \tag{6.26}$$

(3) Convolving the sinusoidal source generates the integral

$$i(t) = \frac{V_0}{L} \int_{-\infty}^{+\infty} u(\tau) \sin(\omega\tau) e^{-\frac{R}{L}(t-\tau)} u(t - \tau) d\tau \tag{6.27}$$

Analyzing the integration limits imposed by the step functions reduces the integral to

$$i(t) = \frac{V_0}{L} e^{-\frac{R}{L}t} \int_0^t \sin(\omega\tau) e^{\frac{R}{L}\tau} d\tau \tag{6.28}$$

Employing one's personal integration method of choice with sinusoids yields the final current:

$$i(t) = V_0 \frac{R}{R^2 + \omega^2 L^2} \left[\sin \omega t + \frac{\omega L}{R} \left(e^{-\frac{R}{L}t} - \cos \omega t \right) \right] u(t) \tag{6.29}$$

These three solutions were generated relatively quickly with convolution. Alternatively, analysis would have required solving three differential equations, which is generally considered slower and less efficient.

Convolution is a powerful tool for analyzing the power distribution for various current demands of a load and has the benefit of being in the time domain, so ripple, droop, and overshoot are immediately apparent. However, other tools are available in the Engineer's Toolbox for efficient analysis as one starts to consider the frequency domain

6.4 Fourier Series

There is a special technique that may be used for steady-state periodic signals. These signals are considered to have *finite power* but *infinite energy* given that they have no beginning or end in time. The approach separates a repeating signal into sinusoids that when summed can recreate the signal of interest. The Fourier series technique can be employed using sines and cosines, or strictly in the complex domain. For this refresher, the complex representation of sinusoids, $e^{j\omega t}$, is reviewed, and the engineering convention of $j = \sqrt{-1}$ and the positive exponent is continued.

An important behavior of sinusoids is necessary to begin this review: orthogonality. Assume that there are two sinusoids that have periods that are related by integers: $Ae^{jm\frac{2\pi}{T}t}$, and $Be^{jn\frac{2\pi}{T}t}$, where m and n are integers (at least one of them nonzero), A and B are scalar multipliers, and T is some shared periodicity. Taking the product of the first sinusoid and the complex conjugate of the second sinusoid and integrating over one period T (*inner product*) yield

$$\left\langle Ae^{jm\frac{2\pi}{T}t}, Be^{-jn\frac{2\pi}{T}t} \right\rangle = \int_0^T ABe^{jm\frac{2\pi}{T}t}e^{-jn\frac{2\pi}{T}t}dt = AB\int_0^T e^{j(m-n)\frac{2\pi}{T}t}dt \quad (6.30)$$

Solving the integral and processing the limits yield

$$\left\langle Ae^{jm\frac{2\pi}{T}t}, Be^{-jn\frac{2\pi}{T}t} \right\rangle = AB\frac{T}{2\pi j(m-n)}(e^{j(m-n)2\pi} - 1) = 0, \quad m \neq n \quad (6.31)$$

What if $m = n$? Then the product would be

$$\left\langle Ae^{jm\frac{2\pi}{T}t}, Be^{-jm\frac{2\pi}{T}t} \right\rangle = \int_0^T ABe^{jm\frac{2\pi}{T}t}e^{-jm\frac{2\pi}{T}t}dt = AB\int_0^T dt = ABT \neq 0$$

$$(6.32)$$

The mathematical solutions described in Eqs. (6.31) and (6.32) result in a relationship termed *orthogonality*. Orthogonality makes it possible to take generic periodic signals and decompose them into a sum of sinusoids.

The first step in determining the efficacy of the Fourier series is to analyze a random periodic signal. Let $x(t)$ be a generic periodic signal of period T. To decompose the signal into a sum of sinusoids, the following is necessary:

$$\sum_{n=-\infty}^{\infty} A_n e^{jn\frac{2\pi}{T}t} = x(t) \tag{6.33}$$

where A_n are scalar multipliers associated with each discrete frequency $n\frac{2\pi}{T}$ $\left(\omega_n = n\frac{2\pi}{T}\right)$. The frequencies involved are simply integer multiples of the fundamental frequency $f_o = \frac{1}{T}$.

The scalar multipliers can be found by employing the orthogonality between the sinusoids. Multiplying Eq. (6.33) by $e^{-jm\frac{2\pi}{T}}$ and integrating over one period yield the following equation:

$$\int_0^T e^{-jm\frac{2\pi}{T}t} \sum_{n=-\infty}^{\infty} A_n e^{jn\frac{2\pi}{T}t} dt = \int_0^T e^{-jm\frac{2\pi}{T}t} x(t) dt \tag{6.34}$$

Taking liberty with the infinite sum, the integral is moved inside of the sum, which yields

$$\sum_{n=-\infty}^{\infty} A_n \int_0^T e^{j(n-m)\frac{2\pi}{T}t} dt = \int_0^T e^{-jm\frac{2\pi}{T}t} x(t) dt \tag{6.35}$$

where the exponents have been combined, and the scalar multiplier A_n has been left outside of the integral. From the discussion on orthogonality, it is understood that the integral on the left-hand side is zero unless $n = m$, and the result of the integral with $n = m$ is simply T. The scalar multiplier is then left as

$$A_m = \frac{1}{T} \int_0^T e^{-jm\frac{2\pi}{T}t} x(t) dt \tag{6.36}$$

The periodic signal can then be described in a series of sinusoids as shown in the following example.

Example 6.3   *Fourier Series of a Square Wave*   Find the Fourier series of a square wave of period T, expressed as:

$$x(t) = \sum_{k=-\infty}^{\infty} V_0 \left(u(t - kT) - u\left(t - \frac{2k+1}{2}T\right) \right) \tag{6.37}$$

Solution:
Looking at the first period, the coefficients for the Fourier series are given by

$$A_m = V_0 \frac{1}{T} \int_0^T e^{-jm\frac{2\pi}{T}t} \left(u(t) - u\left(t - \frac{T}{2}\right) \right) dt = V_0 \frac{1}{T} \int_0^{\frac{T}{2}} e^{-jm\frac{2\pi}{T}t} dt \tag{6.38}$$

Solving the integral yields

$$A_m = -\frac{V_0}{T}\frac{1}{jm\frac{2\pi}{T}}(e^{-jm\pi} - 1) = V_0\frac{1}{jm\pi}\begin{cases} 0 & \text{for } m \text{ even} \\ 1 & \text{for } m \text{ odd} \end{cases}, \quad m \neq 0 \tag{6.39}$$

The DC ($m = 0$) part of the series must be handled separately given that the expression in Eq. (6.39) is undefined for $m = 0$. The DC component is reduced to

$$A_0 = \frac{V_0}{T}\int_0^{\frac{T}{2}} dt = \frac{V_0}{2} \tag{6.40}$$

The final solution can be written with a change of indices to be expressed as

$$x(t) = V_0\left(\frac{1}{2} + \sum_{n=-\infty}^{\infty}\frac{1}{j(2n+1)\pi}e^{j(2n+1)\frac{2\pi}{T}t}\right) \tag{6.41}$$

Intuitively, the results make sense. The square wave oscillates between 0 and V_0 with a 50% duty cycle. The DC (or average value) over one cycle is $\frac{V_0}{2}$, as shown in the result of Eq. (6.40). Furthermore, the conclusion that only m odd indices are nonzero is understandable given that the square wave is odd about the $t = 0$ axis, just like a sine wave. To demonstrate, separate the infinite sum as follows:

$$\sum_{n=-\infty}^{\infty}\frac{1}{j(2n+1)\pi}e^{j(2n+1)\frac{2\pi}{T}t} = \sum_{n=0}^{\infty}\frac{1}{j(2n+1)\pi}e^{j(2n+1)\frac{2\pi}{T}t}$$
$$+ \sum_{n=-\infty}^{-1}\frac{1}{j(2n+1)\pi}e^{-j(2n+1)\frac{2\pi}{T}t} \tag{6.42}$$

Shift the indices in the first sum to $n = k - 1$, and the second sum to $n = -p$, which results in

$$\sum_{n=-\infty}^{\infty}\frac{1}{j(2n+1)\pi}e^{j(2n+1)\frac{2\pi}{T}t} = \sum_{k=1}^{\infty}\frac{1}{j(2k-1)\pi}e^{j(2k-1)\frac{2\pi}{T}t}$$
$$+ \sum_{p=1}^{\infty}\frac{-1}{j(2p-1)\pi}e^{-j(2p-1)\frac{2\pi}{T}t} \tag{6.43}$$

Moving the indices to a common counter of m leaves

$$\sum_{n=-\infty}^{\infty}\frac{1}{j(2n+1)\pi}e^{j(2n+1)\frac{2\pi}{T}t} = \sum_{m=1}^{\infty}\frac{1}{j(2m-1)\pi}\left(e^{j(2m-1)\frac{2\pi}{T}t} - e^{-j(2m-1)\frac{2\pi}{T}t}\right)$$
$$= \sum_{m=1}^{\infty}\frac{2}{(2m-1)\pi}\sin\left((2m-1)\frac{2\pi}{T}t\right) \tag{6.44}$$

Substituting Eq. (6.44) into Eq. (6.41) provides an alternative representation for the solution:

$$x(t) = V_0 \left(\frac{1}{2} + \sum_{m=1}^{\infty} \frac{2}{(2m-1)\pi} \sin\left((2m-1)\frac{2\pi}{T}t\right) \right) \tag{6.45}$$

The odd distribution around the $t = 0$ axis is evident in the expression of Eq. (6.45) with respect to the non-DC part.

The orthogonality of the Fourier series sinusoids results in analyzing the differential equations as algebra equations. For each part of the infinite sum, the problem can be calculated relatively quickly given that each frequency component must have a result at the same frequency. To demonstrate this by example, consider the generic second-order linear differential equation that is the common result of circuit analysis:

$$y_{\text{source}}(t) = C_1 x(t) + C_2 \frac{d}{dt} x(t) + C_3 \frac{d^2}{dt^2} x(t) \tag{6.46}$$

Each frequency component of the source can be associated to a response at that frequency so that a source component of $A_n e^{jn\frac{2\pi}{T}t}$ results in a response component of $B_n e^{jn\frac{2\pi}{T}t}$. Consequently, the components can be analyzed in Eq. (6.46) as

$$A_n e^{jn\frac{2\pi}{T}t} = C_1 \left(B_n e^{jn\frac{2\pi}{T}t} \right) + C_2 \frac{d}{dt} \left(B_n e^{jn\frac{2\pi}{T}t} \right) + C_3 \frac{d^2}{dt^2} \left(B_n e^{jn\frac{2\pi}{T}t} \right) \tag{6.47}$$

where the C_x terms are generic. The differential operations can be performed, resulting in

$$A_n e^{jn\frac{2\pi}{T}t} = C_1 B_n e^{jn\frac{2\pi}{T}t} + C_2 jn\frac{2\pi}{T} B_n e^{jn\frac{2\pi}{T}t} - C_3 \left(n\frac{2\pi}{T}\right)^2 B_n e^{jn\frac{2\pi}{T}t} \tag{6.48}$$

Canceling the like terms, and solving for B_n, yields

$$B_n = \frac{A_n}{C_1 + C_2 jn\frac{2\pi}{T} - C_3 \left(n\frac{2\pi}{T}\right)^2} \tag{6.49}$$

Consequently, the final solution can be given by

$$x(t) = \sum_{n=-\infty}^{\infty} \frac{A_n}{C_1 + C_2 jn\frac{2\pi}{T} - C_3 \left(n\frac{2\pi}{T}\right)^2} e^{jn\frac{2\pi}{T}t} \tag{6.50}$$

Equation (6.50) shows the result of a generic equation and is hopefully instructional for the application of the Fourier series. The reconstruction in the time domain can be less intuitive and more problematic given the involved solution to the patterns and finding a concise form for the sum of the sinusoids. However, for analyzing by frequency with a time-harmonic waveform, it is a relatively easy and fast technique. The next example illustrates the method.

Example 6.4 *Square Wave in an RL Circuit* Find the current in the circuit of Examples 6.1 and 6.2 with the source as the square wave evaluated in Example 6.3.

Solution:

Using the source equation given in Eq. (6.41) and the differential equation given in 6.14 (without the initial-time constraint), the current for each n is given by

$$I_s(n) = \frac{V_0}{j(2n+1)\pi \left[j\frac{(2n+1)}{T} 2\pi L + R \right]} e^{j(2n+1)\frac{2\pi}{T}t} \tag{6.51}$$

The DC portion is much easier and is simply: $I_{s_{DC}} = \frac{V_0}{2R}$. However, Eq. (6.51) does not yield an intuitive picture in its present state. The denominator is a product of two functions, however, which suggests it may be decoupled. Dividing into separate denominators yields:

$$I_s(n) = V_0 \left[\frac{1}{j(2n+1)\pi R} - \frac{2L}{RT \left[j\frac{(2n+1)}{T} 2\pi L + R \right]} \right] e^{j(2n+1)\frac{2\pi}{T}t} \tag{6.52}$$

The first part of the sum is just a square wave as found in Eq. (6.41). However, the second part is not obvious at first glance. Tables of common and basic Fourier series for signals can be found in the suggested textbooks and are available online for easy review, although less common series are left to the engineer. The sum could be handled pictorially to determine the actual waveform. On the other hand, intuition can also yield some expectation. Ultimately, the series can now be recreated to reveal the following equation:

$$i_s(t) = V_0 \left[\sum_{k=-\infty}^{\infty} \frac{1}{R} \left(u(t - kT) - u \left(t - \frac{2k+1}{2} T \right) \right) \right.$$

$$\left. - \sum_{n=-\infty}^{\infty} \frac{2L}{RT \left[j\frac{(2n+1)}{T} 2\pi L + R \right]} e^{j(2n+1)\frac{2\pi}{T}t} \right] \tag{6.53}$$

Although a simplified expression for the second sum (n) is not included herein, from experience, one might surmise that the sum results in exponential changes in the rise and fall of the waveform. Intuitively, this may be gleaned from the convolution example and will further be a tool of experience after the following sections if not already.

The Fourier series is a powerful tool for analyzing periodic signals in power integrity. However, much of power integrity involves the analysis of transients.

For many problems and typically as a first approach, the Fourier transform proves to be a great tool and is covered in the next section.

6.5 Fourier Transform

The Fourier series is very useful for periodic signals, but power integrity engineers, more often than not, are concerned with finite-time transients. Fortunately, the Fourier series approach can be extended to cover finitely bounded signals in the form of the *Fourier transform*. The Fourier transform is a very useful tool for analyzing signals and circuits, although this transform is somewhat limited on utility for truly analyzing the transient response. The Fourier transform will be further extended to the Laplace transform for complete analysis of transients.

The development begins with the equations developed for the Fourier series in the previous section. The full expression for the Fourier series of an arbitrary signal $x(t)$ can be written as

$$x(t) = \sum_{n=-\infty}^{\infty} e^{jn\frac{2\pi}{T}t} \frac{1}{T} \int_{-\frac{T}{2}}^{+\frac{T}{2}} e^{-jn\frac{2\pi}{T}t} x(t) dt \tag{6.54}$$

The limits of the integral have been shifted slightly from the previous equations, but still cover only one period. Let the period be expressed in terms of the fundamental frequency $f_o = \frac{1}{T}$.

$$x(t) = \sum_{n=-\infty}^{\infty} e^{jn2\pi f_o t} f_o \int_{-\frac{T}{2}}^{+\frac{T}{2}} e^{-jn2\pi f_o t} x(t) dt \tag{6.55}$$

The limits have been left in terms of the period for easier understanding of the equation development, but now we can focus on the way the equation looks and what the components of it mean. Practically speaking, the fundamental frequency is a discrete band such that f_o is really Δf. Looking at the frequency that is contained in the Fourier series, the actual frequency of focus is given as $f = n\Delta f$. Accepting these observations, Eq. (6.55) can be rewritten as

$$x(t) = \sum_{n\Delta f=-\infty}^{\infty} e^{j2\pi ft} \Delta f \int_{-\frac{T}{2}}^{+\frac{T}{2}} e^{-j2\pi ft} x(t) dt \tag{6.56}$$

Some mathematical liberty has been taken in the limits of the sum. The index n is still what is incremented in integer form, but for development, this nomenclature makes the process clearer, given the substitution of $f = n\Delta f$ in the associated exponent. Now a new definition is necessary. Let the transformed part of the signal be given by

$$X(n\Delta f = f) = \int_{-\frac{T}{2}}^{+\frac{T}{2}} e^{-j2\pi ft} x(t) dt \tag{6.57}$$

$X(n\Delta f)$ is essentially the frequency expression of the Fourier series. For a finitely integrate able $x(t)$, assume that the period goes to infinity because it does not repeat, and therefore $\Delta f \to 0$. Taking this limit yields the equation for the Fourier transform of $x(t)$:

$$X(f) = \int_{-\infty}^{+\infty} e^{-j2\pi ft} x(t) dt \tag{6.58}$$

The limit approach of $\Delta f \to 0$ can be expressed as $\Delta f = df \to 0$. Substituting Eq. (6.58) into Eq. (6.56) gives

$$x(t) = \lim_{\Delta f \to 0} \sum_{n\Delta f = -\infty}^{\infty} e^{j2\pi ft} \Delta f X(f) \tag{6.59}$$

Taking the limit turns the sum into an integral and provides the equation for the inverse Fourier transform

$$x(t) = \int_{-\infty}^{+\infty} e^{j2\pi ft} X(f) df \tag{6.60}$$

Although perhaps not mathematically rigorous in presentation, Eqs. (6.58) and (6.60) give the Fourier and inverse Fourier transform equations through a logical and intuitive development. $X(f)$ and $x(t)$ are typically called *transform pairs*, and such pairing is given by:

$$x(t) \leftrightarrow X(f) \tag{6.61}$$

Given Eqs. (6.58) and (6.60) as the equations for the Fourier transform, a number of useful theorems can be developed that aid an engineer in a quick and intuitive understanding of signal behavior when evaluating a system response. A virtual compendium of these theorems is available online and in texts that are fully dedicated to signal analysis [6]. However, a number of theorems are developed in the following subsections to aid in the understanding and as a refresher of the Fourier series methodology.

6.5.1 Convolution Theorem

The Convolution theorem is a very powerful tool in the frequency domain. To begin, take the Fourier transform of the convolution shown in Eq. (6.13):

$$\int_{-\infty}^{+\infty} d\, tf(t) e^{-j2\pi ft} = \int_{-\infty}^{+\infty} dt \int_{-\infty}^{+\infty} d\tau\, x(\tau) h(t-\tau) e^{-j2\pi ft} \tag{6.62}$$

Rearranging the equation yields

$$F(f) = \int_{-\infty}^{+\infty} d\tau\, x(\tau) \int_{-\infty}^{+\infty} dt\, h(t-\tau) e^{-j2\pi ft} \tag{6.63}$$

A substitution can be made, letting $u = t - \tau$, resulting in

$$F(f) = \int_{-\infty}^{+\infty} d\tau\, x(\tau) \int_{-\infty}^{+\infty} du\, h(u) e^{-j2\pi f(u+\tau)} \tag{6.64}$$

The exponential can be expanded, and the equation can be further redistributed to be

$$F(f) = \int_{-\infty}^{+\infty} d\tau x(\tau) e^{-j2\pi f\tau} \int_{-\infty}^{+\infty} du\, h(u) e^{-j2\pi fu} \tag{6.65}$$

Each integral is now a Fourier transform. The final result is simply

$$F(f) = X(f)H(f) \tag{6.66}$$

Knowing the impulse response of a system, the result of a generalized excitation is the product of the transformed excitation and the impulse response.

6.5.2 Time-Shift Theorem

Frequently, an understanding of what happens when an excitation is shifted in time is helpful. In particular, such knowledge is very relevant for the inverse Fourier transform to provide an intuitive understanding of what the signal response looks like back in the time domain.

The development is very straightforward. Suppose that a given signal $x(t)$ is shifted to $x(t - t_o)$.

$$x(t - t_o) \leftrightarrow \int_{-\infty}^{+\infty} e^{-j2\pi ft} x(t - t_o) dt \tag{6.67}$$

A variable substitution can be made letting $u = t - t_o$. Consequently,

$$x(t - t_o) \leftrightarrow \int_{-\infty}^{+\infty} e^{-j2\pi f(u+t_o)} x(u) du = e^{-j2\pi ft_o} \int_{-\infty}^{+\infty} e^{-j2\pi fu} x(u) du$$

$$\Rightarrow x(t - t_o) \leftrightarrow e^{-j2\pi ft_o} X(f) \tag{6.68}$$

A shift in time equates to a phase shift in frequency. In contrast, analyzing a frequency response that includes a phase shift can be inverse transformed to include a time shift of a known transform.

6.5.3 Superposition Theorem

Frequently transient responses are decoupled into the sum of multiple standard signals. Given the sum of signals $x(t)$ and $y(t)$ with scalar multipliers a and b, the functions can be transformed into

$$ax(t) + by(t) \leftrightarrow \int_{-\infty}^{+\infty} e^{-j2\pi ft} (ax(t) + by(t)) dt$$

$$= a \int_{-\infty}^{+\infty} e^{-j2\pi ft} x(t) dt + b \int_{-\infty}^{+\infty} e^{-j2\pi ft} y(t) dt$$

$$\Rightarrow ax(t) + by(t) \leftrightarrow aX(f) + bY(f) \tag{6.69}$$

Consequently, the sum of two signals in time is the sum of two signals in frequency.

6.5.4 Duality Theorem

Tools that increase the number of known transform pairs are always useful. Suppose a known frequency equation represents a time signal, $X(t)$.

$$X(t) \leftrightarrow \int_{-\infty}^{+\infty} e^{-j2\pi ft} X(t) dt \tag{6.70}$$

The negative sign can be tied into the frequency variable, resulting in

$$X(t) \leftrightarrow \int_{-\infty}^{+\infty} e^{j2\pi(-f)t} X(t) dt = x(-f) \tag{6.71}$$

Consequently, the frequency response includes a reversal in frequency.

6.5.5 Differentiation Theorem

Sometimes signals of interest can be represented as a differentiation of another common signal. Generically, the differentiation theorem can be evaluated by analyzing the nth differential of the signal of $x(t)$:

$$\frac{d^n x(t)}{dt^n} \leftrightarrow \int_{-\infty}^{+\infty} e^{-j2\pi ft} \frac{d^n x(t)}{dt^n} dt \tag{6.72}$$

Integrating by parts yields

$$\int_{-\infty}^{+\infty} e^{-j2\pi ft} \frac{d^n x(t)}{dt^n} dt = e^{-j2\pi ft} \frac{d^{n-1} x(t)}{dt^{n-1}} \Bigg|_{-\infty}^{+\infty} + j2\pi f \int_{-\infty}^{+\infty} e^{-j2\pi ft} \frac{d^{n-1} x(t)}{dt^{n-1}} dt \tag{6.73}$$

The signal $x(t)$ is finitely integrable and must go to zero at the limits. Similarly, all the derivatives of $x(t)$ are the slope of zero functions and are consequently zero as well. Successively integrating by parts culminates in

$$\int_{-\infty}^{+\infty} e^{-j2\pi ft} \frac{d^n x(t)}{dt^n} dt = (j2\pi f)^n \int_{-\infty}^{+\infty} e^{-j2\pi ft} x(t) dt \tag{6.74}$$

Therefore, the transform pair is given by

$$\frac{d^n x(t)}{dt^n} \leftrightarrow (j2\pi f)^n X(f) \tag{6.75}$$

6.5.6 Integration Theorem

Just as with the differentiation theorem, signals can be the integrals of other common signals and can be more easily transformed with simple rules. Given a signal that is an integral of a generic signal of $x(t)$,

$$\int_{-\infty}^{t} x(\tau) d\tau \leftrightarrow \int_{-\infty}^{+\infty} e^{-j2\pi ft} \left(\int_{-\infty}^{t} x(\tau) d\tau \right) dt \tag{6.76}$$

Before proceeding, however, some elementary transform pairs are needed.

Example 6.5 *Fourier Transform – Dirac Delta Function* Find the Fourier transform of the delta function $\delta(t)$.

Solution:

The transform can be found through direct evaluation.

$$\delta(t) \leftrightarrow \int_{-\infty}^{+\infty} e^{-j2\pi ft} \delta(t) dt \tag{6.77}$$

$$\delta(t) \leftrightarrow 1 \tag{6.78}$$

Example 6.6 *Fourier Transform – Constant* Find the Fourier transform of the constant $x(t) = 1$.

Solution:

The duality theorem and the result of Example 6.5 can be used to give the transform:

$$1 \leftrightarrow \delta(f) \tag{6.79}$$

Example 6.7 *Fourier Transform – sgn(t)* Find the Fourier transform for the sgn(t) function. The sgn(t) function is defined as -1 for $t < 0$, and $+1$ for $t \geq 0$.

Solution:

The sgn(t) function does not appear to have a transform because the function does not go to zero at infinity and should not be finitely integrable. However, it can be evaluated and is found to have a transform *in the limit*. The function can be rewritten as

$$\text{sgn}(t) = \lim_{\alpha \to 0} e^{-\alpha|t|}(u(t) - u(-t)) \tag{6.80}$$

The transform is then given by

$$\begin{aligned}
\text{sgn}(t) &\leftrightarrow \int_{-\infty}^{+\infty} dt \lim_{\alpha \to 0} e^{-\alpha|t|}(u(t) - u(-t)) \\
&= \lim_{\alpha \to 0} \int_{-\infty}^{+\infty} dt\, e^{-\alpha|t|} e^{-j2\pi ft}[u(t) - u(-t)]
\end{aligned} \tag{6.81}$$

The limits can be adjusted to account for the step functions and the exponential decay:

$$\begin{aligned}
\text{sgn}(t) &\leftrightarrow \lim_{\alpha \to 0} \left[\int_{0}^{+\infty} dt\, e^{-(j2\pi f + \alpha)t} - \int_{-\infty}^{0} dt\, e^{-(j2\pi f - \alpha)t} \right] \\
&= \lim_{\alpha \to 0} \left[\frac{1}{j2\pi f + \alpha} + \frac{1}{j2\pi f - \alpha} \right]
\end{aligned} \tag{6.82}$$

Evaluating the limit provides the Fourier transform pair:

$$\text{sgn}(t) \leftrightarrow \frac{1}{j\pi f} \tag{6.83}$$

Example 6.8 *Fourier Transform – Step Function u(t)* Find the Fourier transform for the step function.

Solution:
An accurate evaluation of the Fourier transform of $u(t)$ cannot be done in the limit as was done for $\text{sgn}(t)$. However, the solution for $\text{sgn}(t)$ can be used to determine the transform. The step function can be given as

$$u(t) = \frac{1}{2} + \frac{1}{2}\text{sgn}(t) \tag{6.84}$$

Employing the superposition theorem and the results of Examples 6.6 and 6.7 provides the transform pair:

$$u(t) \leftrightarrow \frac{1}{2}\delta(f) + \frac{1}{j2\pi f} \tag{6.85}$$

Returning to the theorem, the integral of the signal $x(t)$ can be rewritten as

$$\int_{-\infty}^{t} x(\tau)d\tau = \int_{-\infty}^{+\infty} x(\tau)u(t-\tau)d\tau \tag{6.86}$$

The limits have been preserved by using the step function in the argument. Now Eq. (6.76) can be expressed by

$$\int_{-\infty}^{t} x(\tau)d\tau \leftrightarrow \int_{-\infty}^{+\infty} e^{-j2\pi ft}\left(\int_{-\infty}^{+\infty} x(\tau)u(t-\tau)d\tau\right)dt \tag{6.87}$$

The integrals can now be rearranged as:

$$\int_{-\infty}^{t} x(\tau)d\tau \leftrightarrow \int_{-\infty}^{+\infty} d\tau \int_{-\infty}^{+\infty} d\,t e^{-j2\pi ft}x(\tau)u(t-\tau) \tag{6.88}$$

A substitution of variables is made using $\xi = t - \tau$, leaving

$$\int_{-\infty}^{t} x(\tau)d\tau \leftrightarrow \int_{-\infty}^{+\infty} d\tau \int_{-\infty}^{+\infty} d\xi\, e^{-j2\pi f(\xi+\tau)}x(\tau)u(\xi) \tag{6.89}$$

Rearranging one more time yields

$$\int_{-\infty}^{t} x(\tau)d\tau \leftrightarrow \int_{-\infty}^{+\infty} dt\, x(\tau)e^{-j2\pi f\tau}\int_{-\infty}^{+\infty} d\xi\, u(\xi)e^{-j2\pi f\xi} = X(f)U(f) \tag{6.90}$$

Substituting in the result of Example 6.8 provides the final step of the theorem:

$$\int_{-\infty}^{t} x(\tau)d\tau \leftrightarrow X(f)\left[\frac{1}{2}\delta(f) + \frac{1}{j2\pi f}\right] \tag{6.91}$$

6.5.7   Multiplication Theorem

Signals can also be represented as the product of two or more simpler signals. Suppose a signal can be given as the product of two signals $x(t)$ and $y(t)$. The transform pair is then given by

$$x(t)y(t) \leftrightarrow \int_{-\infty}^{+\infty} dt \, x(t)y(t)e^{-j2\pi ft}$$

$$= \int_{-\infty}^{+\infty} dt \, x(t) \left(\int_{-\infty}^{+\infty} d\xi \, Y(\xi)e^{j2\pi\xi t} \right) e^{-j2\pi ft} \tag{6.92}$$

where the inverse Fourier transform has been substituted for the function $y(t)$. Rearranging the integral provides

$$x(t)y(t) \leftrightarrow \int_{-\infty}^{+\infty} d\xi \, Y(\xi) \int_{-\infty}^{+\infty} dt \, x(t) \, e^{-j2\pi(f-\xi)t} \tag{6.93}$$

Evaluating the right-most integral takes the development as far as it can go:

$$x(t)y(t) \leftrightarrow \int_{-\infty}^{+\infty} d\xi \, Y(\xi)X(f - \xi) \tag{6.94}$$

Had the inverse Fourier transform been used to describe $x(t)$ in the original development, the solution would have been only slightly different:

$$x(t)y(t) \leftrightarrow \int_{-\infty}^{+\infty} d\xi \, X(\xi)Y(f - \xi) \tag{6.95}$$

This solution is offered to show the thought process; however, given the theorems already proved, the same result could have been developed by quickly employing the convolution and duality theorems.

6.5.8   Time-Scaling Theorem

During the power converter design or evaluation cycles, it is sometimes fortuitous to consider making the responses faster or slower to help determine the response to the PDN. The time-scaling theorem provides not only an efficient tool for evaluation but also an intuitive understanding that can provide a quick answer without having to go through the analysis.

Let a be a scalar and evaluate the transform pair of $x\left(\frac{t}{a}\right)$:

$$x\left(\frac{t}{a}\right) \leftrightarrow \int_{-\infty}^{+\infty} e^{-j2\pi ft} x\left(\frac{t}{a}\right) dt \tag{6.96}$$

A variable substitution can now be made letting $u = \frac{t}{a}$:

$$x\left(\frac{t}{a}\right) \leftrightarrow |a| \int_{-\infty}^{+\infty} e^{-j2\pi fau} x(u)du \tag{6.97}$$

The end result is:

$$x\left(\frac{t}{a}\right) \leftrightarrow |a|X(af) \tag{6.98}$$

Equation (6.98) indicates that not only does the magnitude of the frequency response change with the time scaling, but the frequency content does as well, resulting in a possibly significant impact.

As a further note, a could be equal to -1, which would result in what is frequently called the Time-Reversal Theorem yielding:

$$x(-t) \leftrightarrow X(-f) = X^*(f) \tag{6.99}$$

where $X^*(f)$ is the complex conjugate of $X(f)$.

6.5.9 Modulation or Frequency-Translation Theorem

The last theorem provided in this refresher involves multiplying a signal by a generic sinusoid $e^{j2\pi f_o t}$. The transform pair is then given by

$$x(t)e^{j2\pi f_o t} \leftrightarrow \int_{-\infty}^{+\infty} e^{-j2\pi ft}x(t)e^{j2\pi f_o t}dt = \int_{-\infty}^{+\infty} e^{-j2\pi(f-f_o)t}x(t)dt \tag{6.100}$$

Therefore, the resulting pair is

$$x(t)e^{j2\pi f_o t} \leftrightarrow X(f-f_o) \tag{6.101}$$

Example 6.9 *Modulation with $\cos(2\pi f_o t)$* A signal $x(t)$ is multiplied by a cosine waveform given by $\cos(2\pi f_o t)$. Determine the Fourier transform pair.

Solution:
The cosine function can be expanded into exponentials as:

$$\cos(2\pi f_o t) = \frac{1}{2}(e^{j2\pi f_o t} + e^{-j2\pi f_o t}) \tag{6.102}$$

Taking advantage of the superposition theorem and the frequency-translation theorem yields

$$x(t)\cos(2\pi f_o t) \leftrightarrow \frac{1}{2}(X(f-f_o) + X(f+f_o)) \tag{6.103}$$

Other theorems are available in various references. This development, however, should give the power integrity engineer a refresher on the Fourier transform and provide the standard nomenclature used herein. Table 6.1 summarizes the results of the theorems for easy reference.

For further reference, Table 6.2 provides a summary of Fourier transform pairs for commonly used signals.

Table 6.1 Summary of Fourier transform theorems

Theorem	Time domain	Fourier transform
Convolution theorem	$\int_{-\infty}^{+\infty} d\tau\, x(\tau)h(t-\tau)$	$X(f)H(f)$
Time-shift theorem	$x(t-t_o)$	$e^{-j2\pi f t_o}X(f)$
Superposition theorem	$ax(t)+by(t)$	$aX(f)+bY(f)$
Duality theorem	$X(t)$	$x(-f)$
Differentiation theorem	$\dfrac{d^n x(t)}{dt^n}$	$(j2\pi f)^n X(f)$
Integration theorem	$\int_{-\infty}^{t} x(\tau)d\tau$	$X(f)\left[\dfrac{1}{2}\delta(f)+\dfrac{1}{j2\pi f}\right]$
Multiplication theorem	$x(t)y(t)$	$\int_{-\infty}^{+\infty} d\xi\, X(\xi)Y(f-\xi)$
Time-scaling theorem	$x\left(\dfrac{t}{a}\right)$	$\lvert a\rvert X(af)$
Time-reversal theorem	$x(-t)$	$X^*(f)$
Modulation or frequency-translation theorem	$x(t)e^{j2\pi f_o t}$	$X(f-f_o)$

Table 6.2 Summary of Fourier transform pairs.

Signal	Time domain	Frequency domain
Dirac delta function	$\delta(t)$	1
Unity constant	1	$\delta(f)$
"Sign" function	$\mathrm{sgn}(t)$	$\dfrac{1}{j\pi f}$
Inverse time	$\dfrac{1}{t}$	$-j\pi\,\mathrm{sgn}(f)$
Step function	$u(t)$	$\dfrac{1}{2}\delta(f)+\dfrac{1}{j2\pi f}$
Exponential sinusoid	$e^{j2\pi f_o t}$	$\delta(f-f_o)$
Cosine function	$\cos(2\pi f_o t)$	$\dfrac{1}{2}[\delta(f-f_o)+\delta(f+f_o)]$
Sine function	$\sin(2\pi f_o t)$	$\dfrac{1}{j2}[\delta(f-f_o)-\delta(f+f_o)]$
Pulse function	$\Pi(t)$	$\mathrm{sinc}(f)$
Sinc function	$\mathrm{sinc}(t)$	$\Pi(f)$
Decaying exponential	$e^{-\alpha t}u(t),\ \alpha>0$	$\dfrac{1}{\alpha+j2\pi f}$

Using the convolution theorem, basic circuit equations become products. The voltage–current relation for a resistor is $V(f) = I(f) R$. Further employing the differentiation theorem, the inductor and capacitor equations become $V(f) = j2\pi fL\, I(f)$, and $I(f) = j2\pi fC\, V(f)$, respectively. These results can be used to quickly evaluate the frequency response of typical circuits.

However, the Fourier transform is not conducive to analyzing transients, which are vitally important to the power integrity engineer. Nonetheless, Fourier analysis can be used judiciously to determine a variety of possible pitfalls in power delivery designs. Arguably, the most important aspect of Fourier analysis is the result of the convolution and superposition theorems. The input impedance of circuits from arbitrary ports can be calculated to determine the probable problem frequencies. The Fourier transform can be likewise calculated for the stimulus at the chosen port to evaluate if significant energy is created at the port in question.

Example 6.10 *Reduced Circuit Analysis for Power Integrity* A power integrity engineer has a meeting in a half hour to discuss the challenges in supplying power to a new processor. The engineer has not yet been given much information for this brainstorming venture but wants to be able to demonstrate the basic concerns. A reduced circuit diagram is made which shows the components of the voltage regulator, the power planes, and the load-current requirement. The quick sketch of the circuit is provided in Figure 6.1. At what frequencies does a significant problem develop?

Solution:
Figure 6.1 is illustrative of the most basic elements. The voltage regulator is modeled as a DC source. The slew rate and interconnect of the regulator is modeled as a source inductor, and the losses in the regulator as well as the connection path to the load are modeled as the source resistance. The system design includes parallel planes, and the planes are modeled simply as a capacitor. Finally, the load is modeled as a current source. The concern is that the voltage at the load may fluctuate.

The elements are given as impedances. Using the superposition theorem, the impedance of the system observed from the load may be calculated. The voltage source is shorted, and the impedance is quickly calculated as

$$\frac{V_L(f)}{I_L(f)} = \frac{R_S + j2\pi fL_S}{1 - (2\pi f)^2 L_S C + j2\pi fR_S C} \tag{6.104}$$

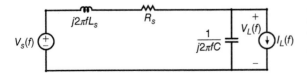

Figure 6.1 A reduced circuit of the power transfer circuit. The load is shown as a current source.

Equation (6.104) is the impedance experienced by the port at the load. The equation shows a high impedance when $f = \frac{1}{2\pi\sqrt{L_S C}}$. The peak is limited by the resistance R_S. If the load includes this frequency in its spectrum (Fourier transform), the current cannot be conducted, the voltage swings significantly, and the load device will not function under that load. Mitigating this possibility is covered in other power integrity chapters, but this simplified example is presented as a reminder of how certain key frequencies can be obtained.

The previous example is very simple, even as a refresher. However, it was chosen purposefully as a vehicle to bring out various features of the Fourier transform that are relevant and important. Employing the Fourier transform to a signal or equation is very straightforward. The power integrity engineer needs to understand the fundamentals of the Fourier transform to best use this powerful tool.

The example could have included a much more involved model for the voltage regulator, a series of decoupling capacitors, the load interconnect, package capacitors, and everything else that can be envisioned for a real power distribution. The circuit would have seemed more realistic for the final product, but it would have taken more space for describing the impedance, and obscured the point: *the equation can be analyzed to determine the problem frequencies and relevant components involved.*

The result of Eq. (6.104) shows a problem frequency: a resonance between the parallel planes and the inductance associated with the voltage regulator. Decoupling capacitors can be employed to effectively boost the localized charge to more easily deliver current and keep the voltage stable, until another resonance is established. The process can continue until the spectrum is such that nothing more can be done on the circuit board, and attention has to be focused on the load device. The process may be successfully completed with circuit simulators. However, without the equations, the target for improvement is often a guess and leads to "solutions" that are detrimental to other disciplines (e.g. electromagnetic compatibility), when mutually beneficial solutions can be achieved.

Of further interest for the PI engineer is the role that loss plays in the design. The loss in Example 6.10 limits the peak of the impedance maximum. However, the loss that limits the impedance peak also burns energy, which reduces the efficiency of the power distribution. Regardless of whether devices are AC or battery powered, efficiency is very important to the end user. As a result, there is a school of PI engineers that look for a target impedance in the profile described in Example 6.10, regardless of the possible excitations. Regardless of the stimulus, the power delivery is supposed to be stable as long as the impedance does not cross the threshold and control is handled as discussed in other sections to limit the loss in the path.

The Fourier transform is indeed a powerful tool for analysis, and this refresher can be informative if the developments are understood. Complete understanding of the circuit behavior includes the initial conditions before a change in stimulus. In order to properly analyze such a situation, a modification in the transform is necessary to extend it to a complex domain.

6.6 Laplace Transform

The Laplace transform is a perturbation of the Fourier transform with some added benefits. As will be seen in the development, the Laplace transform also includes the incorporation of initial conditions. The result is an analysis that truly covers transient behavior, providing a powerful tool to PI engineers for analyzing the system response to changing load conditions. A review of the development and use of Laplace transform follows.

The development begins with the equation for the Fourier transform discussed earlier. An assumption is made that the signal $x(t)$ is zero for $t < 0$. Furthermore, the signal is multiplied by an exponentially decaying signal, where $\sigma > 0$:

$$x(t)e^{-\sigma t} \leftrightarrow e^{-\sigma t}\int_{-\infty}^{+\infty} e^{-j2\pi f t}x(t)dt = \int_{0}^{+\infty} e^{-(\sigma+j2\pi f)t}x(t)dt$$

$$= X(\sigma + j2\pi f) \tag{6.105}$$

To continue the development of the Laplace transform, look at the inverse transform of the final form described in Eq. (6.106):

$$x(t)e^{-\sigma t} = \int_{-\infty}^{+\infty} e^{j2\pi f t}X(\sigma + j2\pi f)df \Rightarrow x(t) = \int_{-\infty}^{+\infty} e^{(\sigma+j2\pi f)t}X(\sigma + j2\pi f)df \tag{6.106}$$

Substituting $s = \sigma + j2\pi f$ into the final part of Eq. (6.105) yields the inverse Laplace transform:

$$x(t) = \frac{1}{j2\pi}\int_{\sigma-j\infty}^{\sigma+j\infty} e^{st}X(s)ds \tag{6.107}$$

Similarly, the substitution for s can be made into Eq. (6.105) to provide the Laplace transform:

$$X(s) = \int_{0}^{+\infty} e^{-st}x(t)dt \tag{6.108}$$

Not all equations are transformable, but signals can be analyzed to determine their ability. Ultimately, a signal that is not transformable will most likely result in an unstable system. With these equations in the toolbox, theorems like the ones for Fourier transforms can be determined to aid in the development of creating Laplace transforms for important signals.

6.6.1 Convolution Theorem

The theorem begins with the Laplace transform of the convolution Eq. (6.13):

$$\int_0^{+\infty} dt\, f(t)e^{-st} = \int_0^{+\infty} dt \int_{-\infty}^{+\infty} d\tau\, x(\tau)h(t-\tau)e^{-st} \tag{6.109}$$

Since $x(\tau)$ is zero for $\tau < 0$, the limits can be changed, and the right-most integrals can be rearranged:

$$F(s) = \int_0^{+\infty} d\tau x(\tau) \int_0^{+\infty} dt\, h(t-\tau)e^{-st} \tag{6.110}$$

A substitution can be made with $\xi = t - \tau$ to yield

$$F(s) = \int_0^{+\infty} d\tau x(\tau) \int_{-\tau}^{+\infty} d\xi\, h(\xi)e^{-s(\xi+\tau)} \tag{6.111}$$

Observing that $h(\xi) = 0$, $\xi < 0$, and $-\tau < 0$ over the limits of τ, and rearranging provides:

$$F(s) = \int_0^{+\infty} d\tau x(\tau)e^{-s\tau} \int_0^{+\infty} d\xi\, h(\xi)e^{-s\xi} \tag{6.112}$$

The final result is consequently

$$F(s) = X(s)H(s) \tag{6.113}$$

As with the Fourier transform, the Laplace transform results in a simple multiplication to understand the response in the s domain.

6.6.2 Time-Shift Theorem

A signal shifted in time results in

$$x(t-t_o) \leftrightarrow \int_0^{+\infty} e^{-st}x(t-t_o)dt \tag{6.114}$$

Substituting $\xi = t - t_o$ yields

$$x(t-t_o) \leftrightarrow \int_{-t_o}^{+\infty} e^{-s(\xi+t_o)}x(\xi)d\xi \tag{6.115}$$

The signal is zero below $\xi < 0$, so the lower limit can be adjusted to zero, providing the final relation

$$x(t - t_0) \leftrightarrow e^{-st_0} X(s) \tag{6.116}$$

6.6.3 Superposition Theorem

The addition and scalar multiplication of signals is a common modification to signals and can be quickly analyzed given the known signals. The Laplace transform applied to such signals ($x(t)$ and $y(t)$) and scalar multipliers a and b gives

$$ax(t) + by(t) \leftrightarrow \int_0^{+\infty} e^{-st}(ax(t) + by(t))dt$$

$$= a \int_0^{+\infty} e^{-st} x(t)dt + b \int_0^{+\infty} e^{-st} y(t)dt$$

$$\Rightarrow ax(t) + by(t) \leftrightarrow aX(s) + bY(s) \tag{6.117}$$

6.6.4 Differentiation Theorem

Signals that are differentials of known signals can be quickly evaluated as the result of this theorem. Taking the transform of a differentiated signal leaves

$$\frac{d^n x(t)}{dt^n} \leftrightarrow \int_0^{+\infty} e^{-st} \frac{d^n x(t)}{dt^n} dt \tag{6.118}$$

Evaluating by integration-by-parts results in the integral

$$\int_0^{+\infty} e^{-st} \frac{d^n x(t)}{dt^n} dt = s \int_0^{+\infty} e^{-st} \frac{d^{n-1} x(t)}{dt^{n-1}} dt - \frac{d^{n-1} x(t)}{dt^{n-1}} \bigg|_{t=0} \tag{6.119}$$

The integral is successively evaluated by integration-by-parts until the known signal $x(t)$ is in the integrand

$$\frac{d^n x(t)}{dt^n} \leftrightarrow s^n X(s) - \sum_{m=0}^{n-1} s^m \frac{d^m x(t)}{dt^m} \bigg|_{t=0} \tag{6.120}$$

The differentiation theorem shows the important difference between the Laplace transforms and Fourier transforms. The initial conditions for the function and the subsequent derivatives (as applicable) must be included in the transform. A simple example may demonstrate the relevance.

Example 6.11 *Laplace Transform of the Inductor Circuit Equations* Show the Laplace transform relating the voltage and current associated with an inductor.

Solution:
The voltage and current relation for an inductor is given as $v(t) = L \frac{di(t)}{dt}$. Applying the Laplace transform to the equation and applying the differentiation theorem yield

$$V(s) = sLI(s) - L\, i(0) \tag{6.121}$$

The initial current is necessary to analyze the ultimate response to a change in stimulus. This change proves important to power integrity when considering how changes in loads and switches influence the power delivery. The last example in this chapter will show the true relevance.

6.6.5 Integration Theorem

Integrated signals may also occur that are reflective of known signals; or similar to the previous example, integration may show up as the most efficacious approach to solving an equation. The approach is not as involved as the Fourier transform theorem given the expectation that the signal is zero below $t < 0$. Start with the integral equation and its transform:

$$\int_0^t x(\tau)d\tau \leftrightarrow \int_0^{+\infty} e^{-st}\left(\int_0^t x(\tau)d\tau\right)dt \tag{6.122}$$

As with the differentiation theorem, integration-by-parts is used to proceed:

$$\int_0^{+\infty} e^{-st}\left(\int_0^t x(\tau)d\tau\right)dt = \frac{X(s)}{s} + \left.\frac{\int_0^t x(\tau)d\tau}{s}\right|_{t=0} \tag{6.123}$$

As a result, if $y(t) = \int_0^t x(\tau)d\tau$, the integration theorem is finally

$$y(t) = \int_0^t x(\tau)d\tau \leftrightarrow \frac{X(s)}{s} + \frac{y(0)}{s} \tag{6.124}$$

The initial condition for the integration theorem looks slightly confusing, but one of its best uses is explained in the next example.

Example 6.12 *Laplace Transform of the Capacitor Circuit Equations*
Show the Laplace transform relating the voltage and current associated with a capacitor.

Solution:
The voltage and current relation were shown using the differentiation theorem in the previous example. However, the relationship can also be described as:

$$i(t) = \frac{1}{L}\int_0^t v(\tau)d\tau \tag{6.125}$$

Applying the Laplace transform and the integration theorem to Eq. (6.123) yields

$$I(s) = \frac{1}{sL} V(s) + \frac{1}{L} \int_0^t v(\tau) d\tau \bigg|_{t=0} = \frac{1}{sL} V(s) + \frac{1}{s} i(0) \tag{6.126}$$

Rearranging the result of Eq. (6.124) gives the same equation as Eq. (6.120). However, the representation in Eq. (6.124) provides another utility for circuit analysis, which will be demonstrated toward the end of this chapter.

6.6.6 Multiplication Theorem

Signals and/or circuits may result in multiplying equations that can be evaluated more quickly by determining the multiplication theorem. As with Fourier transforms, the development begins with observing the transform of the multiplication of $x(t)$ and $y(t)$:

$$x(t)y(t) \leftrightarrow \int_0^{+\infty} dt\, x(t)y(t)e^{-jst} \tag{6.127}$$

Arbitrarily, the inverse Laplace transform of $y(t)$ can be inserted into the equation (the same could be done to $x(t)$ instead) with a dummy variable change to avoid confusion:

$$x(t)y(t) \leftrightarrow \int_0^{+\infty} dt\, e^{-st}x(t)\frac{1}{j2\pi} \int_{\sigma-j\infty}^{\sigma+j\infty} d\xi\, e^{\xi t} Y(\xi) \tag{6.128}$$

The integral can be rearranged to provide

$$x(t)y(t) \leftrightarrow \frac{1}{j2\pi} \int_{\sigma-j\infty}^{\sigma+j\infty} d\xi\, Y(\xi) \int_0^{+\infty} dt\, x(t)e^{-(s-\xi)t} \tag{6.129}$$

The right-most integral is a shifted version of the Laplace transform of $x(t)$. Therefore, the multiplication theorem yields

$$x(t)y(t) \leftrightarrow \frac{1}{j2\pi} \int_{\sigma-j\infty}^{\sigma+j\infty} d\xi\, Y(\xi)X(s-\xi) \tag{6.130}$$

As with the Fourier transform version of this theorem, the development could have chosen to present $x(t)$ as the inverse transform, which would have provided:

$$x(t)y(t) \leftrightarrow \frac{1}{j2\pi} \int_{\sigma-j\infty}^{\sigma+j\infty} d\xi\, Y(s-\xi)X(\xi) \tag{6.131}$$

6.6.7 S-shift Theorem

The S-shift theorem is the result of multiplying a function by an exponential. The exponent c can be complex, but it is assumed that the real part of the

Table 6.3 Summary of Laplace transform theorems.

Theorem	Time domain	Fourier transform
Convolution theorem	$\int_{-\infty}^{+\infty} d\tau\, x(\tau)h(t-\tau)$	$X(s)H(s)$
Time-shift theorem	$x(t-t_o)$	$e^{-st_o}X(s)$
Superposition theorem	$ax(t)+by(t)$	$aX(s)+bY(s)$
Differentiation theorem	$\dfrac{d^n x(t)}{dt^n}$	$s^n X(s) - \displaystyle\sum_{m=0}^{n-1} s^m \dfrac{d^m x(t)}{dt^m}\bigg|_{t=0}$
Integration theorem	$\int_{-\infty}^{t} x(\tau)d\tau$	$\dfrac{X(s)}{s} + \dfrac{\int_0^t x(\tau)d\tau}{s}\bigg|_{t=0}$
Multiplication theorem	$x(t)y(t)$	$\dfrac{1}{j2\pi}\displaystyle\int_{\sigma-j\infty}^{\sigma+j\infty} d\xi\, Y(\xi)X(s-\xi)$
S-shift theorem	$x(t)e^{-ct}$	$X(s+c)$

exponential is decaying in order to ensure that the resulting function is transformable:

$$x(t)e^{-ct} \leftrightarrow \int_0^{+\infty} e^{-st}e^{-ct}x(t)dt = \int_0^{+\infty} e^{-(s+c)t}x(t)dt \qquad (6.132)$$

The final integral with respect to t gives the transform of the s-shifted argument:

$$x(t)e^{-ct} \leftrightarrow X(s+c) \qquad (6.133)$$

Other theorems are available in texts as well as online, but these sections should suffice for this refresher of Laplace transforms. The theorems are summarized in Table 6.3 for later reference.

As with the Fourier transform, a number of basic signals can be analyzed with the Laplace transform to be added to the Engineer's Toolbox. A few such will be calculated below as examples of applying the Laplace transform integral.

Example 6.13 *Laplace Transform of the Dirac Delta Function* Determine the Laplace transform of the impulse function $\delta(t)$.

Solution:
The analysis is a straightforward application of the Laplace transform integral:

$$\delta(t) \leftrightarrow \int_0^{+\infty} e^{-st}\delta(t)dt = 1 \qquad (6.134)$$

Example 6.14 *Laplace Transform of the Step Function* Determine the Laplace transform of the step function $u(t)$.

Solution:
The analysis is straightforward and does not require the same complicated analysis as the Fourier transform given the real part of the s domain.

$$u(t) \leftrightarrow \int_0^{+\infty} e^{-st} u(t) dt = \frac{1}{s} \tag{6.135}$$

Example 6.15 *Laplace Transform of $e^{j2\pi f_o t}$.* Determine the Laplace transform of the sinusoidal exponential $e^{j2\pi f_o t}$.

Solution:
The transform for the sinusoidal exponential can be done quickly by employing the S-shift theorem. However, as another simple example of how to apply the integral, a straightforward application is presented, beginning with

$$e^{j2\pi f_o t} \leftrightarrow \int_0^{+\infty} e^{-st} e^{j2\pi f_o t} dt = \int_0^{+\infty} e^{-(s-j2\pi f_o)t} dt \tag{6.136}$$

Solving the final integral yields

$$e^{j2\pi f_o t} \leftrightarrow \frac{1}{s - j2\pi f_o} \tag{6.137}$$

Equation (6.131) coupled with the superposition theorem can be used to find the transforms of the sine and cosine functions. To continue the example, look at the cosine function expanded it in terms of the exponential:

$$\cos(2\pi f_o t)u(t) \leftrightarrow \int_0^{+\infty} e^{-st} \cos(2\pi f_o t)u(t) dt$$

$$= \int_0^{+\infty} e^{-st} \left[\frac{e^{j2\pi f_o t} + e^{-j2\pi f_o t}}{2} \right] dt$$

$$= \frac{1}{2} \left[\int_0^{+\infty} e^{-st} e^{j2\pi f_o t} dt + \int_0^{+\infty} e^{-st} e^{-j2\pi f_o t} dt \right]$$

$$= \frac{1}{2} \left[\frac{1}{s - j2\pi f_o} + \frac{1}{s + j2\pi f_o} \right] \tag{6.138}$$

Algebraically manipulating the result of Eq. (6.132) yields the transform pair:

$$\cos(2\pi f_o t)u(t) \leftrightarrow \frac{s}{s^2 + (2\pi f_o)^2} \tag{6.139}$$

The same process can be used to find the transform pair for the sine function.

For reference, some transform pairs for commonly used signals are summarized in Table 6.4.

As previously discussed, the Laplace transform is a powerful and useful tool for analyzing transient behavior. Given the theorems, a better understanding of the treatment of the various elements is necessary. A resistor is a linear

Table 6.4 Summary of Laplace transform pairs.

Signal	Time domain	s domain
Dirac delta function	$\delta(t)$	1
Step function	$u(t)$	$\dfrac{1}{s}$
Exponential sinusoid	$e^{j2\pi f_o t}$	$\dfrac{1}{s - j2\pi f_o}$
Cosine function	$\cos(2\pi f_o t)$	$\dfrac{s}{s^2 + (2\pi f_o)^2}$
Sine function	$\sin(2\pi f_o t)$	$\dfrac{2\pi f_o}{s^2 + (2\pi f_o)^2}$
Decaying exponential	$e^{-\alpha t} u(t),\ \alpha > 0$	$\dfrac{1}{\alpha + s}$

element that requires no special treatment for initial conditions, but inductors and capacitors do, and options are available.

Example 6.16 *Laplace Transform of the Voltage and Current Relationship for an Inductor* Calculate the s-domain equivalent circuit for an inductor.

Solution:
The evaluation starts with the time-domain differential equation for the voltage and current relationship across an inductor: $v(t) = L\frac{di(t)}{dt}$. Taking the Laplace transform of both sides and applying the differentiation theorem yield

$$V(s) = sLI(s) - Li(0^-) \tag{6.140}$$

The result of Eq. (6.140) is a Thevenin equivalent circuit of a DC voltage source in series with the inductor s-domain voltage. As a result, the equation can be modified to have a Norton equivalent circuit. This can be done in the typical straightforward fashion from circuit theory, but the concept also provides another useful reminder. Suppose instead of the differential equation governing the inductor behavior, the approach was the integral equation:

$$i(t) = \frac{1}{L} \int_{-\infty}^{t} v(\tau)d\tau \tag{6.141}$$

The lower limit of the integral is necessarily modified to have a lower limit of zero, given the development of the one-sided Laplace transform. Transforming both sides and employing the integration theorem give

$$I(s) = \frac{1}{sL} V(s) + \frac{\int_0^t v(\tau)d\tau}{Ls}\Bigg|_{t=0} = \frac{1}{sL} V(s) + \frac{i(0^-)}{s} \tag{6.142}$$

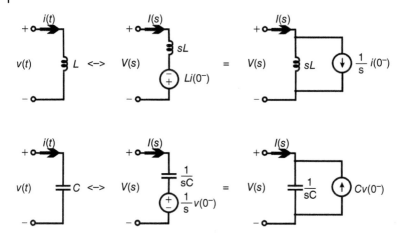

Figure 6.2 Equivalent circuits for inductors and capacitors for evaluation in the s domain. The element parameters are given as s-domain impedances.

Equation (6.142) describes the Norton equivalent circuit for the inductor, which is identical to the direct conversion of the Thevenin equivalent circuit. The same approach can be used to analyze the capacitor equations, yielding the following equivalent circuit equations:

$$V(s) = \frac{I(s)}{sC} + \frac{v(0^-)}{s} \qquad (6.143)$$

$$I(s) = sC\,V(s) - Cv(0^-) \qquad (6.144)$$

Figure 6.2 shows the sign convention that results from the equation development.

The previous developments are meant to serve as a reminder and an opportunity to delve back into the origins of the development. Such understanding is useful in the scrutiny of results developed by the engineer and from software tools. In keeping with such an approach, this review will continue with a Laplace analysis of the Fourier analysis given in Example 6.10. As with the Fourier analysis, the example is very simplified, but it is instructional as a reminder of the tools and approaches necessary for a transient evaluation.

At this point in the refresher, it is necessary to apply all that has been presented. Revisiting the final example in the Fourier transform discussion will serve to demonstrate the utility of the Laplace transform in evaluating transients.

Example 6.17 *Circuit Evaluation Using the Laplace Transform* Revisiting the Fourier transform in Example 6.10, use the Laplace transform to determine the final time-domain solution. The source voltage is a constant V_S. The load current starts at zero and then steps up to I_L ($i_L(t) = I_L u(t)$). Find the time-domain expression for the voltage across the load.

Solution:
The figure has to be modified to include the initial conditions and the s-domain impedances. Figure 6.3 shows the s-domain circuit. The superposition theorem can be used to evaluate the final response. Arbitrarily, the analysis is started with evaluating the response from the DC source, yielding

$$V_{L(1)}(s) = V_S \frac{1}{L_S Cs \left[s^2 + \frac{R_S}{L_S} s + \frac{1}{L_S C} \right]}$$

$$= V_S \frac{1}{s} - V_S \frac{s + \frac{R_S}{L_S}}{\left[\left(s + \frac{R_S}{2L_S} \right)^2 + \left(\frac{1}{L_S C} - \frac{R_S^2}{4L_S^2} \right) \right]} \tag{6.145}$$

Algebraic manipulation is used to make the final summation and subsequent inverse transforms easier.

Next the circuit is analyzed for the initial condition of the capacitor to provide

$$V_{L(2)}(s) = CV_S \frac{R_S + sL_S}{L_S Cs^2 + R_S Cs + 1} = V_S \frac{s + \frac{R_S}{L_S}}{\left[\left(s + \frac{R_S}{2L_S} \right)^2 + \left(\frac{1}{L_S C} - \frac{R_S^2}{4L_S^2} \right) \right]} \tag{6.146}$$

Finally, the load source is evaluated leaving

$$V_{L(3)}(s) = -I_L \frac{R_S + sL_S}{s[L_S Cs^2 + R_S Cs + 1]}$$

$$= -I_L \frac{1}{s} + I_L R_S \frac{s - \left(\frac{1}{R_S C} - \frac{R_S}{L_S} \right)}{\left[\left(s + \frac{R_S}{2L_S} \right)^2 + \left(\frac{1}{L_S C} - \frac{R_S^2}{4L_S^2} \right) \right]} \tag{6.147}$$

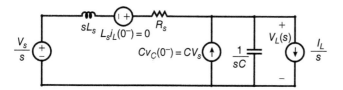

Figure 6.3 Laplace-transformed circuit for evaluation.

The three results can now be combined. Further algebraic manipulation can be done, like terms canceled and arguments adjusted to provide something easily inversely transformed:

$$V_L(s) = V_S \frac{1}{s} - I_L \frac{1}{s} + I_L R_S \frac{s + \frac{R_S}{2L_S}}{\left[\left(s + \frac{R_S}{2L_S}\right)^2 + \left(\frac{1}{LC} - \frac{R_S^2}{4L_S^2}\right)\right]}$$

$$- I_L R_S \frac{\frac{R_S}{2L_S} + \frac{1}{R_S C}}{\sqrt{\frac{1}{L_S C} - \frac{R_S^2}{4L_S^2}}} \frac{\sqrt{\frac{1}{L_S C} - \frac{R_S^2}{4L_S^2}}}{\left[\left(s + \frac{R_S}{2L_S}\right)^2 + \left(\frac{1}{LC} - \frac{R_S^2}{4L_S^2}\right)\right]} \tag{6.148}$$

The theorems and transform pairs discussed can be used to find the time-domain solution. The initial condition of the voltage source is included, and the final solution is given by

$$v_L(t) = V_S - I_L R_S u(t) \left[1 - e^{-\frac{R_S}{2L_S}t} \left(\cos\left(\sqrt{\frac{1}{L_S C} - \frac{R_S^2}{4L_S^2}} t\right) \right.\right.$$

$$\left.\left. - \frac{\frac{R_S}{2L_S} + \frac{1}{R_S C}}{\sqrt{\frac{1}{L_S C} - \frac{R_S^2}{4L_S^2}}} \sin\left(\sqrt{\frac{1}{L_S C} - \frac{R_S^2}{4L_S^2}} t\right) \right) \right] \tag{6.149}$$

The circuit seems simple for such a complicated expression. However, a wealth of information is available in such symbolic calculations. Not uncommonly, engineers rely on circuit simulators to generate results, and when faced with a problem randomly change the circuit to solve the issue without knowing that their solution is the optimum one.

Inspecting Eq. (6.149), one may conclude at first glance that reducing the resistance would eliminate the problem. Such an approach would certainly improve the efficiency. However, looking at the circuit in the limit as the resistance becomes arbitrarily small, the load voltage becomes

$$v_L(t)|_{R_S \to 0} = V_S - I_L u(t) \left[1 + \sqrt{\frac{L_S}{C}} \sin\left(\sqrt{\frac{1}{L_S C}} t\right) \right] \tag{6.150}$$

It may well be that reducing the resistance is the best approach for many reasons, but understanding the ultimate response is very important. Depending on the load increase and the associated inductance and capacitance, the voltage could swing wildly with a very slow decay and cause the load device to stop functioning. Handling this very real response to the power delivery is discussed in other chapters.

6.7 Summary

In this chapter, we provided a review of signal analysis fundamentals. Signal analysis is one of the key toolsets a power integrity engineer must have in order to solve many of the complex problems they will encounter in their career. Although there are many good texts on the subject that delve into significantly more detail than what we have provided here, an introductory review is certainly warranted. Though SPICE is often used for more complex problems with multiple circuits and subsystems, understanding the fundamentals here is key to growing the PI engineer's foundation.

In this chapter, we began with LTI systems (linear time-invariant). Virtually all of the problems a PI engineer will encounter will involve LTI systems and those that are not can usually be approximated as one. This led us to discuss delta functions, convolution, and the Fourier series which are the backbone of many signal analysis techniques and methods.

Because of its importance, the Fourier transform was discussed in some detail. This involved giving the engineer a review of the key theorems, which are important toward analyzing signal and circuit problems. The Fourier transform is suitable for truncated signals common for many power integrity type problems. Specifically, the unit step and the impulse function were discussed along with their transforms to aid the engineer in solving problems analytically.

One of the more common tools used in solving circuit problems is the Laplace transform. Like the Fourier transform, a number of common theorems were discussed to aid the PI engineer. This section ended with a number of examples using the Laplace theorems and transforms including some simple circuit problems.

Problems

6.1 Show mathematically for the following equations which ones are linear and time invariant.

$$y(t) = t\sin(2t + 30), y(t) = 7t + 1, y(t) = 5t + 4t^2, y(t) = 17e^{-i2\pi t}$$

$$(6.151)$$

6.2 Compute the convolution integral for the function

$$h(t) = \frac{1}{L}e^{-\frac{R}{L}(t-k)}u(t - k) \qquad (6.152)$$

6.3 Compute and graph the Fourier series of the square wave result for $n = 5$ in Eq. (6.45).

6.4 Use the time-shift theorem to compute the Fourier transform of the following signal.

$$y(t) = \sin\left(2\pi t - \frac{\pi}{3}\right) u\left(t - \frac{\pi}{3}\right) \tag{6.153}$$

6.5 Compute the Fourier transform using the differentiation theorem.

$$y(t) = \cos\left(2\pi t - \frac{\pi}{3}\right) u\left(t - \frac{\pi}{3}\right) \tag{6.154}$$

6.6 Find the transfer function in Figure 6.1 using Laplace transforms.

6.7 Find the Laplace transform of a capacitor and an inductor in parallel. Add initial conditions to both. What is the resonant frequency? Compute it.

6.8 Determine the Laplace transform for a capacitor and inductor in series. Use initial conditions. Compare the result to the previous problem's solution.

6.9 Find the Laplace transform for the following:

$$\sin(2\pi f_o t)u(t - t_0) \tag{6.155}$$

6.10 Replace the resistor in Figure 6.3 with a capacitor and recompute the time-domain solution as in Example 6.17.

Bibliography

1 Oppenheim, A.V. and Willsky, A.S. (1983). *Signals and Systems*. Prentice Hall.
2 Oppenheim, A.V. and Schafer, R.W. (1975). *Digital Signal Processing*. Prentice Hall.
3 Kunt, M. (1986). *Digital Signal Processing*. Artech House.
4 Pozrikidis, C. (1998). *Numerical Computation in Science and Engineering*. Oxford University Press.
5 Kreyszig, E. (1993). *Advanced Engineering Mathematics*, 7e. Wiley.
6 Ziemer, R., Tranter, W., and Fannin, D.R. (1989). Signals and Systems: Continuous and Discrete. In: *Macmillan Publishing Company*, 2e.

7

Numerical Methods for Power Integrity

Over the past 50 years or more, numerical methods have become a key component toward analyzing the complex technical problems that have faced engineers in modern times. This discipline has slowly crept from the academic arena to now being an industry of its own replete with a plethora of programs available for engineers at all levels of experience. The tools that have been developed have helped to move innovation forward in ways the computer industry could have never imagined. In fact, the results are astonishing if one looks back just 10 years ago to where the electronic markets were and where they are now. It is not a mistake that numerically based programs have been a key enabler of the innovations that have brought us to where we are today.

Though these numerical programs have helped us tremendously in our understanding of the complex problems encountered throughout these past decades, an unfortunate consequence has arisen in both academic and business circles. An inordinate number of engineers have come to rely too heavily on them to do the work that the human brain is significantly more equipped to tackle. This is because too many engineers that use these tools spend the majority of their time learning the user interface and how to manipulate the tool to obtain data, rather than understanding the method to compute the solution *and* the underlying meaning of the resultant data. This has led engineers to rely less on their own critical thinking skills and more on what the tool is telling them. Too often are engineers spending hours upon hours behind a computer screen iterating their analyses instead of understanding the fundamentals and the conditions under which they input the data to get the results (good or bad) they received. In essence, this has led to less analytical thinking and increasingly more time spent becoming a technician of a particular program rather than a thoughtful engineer. Without understanding of the fundamentals, problem-solving becomes a series of mindless simulations, and innovation is slowed to a crawl.

The interesting thing is that this was never the intention, nor the basis for the development of numerical methods. No one envisioned replacing good

Power Integrity for Electrical and Computer Engineers, First Edition. J. Ted Dibene II and David Hockanson.

engineering analysis and practice with the application of any of these methods; in fact, the goal was to *supplement* the analytical process, not thwart it. In the end, the purpose has always been to increase productivity. Though many engineers believe it has made them more productive, in a large number of cases (possibly more than not), it has had the opposite effect.

Given the previous view on numerical tools in general, the goal herein is not to change the behavior of the industry. However, it is important to understand the real purpose of these numerical tools and to use them as they were intended. With respect to power integrity, most of the tools are related to solving fields, mathematical equations, and/or circuits. The math- and circuit-based tools Simulation Program with Integrated Circuit Emphasis (SPICE) typically use the data extracted from the field solvers and thus post-process this information for use in system level modeling. These tools will be utilized and discussed in subsequent chapters, and their limitations are usually more readily understood. Field solving though is the foundation for obtaining many of the parasitic elements that power integrity engineers need when determining interconnections between a particular section of one portion of a power distribution network (PDN) to another. These are mainly inductive and resistive in nature as we have seen and are critical toward gaining an understanding of the system-level effects of the PDN behavior.

In this chapter, the focus will be on understanding some of the more common numerical tools here. The focus, however, will *not* be on utilizing the tools themselves but on the underlying algorithms and methods and limitations behind them. The reason for this emphasis is because it is important to understand how these tools fundamentally operate and where the limits are with respect to accuracy, convergence, and tolerance. There should be less concern in running many of these "black box" programs than in making sure that the data is what engineers expect once an analysis is completed. Moreover, there should be clear objectives set upfront on the actual problem analyzed before one starts to scrutinize the results. This typically starts with some expectation on the result itself in order to bound it so that the iterative process is kept to a minimum.

Because the depth of each subject is large, it is only practical in this text that we touch upon the basic methods and the more common ones at that. This chapter is in no way intended to be all inclusive or even to scratch the surface of even one of these methods. To truly learn one or more of the numerical methods or analytical techniques requires an in-depth study that is far beyond the information provided in this chapter. A serious student should consider taking classes and focus on a technique if he/she wishes to become proficient at it. There are many texts on these subjects (we have provided a brief bibliography at the end of the chapter), and the reader may start the learning process from there or through their own searching. This is encouraged since for many of the programs that engineers use, the code is not exposed to the user, which makes it difficult if not impossible to evaluate the efficacy of the method they are using. For those who are interested in developing their own numerical code, it is imperative that they learn the subject properly. Moreover, this requires one to delve deeply into an area rather than simply take a cursory look at it. The finite element method (FEM), for example, requires one to spend a significant amount of time investigating just the methodology, let alone the details of how to properly code up a given problem; this is true even for the simplest of analyses.

Thus, given the previous statements, this chapter will be organized to cover three main areas. First, we will give an introduction to analytical methods, which are the basis for many of the numerical tools that are used today. Analytical methods have always been the foundation for the growth of numerical methods. The reason is that before computers were invented, analysis was performed with a simple *pen and paper* and many complex problems were solved this way; many even led to the invention of the computer itself. Second, we introduce some of the more common numerical methods used today. Throughout the years, a number of very powerful techniques have been invented (finite difference, finite element, etc.). In addition, variances of each method have been incorporated into many available toolsets. However, once again, our focus here will be to give a limited scope on the ones that are more popular. And third, we provide some of the key fundamental issues with these numerical tools (in general) with the goal of giving the reader a basic understanding of what the results could mean when attempting to solve a particular problem.

It should be reiterated here that a single chapter on numerical methods is not sufficient for the reader to become proficient in any particular methodology. However, the goal here is to help the reader understand the issues with using various tools and thus give them a starting point for understanding what to expect when they set up a particular problem and analyze it using a particular program. Thus, it is our hope that the student will choose a method and then make the effort to become well versed in it. Having a strong foundation in one or more numerical methodologies almost invariably brings with it a deeper understanding of the broader area in which an engineer focuses. Having some understanding of numerical tools, even an introductory knowledge, can only help increase a power integrity engineer's toolset which is often required to be expansive to begin with.

Though it is impossible to teach any one of these subjects within such a limited scope, we do provide some examples as well as a number of programs that the reader can examine to give them an idea of how these methods may be applied. Thus, should the reader decide to write their own code rather than use a particular "black box" program, this may be used as a starting point toward further education, along with the references and many other resources available in the industry and on the web. The code is MATLAB based and may be easily ported into most versions for the student to investigate with. The programs are quite simple and are based upon solving Laplace's equation rather than attempting to solve more complex problems.

In general, as in the previous chapters, we will begin each section with the basics behind each method and then follow the introduction with an example or two. This will be evident in the introduction on analytical methods, leading up to the discussions on the finite difference and FEMs. The chapter will end with a section on errors in results, accuracy, and convergence though these issues will be touched upon along the way as well. For black box–type programs, it is likely that this section may be the most useful since it is intended to give the reader an insight into what to expect from these programs.

A final note about the introduction of numerical methods here: The overall motivation is to present this material keeping in mind what is important for the growth of the power integrity engineer. Thus, many important concepts will be left out that would be critical for a student of numerical methods to learn in order to become proficient. The authors acknowledge this deficiency in the presentation of the following material.

The first section of this chapter will then be to discuss some of the more common analytical methods used today.

7.1   Introduction to Analytical Methods

Before computers were invented and available, scientists and engineers relied on analytical tools to solve their complex technical problems. Through the development of advanced mathematics and with a greater understanding of the physics of the problems, analytical methods were developed to help drive the science and state of the art forward. It is still true, however, that numerical programs have aided in speeding up the solution process in so many ways. However, the foundation was laid with analytical methods, which paved the way for the algorithms in which these numerical methods were based upon. Moreover, in most cases, these analytical tools are still fundamental to the foundation of an engineer's background, particularly when solving electromagnetic (EM) problems. In addition, with the requirement that power integrity requires a reasonably strong background in the physics of electromagnetism, analytical methods should continue to have a place in the PI engineer's tool chest.

Though this chapter is about numerical methods, there are a number of important concepts, which form the basis for the algorithms in which these methods are based upon. In essence, analytical methods are techniques to solve specific equations that result in a closed-form solution. In many instances, once the equations are found, a numerical solution is often still desired which requires changing the parameters within the equation, and thus, the formulae are put into a mathematical solver for further processing. Moreover, the results from these methods can be used for checking more complex solutions obtained from a numerical simulation. In this way, engineers can verify that their modeling efforts have led them down the path that they were striving for.

Another important aspect to learning analytical methods is it strengthens engineers' skills in math and physics while giving them another tool to solve complex problems using *first principles*. Though the time deriving many of these equations and iterating through the math can be tedious, the results are invariably worth the effort particularly when an engineer begins the process of setting up a numerical model. Having already worked through an analytical solution, the engineer usually has a grasp of the problem and knows how to analyze the results which makes for a much more efficient process.

For the electromagnetic problems that are encountered in power integrity, the primary focus will be on finding the inductance and resistance between capacitive components in a PDN. However, to introduce the concepts, we will start with some simple examples and then begin to solve a few of the more common structures that are required for setting up the system simulations and analyses that we will see later in the text. Because the problems that power integrity engineers face are transverse electromagnetic (TEM), the field solutions are invariably found to be static. This simplifies the equations significantly, as well as their solutions. In fact, many of the problems may be solved through setting up Laplace's equation since the first step is often to find the electric potential in the structure. From there, the capacitance and the inductance can be deduced, which is why learning these methods is both insightful and powerful at the same time.

In Section 7.1.1, we will discuss separation of variables and touch upon the other techniques along the way.

7.1.1 Separation of Variables

The method of separation of variables, or sometimes referred to as the *Product* or *Fourier* method, is based upon the notion that there exists a solution to a particular problem where the result is a combination of functions with independent variables multiplied together [1]. The object is to start with a particular function or equation and then find the solution to this based upon the ability to separate out the variables. Thus, for a three-dimensional time-dependent problem, the solution would take the form of the following function:

$$\varphi(x, y, z, t) = F(x)G(y)H(z)K(t) \tag{7.1}$$

where the complete solution is a combination of independent *separable* functions. Because the functions may be separated, the solution to the problem can be broken down into a set of solutions with independent equations whereby the final answer will be similar to the result in Eq. (7.1).

There are three important rules that are required when applying the *separation of variables* method. First, it is based upon the ability to separate the variables in the solution of a particular PDE (partial differential equation) with the goal of finding a known solution. Many PDE's are not separable, and thus the method will break down for equations that have these constraints (see Problem 7.1). It is important to test this before assuming the method may be applied. Second, it is required that the boundary conditions (BCs) be *constant-coordinate* surfaces. For example, if the solution space is *Cartesian*, the boundary conditions must be constant along a rectangular line of a given boundary for the problem definition. As another example, if the solution space is *cylindrical*, the circumference along one of those boundaries would also have a constant value. Finally, the third requirement is that the boundary conditions must not have any partial derivatives associated with them. It is usually best to understand the method by applying it in a real problem.

Referring to Figure 7.1, let us assume we wish to find the solution to the voltage within the region shown based upon a given set of boundary conditions, where eventually we will wish to determine the capacitance of the structure. The length of the box is assumed to be very long (in the z direction), so in the end, we will end up determining the per-unit-length capacitance. Thus, this is another example of an electromagnetics static problem. The boundary conditions are shown in the figure. We let the potential be zero on the two sides where the x and y axis' are shown below (e.g. $V(0 < x < a, 0 < y < b) = 0$).

It will also be assumed for this particular problem that the corners are insulated, but are infinitely close to each other. This will be important for the next step here.

At first glance, the problem appears to be relatively complex since we have two inhomogeneous boundary conditions. This makes it difficult to apply them at the end of the problem when we bring the solution together. Fortunately, we can use the property of superposition to solve this dilemma. Using this fact, we have separated a more complex problem into two simpler ones. The two new problems are shown in Figures 7.2 and 7.3. The complete solution will eventually be of the form

$$V = V_A + V_B \tag{7.2}$$

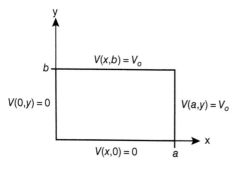

Figure 7.1 Structure for determining voltage.

Figure 7.2 V_A problem setup for Figure 7.1.

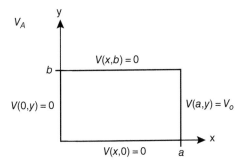

Figure 7.3 V_B problem setup for Figure 7.1.

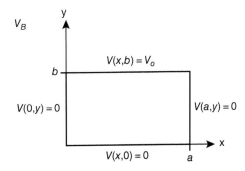

We have defined now two problems where there is only one inhomogeneous boundary condition in each. In this case, we have assigned in Figure 7.2 the potential along the $y = b$ coordinate $V_{xb} = 0$. Also for the problem in Figure 7.3, we have assigned the potential along the $x = a$ coordinate as $V_{ya} = 0$. We will start by solving the problem in Figure 7.2. In this case, the PDE that we wish to solve is Laplace's equation shown here in its general form.

$$\frac{\partial^2 \varphi}{\partial x^2} + \frac{\partial^2 \varphi}{\partial y^2} = 0 \tag{7.3}$$

We will substitute the general variable φ, for the voltage V, or

$$\frac{\partial^2 V}{\partial x^2} + \frac{\partial^2 V}{\partial y^2} = 0 \tag{7.4}$$

The goal is to find the solution to the voltage everywhere inside of the structure and then determine the total capacitance at the end. We start by assuming a solution to the problem where the voltage is a combination of two functions, one in x and the other in y.

$$V(x, y) = F(x)G(y) \tag{7.5}$$

Substituting into Eq. (7.4) we get

$$G\frac{\partial^2 F}{\partial x^2} + F\frac{\partial^2 G}{\partial y^2} = 0 \tag{7.6}$$

Dividing now by the product FG, we end up with two relations, one in F and the other in G

$$\frac{1}{F}\frac{\partial^2 F}{\partial x^2} + \frac{1}{G}\frac{\partial^2 G}{\partial y^2} = 0 \tag{7.7}$$

In order to separate the two relations into two independent equations, we need a separation variable λ such that

$$\frac{1}{F}\frac{\partial^2 F}{\partial x^2} = -\frac{1}{G}\frac{\partial^2 G}{\partial y^2} = \lambda \tag{7.8}$$

This results in two equations, with each of them being a function of one variable

$$\frac{\partial^2 F}{\partial x^2} - \lambda F = 0 \tag{7.9}$$

and

$$\frac{\partial^2 G}{\partial y^2} + \lambda G = 0 \tag{7.10}$$

The goal now is to solve the two equations given the boundary conditions. Going back to Figure 7.2, it is clear that the boundary conditions in terms of y are purely homogeneous. Thus, this should allow us to find a general solution in terms of the equation in y once we apply the boundary conditions. We begin by separating these. The top boundary condition at the $y = b$ boundary is

$$V(x, b) = 0 \rightarrow F(x)G(b) = 0 \rightarrow G(b) = 0 \tag{7.11}$$

Notice that if $F(x) = 0$, this would result in the entire function in x being zero which is the trivial solution. Thus, we have to conclude that $G(b) = 0$ here and $F(x)$ is not. We now examine the other boundary conditions (BC's).

$$V(x, 0) = 0 \rightarrow F(x)G(0) = 0 \rightarrow G(0) = 0 \tag{7.12}$$

and the constant x BCs

$$V(0, y) = 0 \rightarrow F(0)G(y) = 0 \rightarrow F(0) = 0 \tag{7.13}$$

$$V(a, y) = V_0 \rightarrow F(a)G(y) = V_0 \tag{7.14}$$

As can be seen in Eq. (7.14), the last boundary condition is not separable. Thus, we will need to apply this condition at the end after we determine the general solution to the problem.

Starting now with the equation in terms of the variable y, we determine the general solution. We start out by substituting for λ the variable k, or, $\lambda = k^2$.

$$\frac{\partial^2 G}{\partial y^2} + k^2 G = 0 \tag{7.15}$$

There are three possibilities for the variable λ. First, if $\lambda = 0$, then the solution is trivial (see Problem 7.2). Thus, we know that this is not a solution. The next two we will now examine. If $\lambda = k^2$ where $\lambda > 0$, then the solution to Eq. (7.15) is

$$G(y) = c_1 e^{ky} + c_2 e^{-ky} \tag{7.16}$$

or,

$$G(y) = c_3 \sinh(ky) + c_4 \cosh(ky) \tag{7.17}$$

Since either solution is valid, let us apply the boundary conditions to Eq. (7.17). Applying the first BC yields

$$G(0) = 0 \rightarrow c_3 \sinh(0) + c_4 \cosh(0) = 0 \rightarrow c_4 = 0 \tag{7.18}$$

Since the cosh function is never zero at the origin, we will now apply the next boundary condition.

$$G(b) = 0 \rightarrow c_3 \sinh(kb) = 0 \tag{7.19}$$

If c_3 were zero, we would have the trivial solution again; thus, the question is for which values $k > 0$ is the sinh function 0? The answer is, there is none. Thus, $\lambda > 0$ cannot be a solution to the equation. We now examine the solution where $\lambda < 0$. The solution to the equation with this condition now takes the form

$$G(y) = d_1 e^{jky} + d_2 e^{-jky} \tag{7.20}$$

or,

$$G(y) = d_3 \sin(ky) + d_4 \cos(ky) \tag{7.21}$$

Applying the boundary conditions once again, here to Eq. (7.21), yields

$$G(0) = d_3 \sin(0) + d_4 \cos(0) \rightarrow d_4 = 0 \tag{7.22}$$

and,

$$G(b) = 0 \rightarrow d_3 \sin(kb) = 0 \rightarrow kb = n\pi \tag{7.23}$$

or,

$$k = \frac{n\pi}{b}, \quad \text{where } n = 1, 2, 3, \ldots \tag{7.24}$$

Thus, we now have a general solution to the first equation.

$$G_n(y) = \sin\left(\frac{n\pi y}{b}\right) \tag{7.25}$$

where $\lambda = k^2$ are the *eigenvalues* for the *eigen function* in Eq. (7.25). The next step is to solve the second equation given the homogeneous boundary condition. We now know that $\lambda < 0$, gives us the general solution to Eq. (7.9). This has the same form as Eq. (7.17), except now it is in the form of a Fourier series.

$$F_n(x) = a_n \sinh\left(\frac{n\pi x}{b}\right) + b_n \cosh\left(\frac{n\pi x}{b}\right) \tag{7.26}$$

Applying the first boundary condition yields

$$F_n(0) = a_n \sinh(0) + b_n \cosh(0) \rightarrow b_n = 0 \qquad (7.27)$$

We can now put the equation back together and apply the final boundary condition to the problem.

$$V_n(x, y) = F_n(x)G_n(y) = a_n \sinh\left(\frac{n\pi x}{b}\right) \sin\left(\frac{n\pi y}{b}\right) \qquad (7.28)$$

This boundary is *inhomogeneous* and must be applied for the combined final solution. Applying the final boundary condition yields

$$V(a, y) = V_0 = \sum_{n=1}^{\infty} a_n \sinh\left(\frac{n\pi a}{b}\right) \sin\left(\frac{n\pi y}{b}\right) \qquad (7.29)$$

We can determine the coefficient for the Fourier expansion using [2] or from the data in Chapter 6. We recall that the coefficient in a Fourier series for some function r is

$$a_n = \frac{2}{K} \int_0^K f(x) \sin\left(\frac{n\pi r}{K}\right) dr \qquad (7.30)$$

where K is the interval over which the integration of the function takes place. Thus, solving for the coefficient we arrive at

$$a_n = \frac{2V_0}{a \sinh\left(\frac{n\pi a}{b}\right)} \int_0^a \sin\left(\frac{n\pi y}{b}\right) dy \qquad (7.31)$$

The integral for the sine function is

$$\int_0^a \sin\left(\frac{n\pi y}{b}\right) dy = \frac{a}{n\pi}\left[1 - \cos\left(\frac{n\pi a}{b}\right)\right] \qquad (7.32)$$

Thus, for $n = $ even, the coefficient is zero, and for $n = $ odd the solution will be,

$$a_n = \frac{4V_0}{n\pi \sinh\left(\frac{n\pi a}{b}\right)}, \quad n = \text{odd} \qquad (7.33)$$

Substituting in for the coefficient, we get the final solution to the problem.

$$V_A(x, y) = \frac{4V_0}{\pi} \sum_{n=odd}^{\infty} \frac{\sinh\left(\frac{n\pi x}{b}\right) \sin\left(\frac{n\pi y}{b}\right)}{n \sinh\left(\frac{n\pi a}{b}\right)} \qquad (7.34)$$

Using a similar approach, we can solve for V_B where the solution is given in [2].

$$V_B(x, y) = \frac{4V_0}{\pi} \sum_{n=odd}^{\infty} \frac{\sinh\left(\frac{n\pi y}{a}\right) \sin\left(\frac{n\pi x}{a}\right)}{n \sinh\left(\frac{n\pi b}{a}\right)} \qquad (7.35)$$

Table 7.1 Values for plotting the voltage function in Eq. (7.35).

a (cm)	b (cm)	V_0 (V)
2	1	1

Figure 7.4 Grid positions for voltages in Figure 7.1.

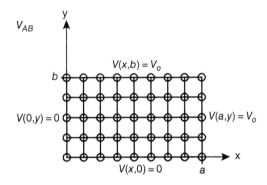

Note the solution to the structure in Figure 7.2 is very similar to the one in Figure 7.3. In fact, if $a = b$, the result *would* be exact. The final solution for the voltage is a linear combination of V_A and V_B now (Eq. (7.2)).

Our next step is to solve for the capacitance of the structure. However, before we do that, it is instructive to plot the function for the voltage to see what the potential looks like. We first break up the region in Figure 7.1 to show discrete points to illustrate the voltage in two dimensions. We may choose values for a, b, and V_0. Table 7.1 gives some specific values so that we may compute the voltage.

Figure 7.4 shows a grid for region in Figure 7.1. The step size is somewhat random for the moment here; however, when we get to solving this structure numerically, we will find that the step size will play an important role in both the accuracy and stability (especially for the finite difference solution). We can generate a matrix of values (row and column) to associate with the grid structure in the figure. For each location in the grid, a value may be computed from Eq. (7.2) where a given value of n is used. A simple MATLAB m-file is created here to compute the first two values as detailed in Code 7.1.

Code 7.1: MATLAB code for Eq. (7.2).

***

```
% This m-file computes the solution the Voltage Potential for the
% Problem in
% Chapter 7 using Separation of Variables. The solution is given for
% VA and VB
```

```
% and the purpose is to compute and plot the results for
  various values
% of n
% in the solution with a series function.


% Set key variables

a=2e-2;
b=1e-2;
V0=1;
ie=9;
je=5;

% Define functions and loop

n=1;

for i= 1:ie,
    for j= 1:je;

    x=(i-1)*(a/8);
    y=(j-1)*(b/4);

    VA(i,j)=(4*V0)/(pi)*( ( sinh( (n*pi*x)/b )*sin( (n*pi*y)/b ) )/
(n*sinh( (n*pi*a/b) ) ) );
    VB(i,j)=(4*V0)/(pi)*( ( sinh( (n*pi*y)/a )*sin( (n*pi*x)/a ) )/
(n*sinh( (n*pi*b/a) ) ) );
    Vn(i,j)=VA(i,j)+VB(i,j);

    end
end

k=3;

for i= 1:ie,
    for j= 1:je;

    x=(i-1)*(a/8);
    y=(j-1)*(b/4);

    VA(i,j)=(4*V0)/(pi)*( ( sinh( (k*pi*x)/b )*sin( (k*pi*y)/b ) )/
(k*sinh( (k*pi*a/b) ) ) );
    VB(i,j)=(4*V0)/(pi)*( ( sinh( (k*pi*y)/a )*sin( (k*pi*x)/a ) )/
(k*sinh( (k*pi*b/a) ) ) );
    Vk(i,j)=VA(i,j)+VB(i,j);

    end
end

V=Vn+Vk;

% end m-file
```

***

Table 7.2 Relative position of voltages.

X → Y↓	1	2	3	4	5	6	7	8	9
1	0	0.8794	1.2004	1.0139	0.8488	1.0139	1.2004	0.8794	0
2	0	0.4348	0.6752	0.7191	0.7217	0.7871	0.8573	0.8708	1.2004
3	0	0.2249	0.379	0.4536	0.4956	0.549	0.6289	0.761	0.8488
4	0	0.0986	0.1733	0.2192	0.2505	0.2872	0.3554	0.5346	1.2004
5	0	0	0	0	0	0	0	0	0

Table 7.2 shows the values for each x/y grid position relative to Figure 7.4. Note that there are some anomalies with respect to the data and their relative positions in the grid. First, one can see that the values along the $y = b$ and $x = a$ lines are not all at 1 V. This is due to the fact that the series was *truncated* when in fact the correct solution should be *infinite*. In addition, the grid size is not infinitely small in size, which would be the case for the exact solution. If we were to increase the size of the number n in the series and decrease the grid size, it would result in a more accurate solution.

The values for each point in the grid may be shown in the surface plot below for $n = 3$. This will give us an idea of what the expected potential would be for the entire region and whether or not the potential in these areas makes sense.

The result is plotted in Figure 7.5. The solution is found by summing the first two components in Eq. (7.28) ($n = 1, 3$). The plot also corresponds with the voltage for the values in the Table 7.2.

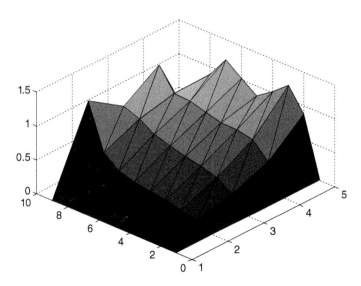

Figure 7.5 Surface plot of voltage.

If we chose a single value in the middle ($V[a/2, b/2]$) and then increase the size of the series (increase n), we should see that value begin to asymptote to the exact solution. This is expected, given the exact result is found by taking the limit of the infinite series for each point in the grid. We can create another small MATLAB m-file to generate the voltage at the middle of the grid as given in Code 7.2. The point we choose is at position $V(ie,je) = V(5,3)$ relative to the x, y coordinates and may be found to be the middle point in the grid by inspection.

Code 7.2: MATLAB script for plotting voltage at center of grid.

***

```
% This m-file computes the solution the Voltage Potential for the
% Problem in Chapter 7 using Separation of Variables at the
   center of
% the grid. The solution is given for VA and VB
% and the purpose is to compute and plot the results for
   various values
% of n in the solution with a series function.

% Set key variables


a=2e-2;
b=1e-2;
V0=1;
ie=5;
je=3;
x=ie;
y=je;

% Loop for number of iterations

n=1;

for k= 1:10,


    n=2*k-1;                     % this makes sure we sum only odd
                                   functions
    x=(ie-1)*(a/8);
    y=(je-1)*(b/4);

    VA(n)=(4*V0)/(pi)*( ( sinh( (n*pi*x)/b )*sin( (n*pi*y)/b ) )/
(n*sinh( (n*pi*a/b) ) ) );
    VB(n)=(4*V0)/(pi)*( ( sinh( (n*pi*y)/a )*sin( (n*pi*x)/a ) )/
(n*sinh( (n*pi*b/a) ) ) );
    V(n)=VA(n)+VB(n);

    Vs=sum(V);                   % Sum of voltage for position ie,je
```

```
Vsp(k)=Vs;                    % We turn Vs into an array for plotting

end

plot(Vsp)

% end m-file
```

***

The graph in Figure 7.6 is a plot of the results for various values of k ($n = 2k - 1$).

As shown in Figure 7.6, we can see clearly that with a few iterations in the series, the function begins to asymptote to the value of $V = 0.5\,\mathrm{V}$ which is the exact value for the voltage potential in the structure. Thus, we have a good check now for how many iterations we may use if we wish to solve this structure numerically for the voltage and achieve a highly accurate result.

Our next goal is to find the capacitance of the structure now. To get the result we are looking for, we need to compute the energy in the system. This may be done by using the relation in Chapter 4 for the energy obtained in a capacitance structure,

$$W = \frac{1}{2}CV^2 \tag{7.36}$$

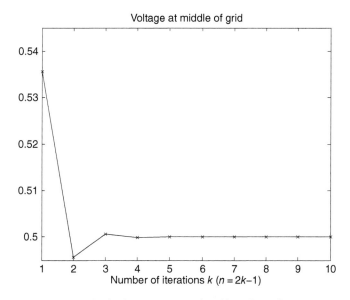

Figure 7.6 Graph of voltage at center of grid for values of k.

The energy stored in the capacitor may also be found from the divergence theorem as discussed in Chapter 4.

$$W = \frac{1}{2} \iiint \epsilon |\mathbf{E}|^2 dv \tag{7.37}$$

where the electric field is related to the voltage.

$$\mathbf{E} = -\nabla V \tag{7.38}$$

Thus, to find the total work *per unit length* for the capacitor, we may substitute for the electric field in Eq. (7.37).

$$W = \frac{1}{2} \iint \epsilon |\nabla V|^2 ds = \frac{1}{2} \iint \epsilon \left(\left(\frac{\partial V}{\partial x} \right)^2 + \left(\frac{\partial V}{\partial y} \right)^2 \right) ds \tag{7.39}$$

Note that the integral is no longer over a volume since we are looking for the per-unit-length capacitance, which means we divided out the z dependence. To get the total work done for a capacitor of length l, we would multiply by the length in Eq. (7.39).

The capacitance is now found from Eq. (7.36).

$$C = \frac{2W}{V_0^2} \tag{7.40}$$

To compute the work done, we can discretize Eq. (7.39). We will do this using finite difference techniques in the next sections. For the problem at hand, we may simply compute the difference in each voltage at a given node and then assume a given change in x and y as we did previously. The writing of the MATLAB script is left as an exercise for the reader. It is instructive to note that once the solution to the voltage is found, finding the gradient with respect to x or y is a simple matter of differentiating the solution found in Eq. (7.2) for the total voltage.

Sometimes, we are given problems that require us to find the optimal solution rather than just one particular result. This is the subject of the next method in which we introduce variational methods.

7.1.2 Introduction to Variational Methods

Variational methods stem from a more broader field of study called the *calculus of variations*. The method is a very powerful one and has been used for optimizing many complex problems over the centuries. As shown in the separation of variables methodology in the previous section, the results may be used to parameterize the overall solution and can be compared with a numerical analysis. In fact, there are techniques to discretize the solution, such as using the *weighted residuals method*, to numerically approximate the result. Thus, variational methods often include a discussion of these additional techniques.

As is the case with most of the numerical and analytical tools discussed herein, the focus will be strictly on an introduction to the subject. There are entire texts dedicated to each of these areas, and the subject of variational methods is no exception. Thus, it is not possible to do any one of them justice within the context of a single chapter (much less a section in a chapter). Nonetheless, our goal here will be to show that methods like this exist and that with a more in-depth study, the student may apply them to solve many of the power integrity problems that they will encounter. As stated at the beginning of this chapter, every engineer should be encouraged to broaden out their background with a focused study of a numerical methodology.

Calculus of variations is indeed a rather old subject and within a few years of the invention of calculus, the great mathematicians of the time started posing problems that could not be solved directly with standard calculus. For reference, Byron and Fuller have done an excellent job of introducing the subject with some historical background in Chapter 2 of their text [4]. We follow closely some of their derivation and then move onto the numerical methods later on in the chapter.

It is believed that variational methods started with a challenge at the end of the seventeenth century; sent out by one of the great mathematicians of the time. The challenge was based on the ability to solve a particularly difficult problem which had been posed a number of years earlier. Though not the first variational problem to be solved, the *brachistochrone* problem may have been the most famous. First proposed by Johann Bernoulli in 1696, it was Bernoulli that issued the challenge to the mathematical community of the time to solve the problem that he himself claimed to have been the first to conquer (this may also have been debatable). In essence, Bernoulli asked what the path would be for a particle to traverse in the shortest time, given the force on the particle was due to gravity alone. This problem is illustrated in Figure 7.7.

Figure 7.7 Brachistochrone problem setup.

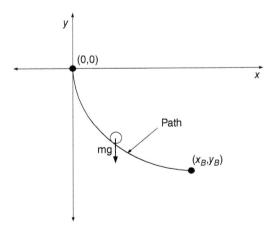

Basically, the object was to find the path between two points that would result in the particle of mass m moving from the origin (or some point in the x–y plane) to the point (x_B, y_B) in the shortest time. Another assumption was that there was no friction in the path (which obviously simplified the problem further). At first glance, the problem appears to be simple. However, at the time (and actually today in fact) it was, and is, quite complex. It was said that it took Sir Isaac Newton over 12 hours to solve. Later, he sent his reply back to Bernoulli anonymously. Bernoulli, however, was not fooled by the anonymous reply and after receiving the manuscript of the solution from Newton was believed to have stated simply, "I knew the lion by his touch."

The term *brachistochrone* comes from the two Greek words, *brachistos* (fastest), and *chronos* (time). Leibniz, who is also credited for inventing the calculus, was said to be the first to have sent in the complete solution. There was speculation in the day that when Johann Bernoulli had sent out the problem (he actually posed two problems to the mathematical community), his goal was to discredit his brother whom he had an ongoing feud with. The older brother (Jacob), whom he had publically called out as being incompetent, was actually able to solve the problem as well. Interestingly, Bernoulli himself did not use calculus of variations to solve the problem [4].

The solution to the Brachistochrone problem was in part the start of a very powerful methodology that involved finding the extremum of a given class of problems. The solution to this problem is found in many texts, and we will show it later on in this section for those who are interested. Many variances of the problem are available on the internet, and a few references are given in the bibliography at the end of this chapter [3].

Most texts introduce the formulation of the variational method with the derivation of the Euler–Lagrange equation. The equation is found by differentiating the function under the integral and setting the entire result to zero. We will start with the simpler case where the problem is in one dimension only. The method may be expanded to multiple variables and functions as needed.

The goal here is to find the function $y(x)$ which is intended to minimize the following integral:

$$I = \int_{x_A}^{x_B} f(x, y, y')dx \tag{7.41}$$

The integral I is called a *functional*. The integrand in the functional (f) is a function of the variables x, y, and y'. By minimizing the integral I, the function y that results will be the solution to the problem that one is looking for. The functional in Eq. (7.41) is subject to two very important constraints:

(1) x_A, x_B, $y(x_A) = y_A$, $y(x_B) = y_B$, and f are all known.
(2) f is twice differentiable in all its arguments.

solving the integral equation for the minimum. The functional will be discussed again when we cover the FEM.

Let us analyze a variance of Eq. (7.48). We recall that the function f does not explicitly depend upon x. Thus, we can use the chain rule. We take the derivative of the equation.

$$\frac{d}{dx}\left[y'\frac{\partial f}{\partial y'}-f\right]=y''\frac{\partial f}{\partial y'}+y'\frac{d}{dx}\left(\frac{\partial f}{\partial y'}\right)-\frac{\partial f}{\partial x}-\frac{\partial f}{\partial y}y'-\frac{\partial f}{\partial y'}y'' \tag{7.49}$$

Simplifying yields

$$y'\frac{d}{dx}\left(\frac{\partial f}{\partial y'}\right)-\frac{\partial f}{\partial x}-\frac{\partial f}{\partial y}y'=-y'\left[\frac{\partial f}{\partial y}-\frac{d}{dx}\left(\frac{\partial f}{\partial y'}\right)\right]-\frac{\partial f}{\partial x} \tag{7.50}$$

But since f does not *explicitly* depend on x and since y is a solution to the Euler–Lagrange equation, we know that

$$\frac{\partial f}{\partial x}=0 \tag{7.51}$$

Thus, finally we arrive at the solution

$$\frac{d}{dx}\left[y'\frac{\partial f}{\partial y'}-f\right]=0 \tag{7.52}$$

Now, coming back to the Brachistochrone problem. As in variational methods, we will be deriving a functional under the integral. To simplify the goal here, we ask the question: How long will it take to traverse the path in Figure 7.7 from the origin to the end point? The velocity is found by determining the distance over the time or

$$v=\frac{ds}{dt} \tag{7.53}$$

where s is the total distance that must be traversed from the origin to the point (x_A, y_A). A differential length of a part of this path, or arc, is found by examining an infinitesimal portion of the arc as shown in Figure 7.9: Thus, a differential time will be

$$dt=\frac{ds}{v}\approx\frac{\Delta s}{v}=\Delta t \tag{7.54}$$

and our integral will then be

$$I=\int_{x_A}^{x_B}dt=\int_{0}^{x_B}\frac{ds}{v} \tag{7.55}$$

The distance ds in terms of the variable y is

$$ds=\sqrt{dx^2+dy^2}=dx\sqrt{1+\left(\frac{dy}{dx}\right)^2}=dx\sqrt{1+(y')^2} \tag{7.56}$$

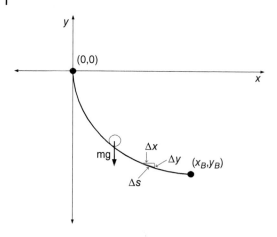

Figure 7.9 Differential path in arc length.

We now need to find the velocity in terms of the variable y. We relate the energy of a moving object to that of one under the influence of gravity to get

$$\frac{1}{2}mv^2 = mgy \tag{7.57}$$

where y is the distance fallen, m is the mass, and g is the acceleration due to gravity. Thus, solving for the velocity yields

$$v = \sqrt{2gy} \tag{7.58}$$

Substituting Eqs. (7.56) and (7.58) into Eq. (7.55) now,

$$I = \int_0^{x_B} \frac{\sqrt{1+(y')^2}}{\sqrt{2gy}}dx = \frac{1}{\sqrt{2g}}\int_0^{x_B}\sqrt{\frac{1+(y')^2}{y}}dx \tag{7.59}$$

The term under the integral is the function that we wish to minimize. In terms of a variational problem, we now need to find the function y that minimizes this functional I overall.

Thus, we know that the integrand is

$$f = \sqrt{\frac{1+(y')^2}{y}} \tag{7.60}$$

We may now use Eq. (7.52) in Eq. (7.60) to get

$$\frac{d}{dx}\left[y'\frac{\partial f}{\partial y'} - f\right] = \frac{d}{dx}\left[\frac{y'(2y')}{\sqrt{y(1+(y')^2)}} - \sqrt{\frac{1+(y')^2}{y}}\right] = 0 \tag{7.61}$$

Simplifying yields

$$\frac{d}{dx}\left[y'\frac{\partial f}{\partial y'} - f\right] = \frac{d}{dx}\left[\frac{-1}{\sqrt{y(1+(y')^2)}}\right] = 0 \tag{7.62}$$

If we square both sides and integrate, we get

$$\frac{1}{y(1 + (y')^2)} = A \tag{7.63}$$

where A is the constant from integration. Solving for y' yields,

$$y' = \frac{dy}{dx} = \sqrt{\frac{\frac{1}{A} - y}{y}} \rightarrow dx = \frac{dy}{y'} = dy\sqrt{\frac{y}{\frac{1}{A} - y}} \tag{7.64}$$

We may integrate this to get

$$\int dx = \int \sqrt{\frac{y}{\frac{1}{A} - y}} dy + D \tag{7.65}$$

To perform the integration, we may make a trigonometric substitution in Eq. (7.65) to solve the integral. This is left as an exercise at the end of this chapter. The solution to the function is a *cycloid*. This was discovered by Bernoulli. A cycloid, interestingly enough, is the path a point on the outer part of a wheel traverses when it rolls on a flat surface without slipping [3]. Thus, the path that an object would traverse would be half of a cycloid.

   Problems such as the Brachistochrone may be solved more readily with variational methods. Unfortunately, most problems that engineers encounter do not result in an elegant closed-form expression. Thus, engineers often pose problems in terms of a variational equation and then discretize it to achieve a solution numerically. The first question one might ask after reading the previous analysis would be, "So how can we apply such a method to solve a complex electromagnetic and power integrity problems?" Variational methods also help form the basis for the FEM (and method of moments), which we will discuss later in this chapter. However, with regard to solving certain problems directly using variational techniques, it is usually best to use an additional numerical technique which makes use of this method. We now introduce a powerful tool for solving such problems, which is a variance of the weighted residual method called the Galerkin method.

7.1.2.1   The Galerkin Method

There are a whole host of numerical methods which can be used to solve variational problems on computers. Many of them are based on the method of weighted residuals, or WRM [1] as it is commonly called. The method assumes a solution of the form whereby an approximation to the exact solution is proposed, resulting in a set of simultaneous equations. There are other techniques, such as the Rayleigh–Ritz method, which may be used to solve variational problems, but there are limitations to this method [2]; thus it is more interesting to discuss approaches which are able to solve the more general problem.

The WRM and its variances may be used to solve some of the more common problems, especially those that do not have well-prescribed ordinary differential-equations (ODEs) or where the functional may not be easily found. Here, we will focus on a particular variance of the WRM, called the Galerkin method. But first, we describe in general the foundation behind the WRM.

The integrand in the functional in Eq. (7.42) falls into a class of problems whereby the integrand may have kth derivatives within it.

$$f(x, y, y^{(1)}, \ldots, y^{(k)}) \tag{7.66}$$

In the previous discussion on the variational method, we were looking for a solution y to minimize the functional. In the weighted residuals method, we instead multiply by a set of *weighting* functions that minimize the integral.

$$\int_a^b f(x, y, y^{(1)}, \ldots, y^{(k)}) \xi_i(x) dx = 0 \tag{7.67}$$

where the function ξ multiplied by the integrand are a set of weighting functions. If the residual vanishes for any value of x and the function y satisfies the boundary conditions, then the exact solution will have been found.

Starting once again with a one-dimensional problem for simplicity, we assume the solution to the problem is of the form

$$y(x) = \sum_{i=1}^{N+1} a_i \phi_i(x) \tag{7.68}$$

The terms under the summation, $a_i \phi_i$, are a set of *basis* functions. The object is to substitute in for y in the integrand of Eq. (7.67) and then solve for the coefficients a_i. In the Galerkin method, we actually set the basis functions equal to the weighting functions. That is

$$\xi_i(x) = \phi_i(x) \tag{7.69}$$

The next question is how do we determine which basis functions to use? In general, the basis functions take the form of the independent variable raised to some power or a sine function, such as

$$\phi_i(x) = x^i \text{ or } \phi_i(x) = \sin(\pi x i) \tag{7.70}$$

Once we plug in for function in the integrand, we may then integrate over the boundary conditions and determine the solution. For a set of $N + 1$ basis functions, the result will give us a set of $N + 1$ simultaneous equations. The best way to illustrate the method is with an example.

Example 7.1   Given the differential equation

$$f(x, y', y) = y'' + y = 0, y(0) = 0, y(1) = 1 \tag{7.71}$$

Find the solution using the Galerkin method.

Solution:
Our first step here is to find the general solution to the differential equation. We have seen from looking at previous ODEs that a general form of the solution is of the following:

$$y(x) = A\cos(x) + B\sin(x) \tag{7.72}$$

Plugging in for the boundary conditions yields

$$y(0) = A(1) + B(0) \rightarrow A = 0 \tag{7.73}$$

Thus, the solution takes the form

$$y(x) = B\sin(x) \tag{7.74}$$

Plugging in now for the other boundary condition yields

$$y(1) = B\sin(1) = 1 \rightarrow B = \frac{1}{\sin(1)} \tag{7.75}$$

Thus, we now have the solution which is

$$y(x) = \frac{\sin(x)}{\sin(1)} \tag{7.76}$$

The next step is to generate an approximate solution to compare against our exact one above. If we start from the premise that we do not know what this solution is as yet, we would begin by generating a set of basis functions which take the form of a power equation.

$$y(x) = a_1 x^1 + a_2 x^2 \tag{7.77}$$

One of the benefits of the Galerkin method is that both boundary conditions need not be prescribed to achieve a solution. Thus, we apply the second boundary condition to the problem.

$$y(1) = a_1(1) + a_2(1) = 1 \rightarrow a_2 = 1 - a_1 \tag{7.78}$$

We now have one unknown to find

$$y(x) = a_1(x - x^2) + x^2 \tag{7.79}$$

Our basis function then takes the form

$$\phi_1(x) = (x - x^2) + x^2 = x \tag{7.80}$$

Thus, we may now substitute into Eq. (7.67) and integrate

$$\int_0^1 f(x, y, y', y'')\xi_1(x)dx = \int_0^1 (a_1(x - x^2) + x^2)x\,dx = 0 \tag{7.81}$$

Solving for the coefficient yields

$$a_1 = \frac{15}{11} \tag{7.82}$$

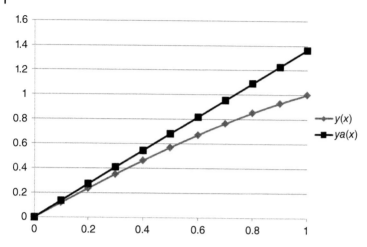

Figure 7.10 Plot of exact and approximate solutions.

Therefore, our approximate solution is

$$y(x) = \frac{15}{11}(x - x^2) + x^2 \tag{7.83}$$

We can plot the results for comparison. This is shown in the graph in Figure 7.10 where $y(x)$ is the exact result and $y_a(x)$ is the approximation. It is not surprising that the result for the approximation deviates somewhat as x approaches the boundary ($x = 1$) since we only chose one basis function for the solution. If we were to increase the number of functions, the accuracy would undoubtedly be better (Figure 7.10).

Our next subject is another analytical method called conformal mapping. We introduce this subject in brief now.

7.1.3 Conformal Mapping

Conformal mapping, like the other techniques discussed, is an analytical method that can be used with or without numerical post processing. As with all of these methods, there is a plethora of books and papers available that illustrate the method [5]. Herein, we will only give a brief snapshot of the method and its capabilities.

Conformal mapping has its applications in many areas, but one in particular has proven quite valuable. And that is the applicability to solve electrostatic problems using Laplace's equation [17]. For our purposes, this is advantageous since many problems that power integrity engineers face involve solving voltage potential problems.

The technique basically entails defining a problem in real number space and then mapping it into the Z or complex plane through a transformation (or through more than one mapping transformation). Once the problem is transformed into this space, the problem is solved (or intended to be solved) more easily, and then the solution is retransformed into the original mapping space. Before we illustrate the method with an example, we will go through a few concepts and then discuss solving a simple voltage potential problem.

Conformal mapping involves using complex variables to solve these particular problems [20, 21]. As a review for the reader, a complex variable is one that is comprised of both a real and a complex portion (or just a complex portion by itself),

$$z = x + jy \tag{7.84}$$

where z is the complex variable in question and x and y are real. The variable may also easily be transformed into polar coordinates where

$$x = r\,\cos\theta, \quad y = r\,\sin\theta \tag{7.85}$$

such that

$$z = r(\cos\theta + j\,\sin\theta) = re^{j\theta} \tag{7.86}$$

Thus, a function $f(z)$ is a function that contains a complex variable. An example of this would be

$$f(z) = \frac{z+1}{z-1} \tag{7.87}$$

This is a special type of mapping function called a *linear fractional transformation* [1]. There are many types of transformations into specific mapping planes though we will focus only on a few [22].

Now suppose we have another complex variable w such that

$$w = u + jv \tag{7.88}$$

This means that for every value of z, there is a corresponding mapping to w. In this way we can define the variable w, as a function of z.

$$w = f(z) \tag{7.89}$$

The derivative of the function w may be defined as we normally define the derivative or

$$\frac{dw}{dz} = \frac{df(z)}{dz} = \lim_{\Delta z \to 0} \frac{f(z + \Delta z) - f(\Delta z)}{\Delta z} \tag{7.90}$$

If the function has a derivative that exists and is unique, we say the function is *analytic*. Analytic functions are unique in that the function is infinitely differentiable in its Taylor series around a particular point while also converging around that point [1]. Going back now to Eq. (7.89), we may differentiate it with respect to the variables x and jy or

$$\frac{\partial w}{\partial x} = \frac{\partial u}{\partial x} + j\frac{\partial v}{\partial x} \tag{7.91}$$

and

$$\frac{\partial w}{\partial jy} = \frac{\partial v}{\partial y} - j\frac{\partial u}{\partial y} \tag{7.92}$$

If the real and imaginary parts of the previous two equations are equal, that is, if

$$\frac{\partial u}{\partial x} = \frac{\partial v}{\partial y} \quad \text{and} \quad \frac{\partial v}{\partial x} = -\frac{\partial u}{\partial y} \tag{7.93}$$

Then the function has a complex derivative that is continuously differentiable. The previous equations are called the Cauchy–Riemann equations. These conditions are a necessary condition for the derivative to be unique at a point and the function f to be analytic.

Essentially, conformal mapping is the method of mapping the solution region and the equations into another complex space in order to solve the original problem in an easier way [24]. After finding a solution in the new space, the result is then remapped into its original space. The following simple example illustrates the method.

Example 7.2 Determine the capacitance, inductance, and impedance between two concentric circles as in the coaxial circuit in Figure 7.11 using conformal mapping techniques. Assume the length is very long relative to the other dimensions.

Solution:
Figure 7.12 shows the transformation that we will use. First, we will assume that we want to map onto the complex space for w. The mapping to the z space is then simply

$$w = \ln(z), \quad z = e^w = e^{u+jv} \tag{7.94}$$

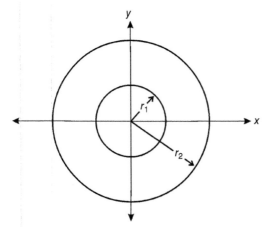

Figure 7.11 Coaxial structure.

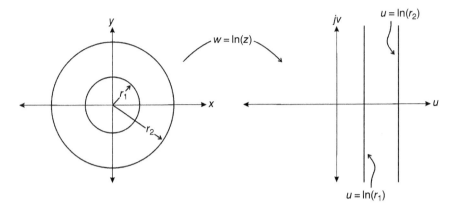

Figure 7.12 Transformation for conformal mapping.

Plugging in the complex representations and using Euler's formula yields

$$z = x + jy = e^{u+jv} = e^u(\cos v + j \sin v) \tag{7.95}$$

Now, the equation for a circle is

$$x^2 + y^2 = r^2 \tag{7.96}$$

Thus, using Eq. (7.95)

$$x^2 + y^2 = (e^u \cos v)^2 + (e^u \sin v)^2 = e^{2u} \tag{7.97}$$

Thus, for each circle, we get

$$x_1^2 + y_1^2 = r_1^2 = e^{2u_1} \rightarrow r_1 = e^u \rightarrow u_1 = \ln(r_1) \tag{7.98}$$

and, subsequently

$$u_2 = \ln(r_2) \tag{7.99}$$

Now, we recall for a simple capacitor that the capacitance is

$$C = \frac{\epsilon_0 \epsilon_r A}{d} \tag{7.100}$$

where A is the area and d is the separation. We assume that the structure is very long (essentially infinite), so we will compute the per-unit-length parameters. Thus, plugging in from Figure 7.12 we get

$$C = \frac{\epsilon_0 \epsilon_r 2\pi}{\ln(r_2) - \ln(r_1)} = \frac{2\pi\epsilon_0\epsilon_r}{\ln\left(\frac{r_2}{r_1}\right)} \tag{7.101}$$

The inductance may be found by recalling that the relationship between the phase velocity and the inductance is

$$LC = \mu\epsilon_0\epsilon_r \rightarrow L = \frac{\mu\epsilon_0\epsilon_r}{C} = \frac{\mu}{2\pi} \ln\left(\frac{r_2}{r_1}\right) \tag{7.102}$$

Finally, the impedance may be easily computed as

$$Z = \sqrt{\frac{L}{C}} = \sqrt{\frac{\mu}{\epsilon_0 \epsilon_r}} \ln\left(\frac{r_2}{r_1}\right) \tag{7.103}$$

7.2 Numerical Methods

We now turn the discussion to the main subject of this chapter: numerical methodologies. In this case, we will focus on essentially the two most popular methods: finite difference and finite element. When the element of time is added, the first one is termed the finite difference time domain method (FDTD). Because our focus is on static fields for power integrity problems, we will forego the discussion on time dependence.

With respect to the previous sections, the main difference between the analytical and the numerical methods is that numerical tools are geared toward solving problems using computers. Though one can certainly use them to solve a problem analytically, it would be a laborious and long proposition for even many of the simpler problems. This is where numerical tools and computers really do become necessary and advantageous. Once the problem is well posed and set up properly to be run with a program on a computer, the efficiency of a computational engine is clearly infinitely more expedient. This is particularly true for problems that require a large computational memory and/or involve many complex computations.

Through the introduction of analytical methods, we saw how to set up and solve a problem and then start the process of bridging the gap toward numerically computing the result. The next step then is to introduce these two methods and then discuss some of the salient issues, which come about when generating data with a numerically based program. This is a critical aspect toward understanding the limits of numerical methods. It is particularly true when trying to solve boundary value problems. Most problems usually require one to create boundaries that do not perfectly represent those of the original problem, which can lead to errors that may not be evident until it is too late.

First, we will examine the finite difference method and discuss how to solve a problem numerically using this technique.

7.2.1 The Finite Difference Method

One of the more powerful numerical methods that is broadly used today is the finite difference method. Referred to as the FDTD method when the variable of time is added, the method is based on basic differential calculus and has the advantage over other methods in that it is relatively easy to code and does not rely on an explicit mesh to determine a solution. Since the time dependence is simply an extension of the method and because our focus is on static computations for component extractions in power integrity, we will not discuss the *FDTD* method. Some excellent references are provided at the end of the chapter and the reader is encouraged to review them [6].

Compared to other numerical techniques, finite difference schemes are simpler to understand and are usually straightforward to set up. The method involves three simple steps: develop a grid for the solution (implied with respect to the code), develop an approximation to the equations using a difference technique, and finally, solve the equations given a prescribed set of boundary conditions.

As mentioned earlier, the grid in finite difference implementations are *implied*; that is, the grid is typically implemented within the code and the formulation of the equations. For two-dimensional problems, one simply partitions the problem space with a grid and then determines the distance between each grid point (e.g. for two dimensional problems, $dS = dx\,dy \cong \Delta x \Delta y$). Depending upon the problem, the grid spacing may be chosen to be the same in both the x and y directions for ease in coding. Because difference methods approximate the derivative of the function, at some point, there will be an error associated with the result. Figure 7.13 illustrates this. The function that we are interested in approximating with a difference formula is $f(x)$. This is the *exact* function. The approximations for the point $[f(x_0), x_0]$ are shown in the figure as well. It can be seen that if one were to estimate with a difference formula any point along the curve $f(x)$, there would be an error between the approximation and the actual function. This will be more evident when we derive the difference equations in a moment.

There are three variations to the derivation of the difference equations, which programmers rely on: the *forward*, *backward*, and *central* difference formulations. The main variance between the three is where the difference computation takes place on the grid.

Figure 7.13 Approximation of function for finite difference method.

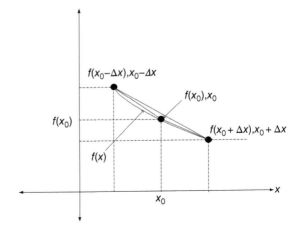

The forward difference equation may be derived through a truncation of a Taylor series (as may all of the difference formulations). To find the derivative of a function in x, we expand it around a small change Δx with the objective of solving for the first derivative with a Taylor series.

$$f(x + \Delta x) = f(x) + \frac{f'(x)}{1!}\Delta x + \frac{f''(x)}{2!}(\Delta x)^2 + \ldots \tag{7.104}$$

This expansion is exact if we expand the terms out to infinity. However, for our purposes, everything with a second derivative or higher is assumed to be small relative to the first two terms; that is, the *error* associated with the additional terms compared to the exact value may be ignored in the formulation. This is the key to using a finite difference formula. If we approximate the function now by combining all terms of second order and higher into one function, we get the following result:

$$f(x + \Delta x) = f(x) + \frac{f'(x)}{1!}\Delta x + O(\Delta x)^2 \tag{7.105}$$

The last element in the equation $(O[\Delta x^2])$ above is the *error* associated with the other components in the expansion. As stated previously, if we assume these are negligible, we may truncate the expansion. Solving for the first derivative and setting the error O to zero yields the solution for the first derivative that we are looking for.

$$f'(x) \cong \frac{f(x + \Delta x) - f(x)}{\Delta x} \tag{7.106}$$

The backward difference and central difference formulas may be similarly derived (see Problems 7.4 and 7.5).

$$f'(x) \cong \frac{f(x) - f(x - \Delta x)}{\Delta x} \tag{7.107}$$

$$f'(x) \cong \frac{f(x + \Delta x) - f(x - \Delta x)}{2\Delta x} \tag{7.108}$$

To find the *second* derivative of a function using the forward difference approximation, we take the derivative of Eq. (7.106).

$$f'(x + \Delta x) = f'(x) + \frac{f''(x)}{1!}\Delta x + O(\Delta x)^2 \tag{7.109}$$

Solving for the second derivative, we arrive at

$$f''(x) \cong \frac{f'(x + \Delta x) - f'(x)}{\Delta x} \tag{7.110}$$

We can use Eq. (7.106) and substitute for the functions there to put the terms on the right-hand side for the nonderivative values.

$$f'(x + \Delta x) \cong \frac{f(x + 2\Delta x) - f(x - \Delta x)}{\Delta x} \tag{7.111}$$

Substituting back in Eq. (7.110) yields

$$f''(x) \cong \frac{\dfrac{f(x + 2\Delta x) - f(x - \Delta x)}{\Delta x} - \dfrac{f(x + \Delta x) - f(x)}{\Delta x}}{\Delta x} \tag{7.112}$$

and simplifying results in

$$f''(x) \cong \frac{f(x + 2\Delta x) - f(x - \Delta x) - f(x + \Delta x) + f(x)}{(\Delta x)^2} \tag{7.113}$$

This is in the form of a *forward difference* approximation for the second derivative of a function. Depending upon the code and the programmer's choice of method to solve the problem, it could easily have been put into a *central difference* or *backward difference* formulation.

Some functions, such as parabolic PDEs (the heat equation is an example), involve a few different approaches to solving finite difference problems. The first is called the *explicit* method. The second is the *implicit* method. The different methods occur because there are aspects concerning the stability of the finite difference formulation. This has to do with the type of equation one is solving and the choice of grids for the problem. Stability and convergence are important for any numerical methodology but is critically important when using finite difference methods. For details concerning these equations, the readers are referred to the references at the end of this chapter [23, 28].

An elliptical PDE (such as Poisson's equation) may be solved by formulating the function in terms of a grid-based structure and then creating a molecule that steps through the grid as the problem is solved numerically [23]. We can illustrate the method with a simple example of solving Laplace's equation (a specific variant of Poisson's equation) for the voltage of a capacitance structure as was done in Section 7.1.1. In this case, we set up the initial boundary value problem and then create the molecule through formulation of the equation.

The first step is to set up the difference equation. Laplace's equation for the voltage was found from Eq. (7.3) in Section 7.1.1. We may approximate the equation in two dimensions as we did previously.

$$\frac{\partial^2 \varphi}{\partial x^2} + \frac{\partial^2 \varphi}{\partial y^2} = 0 \tag{7.114}$$

Let us first find the central difference formula for the second derivative in x; it will be identical in terms of the variable y. We may start with expanding the Taylor series around Δx as we did in Eq. (7.106), but this time we will add some additional terms.

$$f(x + \Delta x) = f(x) + \frac{f'(x)}{1!}\Delta x + \frac{f''(x)}{2!}(\Delta x)^2 + \cdots \tag{7.115}$$

Expanding for $-\Delta x$ yields

$$f(x - \Delta x) = f(x) - \frac{f'(x)}{1!}\Delta x + \frac{f''(x)}{2!}(\Delta x)^2 - \cdots \tag{7.116}$$

j

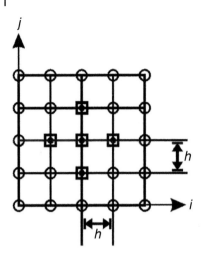

Figure 7.14 Grid region for computation molecule.

h

i

h

Adding Eqs. (7.115) and (7.116),

$$f(x + \Delta x) + f(x - \Delta x) = 2f(x) + \frac{2f''(x)}{2!}(\Delta x)^2 + O(\Delta x)^2 \tag{7.117}$$

Solving for the second derivative yields

$$f''(x) \cong \frac{f(x + \Delta x) - 2f(x) + f(x - \Delta x)}{(\Delta x)^2} \tag{7.118}$$

We want to now discretize this function such that we can create an equation that is solvable using numerical techniques. First, we create a grid for the solution space. This is similar to what we did in Section 7.1.1. By doing so, we can then discretize the function to represent points on the grid. A simple graph of the grid region is shown in Figure 7.14.

The region in the center with the squares and solid dots is called the *computation molecule*. These are nodes in the grid where the equation is defined and the function is computed. This may be better seen through analyzing the function.

Discretizing this function for the variable φ using Eq. (7.118) now gives

$$\frac{\partial^2 \varphi}{\partial x^2} \cong \frac{\varphi(i + 1, j) - 2\varphi(i, j) + \varphi(i - 1, j)}{h^2} \tag{7.119}$$

Note that we have replaced the variable x with its discrete position in the grid with the variable i. Also, we have replaced the differential for Δx with h. For simplicity, we will use the same spacing for the equation in y.

$$\frac{\partial^2 \varphi}{\partial y^2} \cong \frac{\varphi(i, j + 1) - 2\varphi(i, j) + \varphi(i, j - 1)}{h^2} \tag{7.120}$$

The two equations now represent the final equation to be solved numerically. Note that for the grid size here, the grid spacing (h) does not come into play since we can multiply both sides of the equation by h^2 to remove it in the final discretization of Laplace's equation. Thus, if we solve for the (i, j) component in the center of the grid, we get the result

$$\varphi(i,j) = \frac{1}{4}[\varphi(i+1,j) + \varphi(i,j+1) + \varphi(i-1,j) + \varphi(i,j-1)] \qquad (7.121)$$

We now change the generic variable to the voltage potential for clarity as we did in Section 7.1.1. This gives us the function we wish to solve for.

$$V(i,j) = \frac{1}{4}[V(i+1,j) + V(i,j+1) + V(i-1,j) + V(i,j-1)] \qquad (7.122)$$

This is the main equation we will use for setting up a finite difference code to solve for the voltage in a given structure. Let us apply this to the example where we solved for the exact solution in the previous problem using separation of variables.

Example 7.3 Using a finite difference technique, determine the voltage for the structure in Figure 7.6. Compare the results using the same grid size. Then shrink the grid size to 1/4 of the size and compare the results again. (In essence, this will be the same as increasing the number of points in the computational space by a factor of 4 in each direction.)

Solution:
The setup for the problem requires us to apply the initial boundary conditions and then solve the equation in a loop. There are multiple ways to do this of course. MATLAB allows us to create vectors where we can define columns and vectors and simply compute them in a loop. This is more efficient (usually) computationally but does not always illustrate how to initialize the variables and how the equations are formulated. We will instead step the equation in a loop stepping through each direction iteratively to illustrate the method. We will also apply an error term to the equation so that it will limit the number of iterations to achieve a solution.

The first thing we want to do is initialize the key boundary elements. We can set the size of the region to whatever we wish. Let this be the variable "num" in the code below. To make it consistent with the "separation of variables" problem, we keep the ratio of length to height the same. The values for the width and length of the region are "*ie*" and "*je*," respectively, in the code. These are the number of grids in each direction for the solution space. After initializing the boundary conditions and region inside of the boundaries, we compute an error. To do this, we choose a node value that we can iterate on to determine if it asymptotes to a specific expected value. Once again, as in the example in the earlier section, we need to compute the value we wish to achieve a certain error for. In this case, however, we are *specifying* the error in the result for a specific value rather than *computing* the error by changing the number of iterations or (as in the previous example) changing the number of terms in the summation. We chose this node because upon the first iteration in the loop, we knew we would get a value from the molecule based on its position in the grid. Also, note that we created an implied grid size h based on the x and y lengths, in this case 16 and 9, respectively. This is reasonable since we are essentially taking the average value of the four nodal voltages around the center each time we compute the voltage. Once the computation is complete, we will determine the number of iterations it took to achieve that error and then plot the number of iterations as a function of the value we achieved. This will give us a good idea if the node value started to asymptote to a given static value. The code for the problem is shown in Code 7.3.

Code 7.3: Computation of voltage potential using FD method.

```
                                    ***
%   This Matlab program calculates the voltage for a parallel
%   plate capacitor using the Finite Difference Method
%
%   Author: J.Ted DiBene II
%
%   **** inputs ****
%   ie = number of spatial cells in 'x' direction - #columns
%   je = number of spatial cells in 'y' direction - # rows
%
%   **** outputs ****
%   V = static potential in region
%

% Initialize input variables

num = 16;                % number of grid points in x direction
ie = num;                % set number of grid points in x direction
je = num/2 + 1;          % set number of grid points in y direction
nsteps=1;                % Initialize number of steps

% Initialize boundaries in loops

for i=1:ie;
    for j=1:je;
        V(i,j) = 0;      % initialize the whole region to zero
    end;
end;

for i=1:ie;
V(i,je) = 1;             % set y=b boundary to 1V
end;

for j=1:je;
    V(ie,j)=1;           % set x=a boundary to 1V
```

```
end;

%  We want to minimize the error so we set an initial value
%  We note that it takes a few iterations to fill up the nodes
%  With values so we chose one that should have a value which is
%  in the corner next to the furthest node.

V(ie-1,je-1)=1;

% We now generate a loop and go thru until error is small.

error=1;                  % Initial error in loop
error_set = 0.0001;      % setting error
ii=1;                     % loop counter for error
nsteps=1;                 % loop counter for number of itera-
tions in main loop

% Main Loop for computing potential

while error >= error_set;  % The error can be varied to get
 iterations and accuracy.

    vold=V(ie-1,je-1);

% Loop for Calculating Voltage Equations

for i=2:ie-1;
    for j=2:je-1;
        V(i,j)=0.25*( V(i+1,j) + V(i,j+1) + V(i-1,j) + V(i,j-1) );
    end;
end;

    V(1:ie,je)=1;          % reset x boundary condition
    V(ie,1:je)=1;          % reset y boundary condition

    vnew=V(ie-1,je-1);  % new value for computing error
    errorp(ii)=abs( vnew-vold );  % compute error from new to old
                                      value
    error=errorp(ii);

    ii=ii+1;
    nsteps=nsteps+1;

end;
%  Final output Variables

V
nsteps
```

***

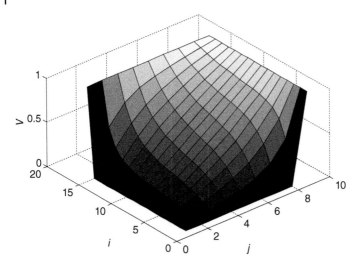

Figure 7.15 Surface plot of voltage.

The plot of the voltage is shown in Figure 7.15. As in the separation of variables plot, we show the shades of gray representing the scale. Because the voltage falls off from the corners from 1 to 0 V, the gray scale interpolator represents that side as the color black, which is actually intended to be at 0 V. However, by looking at the middle of the plot, one can see that the perspective view shows the voltage falling off to zero into the lowest value nodes for i and j as expected. Thus, one can see that on the ends where the voltage potential is at 1 V and the nearest node is at 0 V, the numerical analysis must assume one of those grid points is at either 0 or 1 depending upon how the boundary conditions are set.

The next question we want to answer is on convergence. Did the node we chose converge to a value that appears to be approximately constant with the number of iterations (and error)? That is, does the value for that node start to flatten out to a particular voltage potential that we expected to see based on theory? To see this, we can plot the number of iterations as a function of the voltage for this node. We can see by the plot in Figure 7.16 that the voltage indeed is beginning to asymptote to a single value.

Because the voltage for this particular node appears to be approaching a limit in the grid as we iterate through the loop, it appears that the simulation is yielding something that we can check on to determine if it indeed was the correct value.

However, another check is to run the simulation for a larger number of grid points and see if there is a change as compared with the smaller number of grid points. If we increase the grid size by, say, a factor of 4 that of the previous simulation (as was requested in the problem), we should see a finer mesh structure

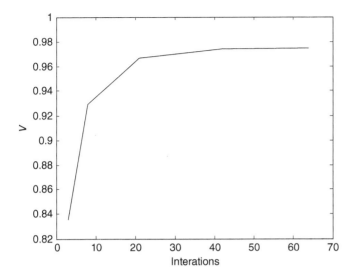

Figure 7.16 Node $V(15,8)$ voltage as a function of iterations through loop.

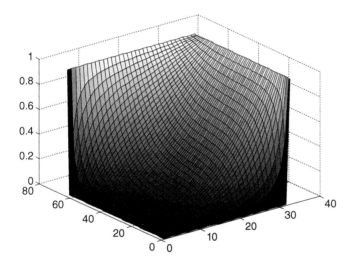

Figure 7.17 Surface plot with 4× finer mesh.

at least. We do this by changing the values for *ie* and *je* in the code. This is shown in Figure 7.17.

Note that not only did we see an increase in the number of mesh points but we also saw an increase in the number of iterations through the loop due to this (why?). The interesting thing about increasing the number of grids is that for this particular problem, we would not expect to see an increase (necessarily) in the accuracy of the solution space on a per nodal basis. However, because there are more points generated in the same surface, the result will certainly show a smoother transition in the overall plot with the potential since we have more data per unit area in which to examine our results. This is evident in the plot in Figure 7.17.

The important thing to note here is that it is not always advantageous to increase the number of grid points just to achieve better accuracy. There is a trade-off between the spatial region under which the simulation is performed and the expected accuracy, and every problem must be approached with these issues in mind.

We also have the value of the node we are looking for printed out as a function of *nsteps*, the number of iterations through the loop, for a point similar to the 16×9 grid (Figure 7.18).

Note that the number of iterations through the loop tends to *increase* with the grid size, but the value at the node appears to be not quite flattening with the number of iterations. This is expected to some extent since there are many more computations to make in each equation, and it is likely that there is still a significant error between the "exact" settled value for this node and the one that occurs as a function of the number of iterations through the loop.

This program was illustrative on how to use the finite difference (FD) method for computing the voltage potential. We now examine how to use a similar code to compute a practical parasitic value for a PDN.

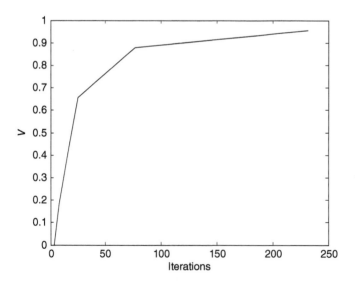

Figure 7.18 Voltage for node position $V(60,29)$ on grid.

The previous example showed how we could develop a simple FD code to determine the potential by using Laplace's equation. However, what if we want to compute the inductance and capacitance of a more practical structure using a similar method? Because of our work from Chapter 6 on transmission lines, we learned that for static fields (TEM approximation), we may eliminate the time dependence in our computations. Thus, we can make use of the fact that for a given structure, there will be *no change* to electric or magnetic fields down the length of a particular structure for a PDN parasitic. There *will* be a change to the *losses* of course if we assume a finite resistance. However, in general, we may start with a perfect conductor with a potential on it and solve for the fields.

One of the more common structures we will encounter will be a strip of finite width and height that extends between one point in the PDN and another. We assume that this structure does not change in the z dimension as well (no bends). Thus, we can use our two-dimensional approximations and determine the per-unit-length parameters for this structure. The structure that we chose here is what we became familiar with in Chapter 6 which was the stripline. This is a structure that is bounded on the top and the bottom by a ground plane and in which there is an electric potential placed on the center conductor. We can choose to apply different boundary conditions to the sides if we wish to represent the actual problem more accurately. In a real parasitic extraction, we would obviously want to represent the boundary conditions as accurately as possible. However, to simplify the code, we will assume for now that the entire boundary is at ground for the moment. This is a reasonable approximation for getting an approximate result. The structure is shown in Figure 7.19.

For the center conductor, we apply a 1 V potential again. The dimensions of the structure depend upon the problem in which we wish to solve. Fortunately,

Figure 7.19 Spatial region for parasitic analysis with FD method.

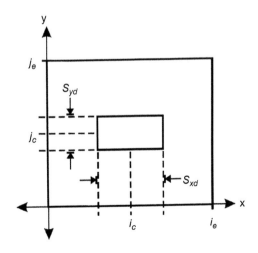

we can change these values to whatever we like in the code (within reason of course). For now, we will keep the geometry relative to Figure 7.19.

Because of the fact that we have assumed *quasi-static* conditions, we can solve once again for the voltage using Laplace's equation. However, if we wish to solve for the capacitance, we will need to do a bit more work. Using Gauss's law, we can compute the total charge on the center conductor by integrating around the loop. We use Stokes theorem and assume that we are after the per-unit-length capacitance, which allows us to divide by the component in the z dimension.

$$\frac{Q_E}{l} = \oint_L \vec{D} \cdot d\vec{l} = \oint_L \varepsilon \vec{E} \cdot d\vec{l} = \oint_L \varepsilon \frac{\partial \vec{V}}{\partial n} \cdot d\vec{l} \qquad (7.123)$$

The gradient of the voltage is essentially the electric field around a given region. Thus, the charge per-unit-length is equivalent to the capacitance per unit length times the voltage.

$$\frac{Q_E}{l} = \frac{C}{l} V \qquad (7.124)$$

By taking the derivative of the voltage around center conductor using a difference formula, we can determine the charge per unit length with and without the dielectric in place. We know from basic transmission line analysis that

$$v_0 = \frac{1}{\sqrt{LC_0}} \qquad (7.125)$$

where C_0 is the capacitance with air as the dielectric. Adding in the dielectric, we get,

$$v_D = \frac{1}{\sqrt{LC_D}} \qquad (7.126)$$

Substituting for the inductance yields

$$v_D = \sqrt{\frac{v_0^2 C_D}{C_0}} = v_0 \sqrt{\frac{C_D}{C_0}} \qquad (7.127)$$

where v_0 is the speed of light. Since we now have both the capacitance with and without the dielectric in place, we may solve for the inductance as well (along with the impedance).

The next example computes the parasitic capacitance and inductance for the structure in Figure 7.19.

Example 7.4 Compute the capacitance and inductance for the structure in Figure 7.19. Using the finite difference method, build the code and compute the results.

Table 7.3 Comparison of closed form vs. FD analysis.

Parameter	Closed form	Finite difference	% Difference
Z_0	$32.18\,\Omega$	35.32	8.9
C_0	$7.12\,\text{pF}$	$4.79\,\text{pF}$	32.7
L_0	$7.4\,\text{nH}$	$5.98\,\text{nH}$	24

Solution:
Similar to the previous example, we can use the methodology already derived to compute the values we wish. Our first step is to create a code and set the boundary conditions for the problem.

In this example, we use MATLAB notation to initialize the boundary conditions and to compute the equations in the loop.

The difference in this computation is that we will compute the total charge on the center conductor as well. This allows us to compute the capacitance and the inductance using the equations discussed earlier.

As we did in the simple capacitor example, we compute the voltage potential in the loop and iterate using a fixed number of steps in a "for" loop.

We generate the values for the impedance, velocity of propagation, capacitance and inductance (per unit length) of the structure. This data is tabulated in Table 7.3. In this example, we use MATLAB code to initialize the boundary conditions and to compute the equations in the loop. The program is given in Code 7.4.

Code 7.4: Computation of voltage potential using FD method.

★ ★ ★

```
% This Matlab program calculates the static electric
% fields and Voltage for a rectangular coaxial waveguide which
% has the dielectric region half-filled
% It also computes the capacitance and inductance for the structure
%
% An alternate way to call this is with a function
%
% function[V_final,V0_final,CD_final,C0_final,ZS_final]...
%    = prog_cap(ie,je,sc,nsteps)
%
% Author: J.Ted DiBene II
%
% **** parameters of Structure *****
%
% ic = middle of x-soln region
% jc = middle of y-soln region
% sd = 1/2 length of one side of center conductor
% sda = 'y' distance to first edge of center conductor in y-direction
```

```
% sdb = 'y' distance to second edge of center conductor
% sea = 'x' distance to first edge of center conductor
% seb = 'x' distance to second edge of center conductor
%
% **** inputs ****
% ie = number of spatial cells in 'x' direction - #columns
% je = number of spatial cells in 'y' direction - # rows
% sc = number of spatial cells or size of
% one side of center conductor
%
% **** outputs ****
% V = static potential in region
%
% These are the input variables for file

nsteps=50;          % number of steps thru the loop
ie = 40;            % number of grid points or 'size' in 'x'
                       direction
je = 40;            % number of grid points or 'size' in 'y'
                       direction
wc = 20;            % number of grid points or 'size' of center
                       conductor in 'x' direction
hc = 10;            % number of grid points or 'size' of center
                       conductor in 'y' direction


% Initialize input variables

ic = ie/2;          % this is the center of the region in the 'x'
                       direction
jc = je/2;          % this is the center of the region in the 'y'
                       direction
sdw = wc/2;         % this is half the size of the center conductor
                       in 'x' direction
sdh = hc/2;         % this is half the size of the center conductor
                       in 'y' direction
sda = ic - sdw;     % this will be the number of rows to the first
                       edge of the center conductor
sdb = jc - sdh;     % this will be the number of columns to the
                       first edge of the center conductor

xs = sda -4;        %
xe = sdb+hc+4;
ys = sdb -4;
ye = sda+wc+4;

% Initialize boundary conditions - we use Matlab matrix nota-
tion to initialize

V_bot = zeros(sdb,je);  % this initializes the region below the
center conductor to zero
```

```
V_mid1= zeros(hc,sda);
V_cond= ones(hc,wc);
V_mid3= zeros(hc,sda);
V_top = zeros(sdb,je);
```

% The following line concatenates the matrices to initialize the
  region properly

```
V(1:je,1:ie)=[V_bot;V_mid1 V_cond V_mid3;V_top];
V0(1:je,1:ie)=[V_bot;V_mid1 V_cond V_mid3;V_top];
```

% We now initialize the variables used in the computations

```
ep1 = 8.854e-12;        % Dielectric in air
ep2 = 4*(8.854e-12);    % Dielectric with permittivity of 4.0
U0=3.0e8;               % velocity per inch in air.
```

% mu = pi*4e-7; % permeability constant

% ***

% We initialize the values that we will compute for.
% These will be the charge, capacitance, and impedance

```
QDV=0;
QDH=0;
QD=0;
CD=0;
C0=0;
ZS=0;
```

% Loop for Calculating Voltage Equations using Matlab Notation

```
for n=1:nsteps;
```

% Main Equation; grad^2(V)=0

```
V(2:ie-1,2:je-1)=0.25*(V(2:ie-1,3:je)   ...
   + V(3:ie,2:je-1) + V(1:ie-2,2:je-1)...
   + V(2:ie-1,1:je-2));
```

% Reset Boundary Conditions in Loop

```
V(1,1:ie)=0;                % This is the bottom of outer region
V(2:je-1,1)=0;              % This is left side of outer region
V(je,2:ie-1)=0;            % This is the right side of outer region
V(2:je-1,ie)=0;            % This is the top side of outer region
V(sdb:(sdb+hc),sda:(sda+wc))=1;   % This is the center conductor
```

```
end;
```

```
% Loop for Calculating Second Voltage Equations

for n=1:nsteps;

% Main Equation; grad^2(V)=0

V0(2:ie-1,2:je-1)=0.25*(V0(2:ie-1,3:je)   ...
    + V0(3:ie,2:je-1) + V0(1:ie-2,2:je-1)...
    + V0(2:ie-1,1:je-2));

% Reset Boundary Conditions in Loop

V0(1,1:ie)=0;
V0(2:je-1,1)=0;
V0(je,2:ie-1)=0;
V0(2:je-1,ie)=0;
V0(sdb:(sdb+hc),sda:(sda+wc))=1;

end;

% Calculate the Charge in the loop with Dielectric
% Use Matlab notation for vectors and then sum up
% elements to get total charge - remember you must
% exclude corner elements of outside corners and
% count inner corners twice.

QDV = ep2*(-V(xe,ys+1:ye-1)+V(xe-1,ys+1:ye-1))...
    + ep2*(-V(xs,ys+1:ye-1)+V(xs+1,ys+1:ye-1));
QDH = ep2*(-V(xs+1:xe-1,ye)+V(xs+1:xe-1,ye-1))...
    + ep2*(-V(xs+1:xe-1,ys)+V(xs+1:xe-1,ys+1));
QD = sum(QDV) +sum(QDH);

% Now Calculate Capacitance with Dielectric in place

CD = QD; % remember V_cond = 1, also this is
% capacitance per inch

% Now Calculate Charge without Dielectric in place

Q0V = ep1*(-V(xe,ys+1:ye-1)+V(xe-1,ys+1:ye-1))...
    + ep1*(-V(xs,ys+1:ye-1)+V(xs+1,ys+1:ye-1));
Q0H = ep1*(-V(xs+1:xe-1,ye)+V(xs+1:xe-1,ye-1))...
    + ep1*(-V(xs+1:xe-1,ys)+V(xs+1:xe-1,ys+1));
Q0 = sum(Q0V) +sum(Q0H);

C0 = Q0; % capacitance per inch

% Calculate Phase Velocity
U = U0*sqrt(C0/CD);

% Calculate Impedance
```

```
ZS = 1/(U0*sqrt(C0*CD));

% Calculate Inductance
L0 = ZS*ZS*CD;

% Final output Variables

V_final  = V;
V0_final = V0;
CD_final = CD;
C0_final = C0;
ZS_final = ZS;
L0_final = L0;
```

***

The first thing we want to examine is the surface plot of the voltage to make sure it looks reasonable. The plot of the voltage is shown in Figure 7.20. As in the previous plot, we show the shades of gray representing the scale. As a first check, the voltage potential over the region looks reasonable. We expect it to fall off symmetrically since the outer boundaries are at ground. Also, we see that the voltage in the center is still 1 V which is correct. Note the grids in the center conductor. This is a good check to see if the proper number of grids were assigned in the code as well.

We now want to examine if we computed a reasonable value for the capacitance and inductance per unit length. To do this, we compare the obtained result against a known formula with the same dimensions and see how close

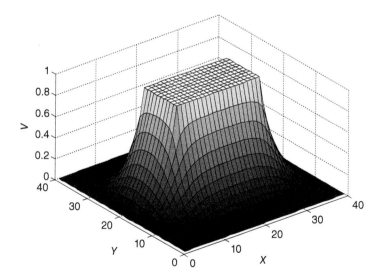

Figure 7.20 Surface plot of voltage for center conductor problem.

to this value did we get. Using a formula found in [7] for a stripline, we may compute the capacitance and inductance per unit length along with the impedance. Using a formula from [7], we get the estimated impedance

$$Z_0 \cong \frac{60}{\sqrt{\varepsilon_r}} \ln \left[\frac{1.9(2H + T)}{(0.8W + T)} \right] \tag{7.128}$$

The capacitance per unit length is then

$$C_0 \cong \varepsilon_r \frac{5.55 \times 10^{-11}}{\ln \left[\frac{3.81H}{(0.8W + T)} \right]} \tag{7.129}$$

The inductance may be computed from the previous two formulae using the following:

$$L_0 \cong Z_0^2 C_0 \tag{7.130}$$

Using these formulas, we may compute the values and compare against the FD code computations. This is shown in Table 7.3.

The first thing that stands out in the table is the large discrepancy between the parasitic values and the FD solution. We assume that the closed-form expression is accurate to within 5% of the exact value. However, we see that the parasitic values are off by more than 20%. This is actually expected and can be explained by looking at the boundary conditions on the sides. First, by observing Figure 7.19, it is clear that we assigned boundary conditions to the sides where we grounded the vertical edges. This was the first approximation that was not correct. However, the more important discrepancy is in the dimensions. In the actual closed formula, the distance to the edge of the boundaries on the sides is typically assumed to be much longer than the width of the conductor. Thus, if we extended the dimensions in the x direction, holding the other dimensions in the y direction constant, we should see the error between them diminish. The reason is that the coupling from the center conductor to the vertical boundaries will have less of an effect. This is left as an exercise for the reader at the end of the chapter.

We could also run more iterations to see if the solution is converging. This would in all likelihood result in a smaller error between the closed-form expression and the FD code but would not make up enough difference to explain the main error just discussed.

This example illustrated that boundary conditions and dimensions *do* matter when one is trying to get a more accurate solution. However, once again, what is considered as "accurate" is highly dependent upon the original objective of the original analysis. It is not surprising to find out that a solution of 20–30% may be perfectly acceptable for some problems and unacceptable for others. When setting up problems in "black box" programs, the boundary conditions and dimensions do matter and considerations of the setup will lend itself to the accuracy of the result.

Section 7.2.2 introduces one of the more common methods used in numerical analysis. This is the FEM. It is likely the most popular method used for problems involving field solving today.

7.2.2 The Finite Element Method

The FEM is probably one of the most popular techniques developed for solving electromagnetic fields in electrical engineering (among other problems) in the industry. FEM methodologies were first introduced to decipher thermal problems but were soon adapted by other disciplines due to its ability to compute solutions for complex geometries. One of the major advantages of the method is that for boundaries where the fields become very large or infinite relative to the other regions of the problem space, FEM techniques can coalesce the elements more easily in these areas with minimum compromise to the number of computations required to post-process the equations. This is in contrast to other numerical processes, such as the finite difference method, which does not scale the grid region as easily. Note that in the previous sections for the problems that we solved, the grid was kept constant. However, what if we had wanted to concentrate the grids in a region around the center conductor, say, for the stripline problem? This would have required us to build in an algorithm to handle different regions and match up different grid dimensions to compute the equations within those regions. Moreover, the boundaries between the different grids would need to be handled in a special way since any change to the grid size necessitates a change to the variables that are being solved (e.g. fields). There are methods that allow us to do this, but they are more cumbersome and difficult to program than using FEM-based tools.

However, the FEM is a bit more complicated to understand and implement in comparison to other methods. It is based upon variational techniques, and there exists a number of approaches to solve or post-process the equations that are generated by the method.

The FEM usually starts out with a definition for an explicit grid, which may be constructed from elements that are of different shapes [19]. The most popular shape for these elements is triangle. Thus, we will focus our attention on these from this point onward. Other shapes that exist include quadrilaterals for two-dimensional problems and tetrahedrons for three-dimensional spaces among others.

We will *partially* solve the same equation that we did in the previous examples but will use the FEM instead. The reason for not completely solving the problem is that fully illustrating the FEM is beyond the scope of this text, and our objective is to only introduce a few salient details so that we may emphasize what is important when using FEM-based tools using standard programs, which is normally the case for power integrity engineers. For a more thorough review of FEM-based methods and numerical methods for EM problems in general, Sadiku [2] has a very good simple description in his text that solves Laplace's equation in two-dimensional space. We will follow a similar path in how we develop our overview of the procedure. Another excellent reference is found in Reddy [8] which covers the FEM in detail.

It should be noted here that the reason we are focusing on Laplace's equation is because it is relatively straightforward to understand in the context of solving FEM problems. Delving into other more advanced equations would require an in-depth study of the FEM methodology.

To illustrate the method, we can use a small computational space to begin with. Once this is understood, moving to larger grid sizes and automating the algorithms can be handled separately. Let us assume to start that we have a small group of triangles and that we wish to solve for the voltage potential inside of each one as shown in Figure 7.21.

Each node of our group of triangles are labeled as well as the triangle elements. Here, we have three triangles with five nodes. Our goal is to find an approximation to the voltage inside of each triangle element. Written mathematically, this becomes,

$$V(x, y) \cong \sum_{e=1}^{3} V_e(x, y) \qquad (7.131)$$

We may approximate the voltage inside of each element using a polynomial expression. As was the case for the finite difference method, the equation is once again not an exact match to the exact value. However, using a simple function, we can establish a reasonable approximation. In most cases, it is not

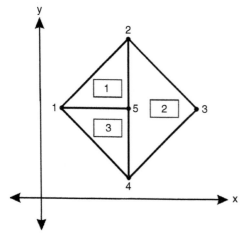

Figure 7.21 Simple grid of three triangles for mesh.

Figure 7.22 Single triangle element.

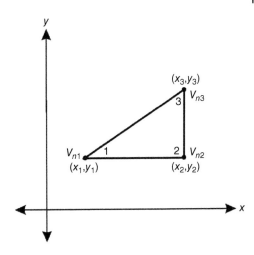

required to estimate the voltage with a complex mathematical expression. In this case, we may use a simple linear equation such that,

$$V_e(x, y) = a + bx + cy \tag{7.132}$$

Our goal now is to solve for the coefficients for each triangle within the solution space. Let us take a single triangle element for the moment and analyze it. Here, we have three nodes indicated by V_{n1}, V_{n2}, and V_{n3}. Internal to the element, we have three local nodes which are numbered as local nodes 1, 2, and 3. The direction here is important, and local numbering is done in the counterclockwise direction (Figure 7.22).We have, in essence, three equations with three unknowns:

$$V_{n1}(x, y) = a + bx_1 + cy_1 \tag{7.133}$$

$$V_{n2}(x, y) = a + bx_2 + cy_2 \tag{7.134}$$

$$V_{n3}(x, y) = a + bx_3 + cy_3 \tag{7.135}$$

We may put this into matrix form and then solve for the coefficients.

$$\begin{bmatrix} V_{n1} \\ V_{n2} \\ V_{n3} \end{bmatrix} = \begin{bmatrix} 1 & x_1 & y_1 \\ 1 & x_2 & y_2 \\ 1 & x_3 & y_3 \end{bmatrix} \begin{bmatrix} a \\ b \\ c \end{bmatrix} \tag{7.136}$$

If we invert the matrix in the middle, we can solve for each coefficient

$$\begin{bmatrix} a \\ b \\ c \end{bmatrix} = \begin{bmatrix} 1 & x_1 & y_1 \\ 1 & x_2 & y_2 \\ 1 & x_3 & y_3 \end{bmatrix}^{-1} \begin{bmatrix} V_{n1} \\ V_{n2} \\ V_{n3} \end{bmatrix} \tag{7.137}$$

or that,

$$\vec{c} = M^{-1}\vec{v} \tag{7.138}$$

where $\vec{c}$ and $\vec{v}$ are vectors and M is 3×3 matrix. We recall from basic linear algebra that

$$M^{-1}M = I \tag{7.139}$$

Thus, the inverse may be found through Cramer's rule or through Gauss elimination. In this case, we can solve for each coefficient (fairly elegantly) by taking the determinants for each element where

$$a = \frac{M_x}{M} = \frac{\begin{vmatrix} a & x_1 & y_1 \\ b & x_2 & y_2 \\ c & x_3 & y_3 \end{vmatrix}}{\begin{vmatrix} 1 & x_1 & y_1 \\ 1 & x_2 & y_2 \\ 1 & x_3 & y_3 \end{vmatrix}} \tag{7.140}$$

where we find the solution to each constant by replacing the column of the determinant by the vector which is the solution. The result is then

$$\begin{bmatrix} a \\ b \\ c \end{bmatrix} = \frac{1}{|M|} \begin{bmatrix} (x_2y_3 - x_3y_2) & (x_3y_1 - x_1y_3) & (x_1y_2 - x_2y_1) \\ (y_2 - y_3) & (y_3 - y_1) & (y_1 - y_2) \\ (x_3 - x_2) & (x_1 - x_3) & (x_2 - x_1) \end{bmatrix} \begin{bmatrix} V_{n1} \\ V_{n2} \\ V_{n3} \end{bmatrix} \tag{7.141}$$

Now there is an interesting aspect to the determinant M here. By examining the triangular element above and looking at the equation, it can be observed that the determinant is actually twice the area of the triangle. Thus, we can substitute in for the coefficients for V_n in Eq. (7.132) and then rearrange to solve for the voltage for that element

$$V_n = [1 \quad x \quad y] \frac{K}{2|A|} \begin{bmatrix} V_{n1} \\ V_{n2} \\ V_{n3} \end{bmatrix} \tag{7.142}$$

where K is the matrix in Eq. (7.136) and A is the area of the triangle. The matrix and determinant in the middle of the matrix equation correspond to what are called the *shape* functions for the element. The shape functions are the coefficients for the solution of the voltage within that element. Thus, we may give the solution as

$$V_n = \sum_{i=1}^{3} \alpha_i(x, y) V_{ni} \tag{7.143}$$

where the αs are simply the multiplication of the components in Eq. (7.143). For example, the first α is

$$\alpha_1(x, y) = \frac{1}{2|A|}[(x_2 y_3 - x_3 y_2) + x(x_3 y_1 - x_1 y_3) + y(x_1 y_2 - x_2 y_1)] \quad (7.144)$$

The other shape functions may be similarly derived. Recall from Section 7.1.2 that we used variational methods to determine the functional for a given function. The functional may be developed using the technique outlined in Sadiku [2]. As a matter of fact, the functional here is equivalent to the work function for the system or

$$W = \frac{1}{2} \int \epsilon |\nabla V|^2 dA = \frac{1}{2} \epsilon \mathbf{V}^T \mathbf{C} \mathbf{V} \quad (7.145)$$

where $\mathbf{V}$ is a vector and $\mathbf{C}$ is a matrix. Solution to the previous equation may be found in brief in the Appendix. For readers that wish to understand the FEM method in more detail, a good starting place would be to review some of the references at the end of this chapter.

For power integrity engineers who use black box programs primarily, the key issues are around accuracy, errors, and convergence. We discuss these issues in Section 7.3 for both the finite difference and FEMs.

7.3 Error and Convergence

At this point, it is a good idea to examine a few of the critical concerns with respect to the bounds on the solution space and the computation itself for the numerical methods discussed in Section 7.2. These are the error and convergence of the computation. We will break these items out one at a time here.

Though there are numerous sources of potential errors when using numerical tools and though most of them can be made to be readily small and when the tools are essentially *black box programs*, the code inside attempts to minimize these errors as part of their algorithms. For individuals that write their own code, the errors are usually readily apparent in the setup of the problem (typically), though it requires more diligence in how they arrive at their solutions.

Error in numerical computation is different from accuracy in that it is considered the true deviation from the exact result. This is not to be confused with the *accuracy* of the solution. Accuracy, on the other hand, is the measure of how good one expects the result to be. In this case, error is simply the accumulation of different components that contribute to the result, and understanding the components here is the goal to help minimize them. The accuracy of the result in comparison is dependent upon many things, most importantly, a qualitative assessment by the user of how good the result needs to be. This is where individuals trade off different metrics in the tools to achieve the final goal.

Convergence is related to how the computer crunches through the code to arrive at the solution and under what conditions the code eventually settles on a value or set of values. Error, accuracy, and convergence are always interrelated as we saw earlier in our finite difference codes [29, 30].

Accuracy is normally the most important aspect for PI engineers when they are attempting to solve a particular problem. This is true whether it is a SPICE simulation or an extraction of a parasitic element. This where our previous discussions of first principles, and the use of other approximations, help to bound the solution. Often engineers have an idea of how accurate the result should be prior to the start of an analysis (if they are not, then they should have). If they are embarking on a numerical solution in the first place, it usually means that whatever result they have or expect is not sufficiently accurate (whether a hand computation or even another numerical analysis). The goal, however, is to achieve a certain result that satisfies the requirement.

The next sections delve into these topics in additional detail.

7.3.1 Errors in Numerical Analysis

Errors in computational analyses are evident in all aspects of numerical simulations. However, users of these programs (such as field solvers) rarely are exposed to the errors inside. In some cases, the program can indicate where to improve the result. However, the sources of errors from these programs, in general, are still not evident. It is thus important to have a basic understanding of these errors so that when an engineer is setting up an analysis, they are not caught off guard when the result ends up being much different than they anticipated.

The *error*, by definition, is of course the difference between the exact (possibly measurable) value and the one the numerical computation puts out. In the case of most system-level problems, including power integrity analyses, it is rare that the engineer has the exact solution of the problem handy to compare against their numerical model. This is where having a first principles, measurement, canonical, or reasonably close baseline solution (sometimes from a previous, but similar, analysis) is very important. Having something to compare against is one of the first steps an engineer should embark upon before performing an analysis.

The error in the solution space takes on many forms. Moreover, very few engineers consider these errors when starting an analysis. This is even truer when using *black box* programs. It should be noted that these types of tools are typically not any worse or better with respect to error generation than any other numerical method. In fact, many advanced field solvers are very sophisticated and quite well developed. However, there is a reason why commercial codes are considered black box; they typically do not reveal the details of their code or algorithms to the user. In comparison, when developers create their own code, they end up having a deeper insight into the issues.

However, it is not always clear where the errors will manifest themselves in either a homegrown code or an existing program. It is though evident that they all share the same types of sources. Thus, it is sometimes instructive to discuss these error types and when to examine them in the construct of a real analysis. The following list shows some of the key errors in a numerical analysis.

- global errors
- truncation errors
- round-off errors
- localization in grid errors

We use the finite difference formulation in one dimension here as our baseline to illustrate. The FEM analysis is similar in concept. A brief discussion on each of the types of errors follows.

Given the result is in the form of a set of data points at each grid node, we may represent the **global error** as the difference between the exact result and the actual result.

If we examine the values of the solution at the node points as a set of vectors across the computational space, the error may be represented globally as,

$$E = |U - U_A| \tag{7.146}$$

where U is the exact solution and U_A is the approximation. If h is the general grid dimension and e_i is the error at each grid point (for one dimension), then the magnitude, or *norm*, of the global error would be [11]

$$\|E\| = \sum_{i=1}^{N} h|e_i| \tag{7.147}$$

A finite difference formulation is said to converge if

$$\lim_{h \to 0} \|E\| = 0 \tag{7.148}$$

In Example 7.3, we would see that the result would take the form of the voltage at each node point. We also noted that as we increased the number of grid points (shrinking the value of h), the accuracy increased. The difficulty with complex problems is that checking to see if the numerical approximation has a reasonable error requires knowing the solution ahead of time. Unfortunately, this is rarely if ever the case. Thus, performing a similar analysis on a canonical problem is sometimes warranted to check the efficacy of the platform and solution space. Often, the black box programs have examples where this has been done (e.g. a simple planar capacitor, coaxial structure). As a practical matter, the global error in a formulation is not sufficient to indicate whether or not the result satisfies the user's requirement. It is, in the end, up to the PI engineer to have a goal in mind with the final result at multiple data points (usually) to ensure that the error is adequate.

The **Truncation Error** is simply the portion of the equation, which is a result of truncating the function itself. Thus, if the equation to be solved, for example, is a differential equation, one can clearly see from the start that simply because the equation is approximated, an error will exist. However, depending upon what result the user is looking for, this error could be quite small. As we saw in the finite difference approximation, one can clearly see that when using the difference relation, there is an obvious error due to the truncation of the Taylor series expansion. This was shown in Eq. (7.105) for a single variable finite difference function.

$$f(x + \Delta x) = f(x) + \frac{f'(x)}{1!}\Delta x + O(\Delta x)^2 \tag{7.149}$$

Applying this truncation to a PDE, one can then determine the order of the error. This error is due to the square of the differential change or, in the FD approximation, the grid size. Thus, the error propagates as the square of the difference in the grid change. For most numerical problems, the number of grid points, relative to the analytical region of interest, is quite small. In the previous examples, we noted that we were able to get reasonably accurate results by creating a grid size that was 1/8th the size of one side of the computational space. Though this is highly dependent upon the problem posed, it does indicate that choosing the correct grid size is important. However, as noted in the previous sections, this can affect both computational time and convergence.

Round-off errors are due to approximation of the number of digits that are stored in the computer for the actual value in question. One can easily see this when performing a simple Excel calculation in a spreadsheet as an example. Numerically, one normally can never get the exact value of a particular solution for a given problem. This is because the number is rounded off during the numerical computation. For single precision, the machine accuracy is typically 10^{-8} [31]. If the number is 32 bits long, the bits are stored as follows: sign = 1 bit, exponent = 8 bits, mantissa = 23 bits. Thus, the largest positive number that can be stored (in binary) will be 1.701×10^{38}. This is rather large considering that often double precision is used by most programs (twice the bit length). Thus, even with multiple iterations through a program, as long as the program itself does not round off the numbers prematurely, the overall round-off error should be quite small.

Localization or grid-related geometry errors are better understood with a simple visual. Let us assume a PI engineer is examining a rectangular structure and wants to know the electric field near a corner of the structure where the voltage source is. By definition, this field should go to infinity as it nears the corner of the metallic structure. However, the machine cannot store infinity. Thus, it is severely rounded off as the fields approach this point. For larger grids in this vicinity, it is typical to get gross errors in the field strength and sometimes the overall result. The FEM has algorithms to shrink the grid size (most programs do) to reduce this error though the finite difference method typically has a single grid size. Thus, for complex geometries, engineers often find the FEM method more accurate. However, even with triangular adjustable gridding, the errors may be large. Often, programs have what are called *grid controls* in regions where this occurs; that is, the grid density may be adjusted in small regions where a corner is present and then increased in more open areas where field changes are small.

No matter what the error is, for black box programs at least, it is best to set up the initial conditions of the problem such that the errors are minimized prior to beginning the analysis. Section 7.3.2 discusses convergence and accuracy in brief.

7.3.2 Convergence and Accuracy

Convergence to a solution numerically is related to many aspects about how the problem is posed mathematically as well as how one sets up the program or code. Normally, we think of convergence as the code in a program completing its output and the result being meaningful. This, in essence, is for the most part, what engineers are only interested in. That being said, the key to convergence is not that the code yielded a result, but that it in fact converged on a solution that is accurate relative to the problem posed.

Some problems take some time to set up and can even take many hours or even days to complete once the analysis is started. However, today, the computational abilities of even small laptops is phenomenal, and even reasonably complex problems can converge quickly to a solution. However, this is not always the case, and when an engineer has set up a particularly large problem with many data points, where engineers are often left starting a job on a workstation and end up leaving their post for a cup of coffee or another task (hopefully). This is particularly true when the number of calculations or simultaneous equations is large or if the grid size is small and the number of grid elements is quite large. Though time is always constrained, it is worthwhile to understand that when posing a particular problem, the expectation is that the results at the end will indeed be meaningful.

Determining whether or not a particular piece of code will converge based on simply looking at the test structure is difficult to nearly impossible; this is particularly true with black box programs. One way to determine if a particular code is converging is to examine the error based on mesh size changes; that is, examine a data point (or average of a set of data points) and then plot the error as a function of effective mesh size change. This is similar to Richardson's extrapolation which examines the error based on changes in base equation between runs. A similar equation was shown in the previous section when we discussed global errors. Mathematically, this looks like the following:

$$\Delta f_{err} = |f(M_{i+1}) - f(M_i)| \tag{7.150}$$

where M is the mesh size for a given run and f is the result at a given data point or set of data points. If the error continues to decrease as a function of changes in the mesh (decreasing size), then this normally indicates the solution is converging. That is if

$$\lim_{\Delta h \to 0} \frac{\Delta f_{err}}{\Delta h} = 0 \tag{7.151}$$

Then the code is converging on a solution.

There are times, however, when the problem is not well posed or the code is unstable with respect to the problem setup. In this case the analysis may be unstable. The previous function will normally show this as well. The change in the error may increase as a function of decreasing mesh size which would indicate a problem with the setup. In that case, it is best to examine the initial boundary value problem and mesh choice to ensure stability is achieved.

For extracting parasitics for a PDN, even small structures can cause issues that do not allow the solution to converge. Often, this is related to the setup of the problem and how the boundary conditions are applied. An example of this would be a simple two-dimensional extraction of a series circuit element. Suppose an engineer has the goal of extracting the inductance of a structure in two dimensions and extending it linearly into the page to get the per-unit-length inductance. As a first step, it is clear that one of the boundaries is ground which is straightforward. This is called a *Dirichlet* boundary condition which essentially means that the boundary does not vary across it spatially ($dr/dt = 0$). However, the region above the main conductor may not be bounded. It has an air region that extends essentially to infinity. Now it is unreasonable to create an infinite space above this conductor to solve the problem (then certainly the solution would not converge, at least in our lifetime!). However, usually, this is handled by creating an artificial boundary condition that spatially changes across its interface. That means the derivative across the boundary is not zero. This is called a *Neumann* boundary condition. Though many black box programs set up these boundary conditions quite well, there are often issues creating them, which can cause spurious variations in the solution. For the FD method, and such structures, a PML or *perfectly matched layer* condition is then applied that actually assumes the fields decay over a number of grid points across this boundary. In reality, the fields will decay into space; the only question is when and how to properly represent them. In many commercial codes, this is handled internally, and the user does not know how this is achieved (nor do they usually want to know). However, if the algorithm to handle the PML is not done correctly, the program can run for extended lengths and not converge on a solution. Also, a poorly posed boundary condition can also result in an inaccurate solution.

In all cases, before an individual embarks on solving a problem numerically, it is critical that a bounded solution to the problem should be known first. This usually comes in the form of some first principles computation, or as described in the previous sections, an analytical solution that gives a result that is either exact or closely approximates the actual solution.

A common question arises when engineers begin the process of solving a complex structure numerically: What is a good approximation to the problem they are solving? Invariably, the answer is usually one that yields a result that is known for a similar structure but takes a shorter amount of time to compute and verify. Most structures for PDN interconnects that PI engineers encounter are limited in construction, and their solution has already been computed by a closed-form equation, such as those found in Chapter 4. Thus, once the geometry is known, one may start the process of solving a given problem with checking the result against a first principles equation.

7.4 Summary

In this chapter, we discussed numerical methods very briefly as they related to power integrity analyses. Though brief, this introduction is important when considering circuit extraction as well as for using general analytical tools, such as black box programs.

We began the discussion by discussing analytical methods. These form the basis for nearly all numerical methods and often may also be converted into a numerical format for use on the computer (though may be less efficient). Separation of variables is a technique that we discussed which gives one an excellent starting point for solving complex problems analytically without resorting to numerical tools. This is especially true for many boundary value problems that may be posed in electromagnetics. We then discussed variational methods which is the basis for the FEM. Variational methods have their origin in the late 1600s and have since been used to solve many complex problems including optimization of integral equations. Conformal mapping was briefly introduced as well and, though somewhat complex, is useful in solving many problems with angles in the geometry of the boundary value problem.

We then discussed two of the most common numerical methods used today: the finite difference and the finite element. The FD method clearly is easier to understand and also much easier to set up being based on an implicit single grid size. The math is also straightforward as it is based on using a truncated Taylor series for estimating the equations. The FEM is more complex but lends itself to complex geometries much more readily. Most FEM-based tools use triangular grids which may be varied within regions of the analytical space (often) to help the program converge to a solution that is also accurate.

Errors and convergence are two important areas which need consideration when using numerical tools. There are a number of potential errors that may spring up in a numerical analysis and understanding how to bound them is critical. Global, truncation, and round-off errors are common. Also, engineers need to consider the geometry of the structure and when fields get large within certain regions of the analytical space, resulting in large errors due to grid size and truncation.

Convergence and accuracy are also important where it is critical that the code converges into a solution where the result is viable. Accuracy in the case of a PI problem is related to the expectation of the result being within the bounds of what the engineer decided.

Though numerical tools have a very important place in the study of power integrity, they are, again, just tools and should not be fully relied upon to give an engineer all of the answers they are looking for. As stated throughout this text, having a foundation and breadth of knowledge in a number of areas not only helps the PI engineer become more efficient but also makes them more valuable to the organization they work for.

Problems

7.1 For the structure depicted in Figure 7.23, solve Laplace's equation with the given boundary conditions using separation of variables. Plot the voltage potential.

7.2 Show mathematically why the trivial solution ($\lambda = 0$) is not a solution for the previous problem.

7.3 Integrate Eq. (7.65) using a trigonometric substitution and solve.

7.4 Derive the central difference and backward difference relations. What is the order of the error?

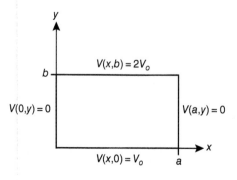

Figure 7.23 Separation of variables problem.

y

$V(x,b) = 2V_o$

b

$V(0,y) = 0$

$V(a,y) = 0$

$V(x,0) = V_o$

a

x

7.5 Derive the central difference formulation for the second derivative. What is the order of the error?

7.6 Using the code for the stripline problem, recompute the impedance by extending the boundaries on the sides out by a factor of 4 on each side. What happens to the values as compared with the previous analysis? Compare with the closed-form expressions for the impedance, inductance, and capacitance that are given in Table 7.3.

7.7 Write a MATLAB script to compute the work to find the capacitance in Example 7.1.

7.8 Using a finite difference relation (implicit method), solve the problem in Example 7.3 using a $dx = K$ (where $K = 1/40$th of the region size). Compare your results with the analytical solution. What is the error? Now cut dx to $0.25\,K$. What is the error now? Is this what you expected?

7.9 Find the inductance in Example 7.4 using a central difference formulation. Compare your results with Table 7.3. What is the error between them? How does your result compare to the analytical solution?

7.10 Solve for the capacitance in Figure 7.2. A few hints on this one. The function for the voltage has already been solved for. Take the derivative of the function and then integrate over the area. This gives the total energy. Once this is computed, you can check the result for a simple parallel plate capacitor. They should be quite close.

Bibliography

1 Kreyzig, E. (1993). *Advanced Engineering Mathematics*, 7e. Wiley.
2 Sadiku, N.O. (1992). *Numerical Techniques in Electromagnetics*. CRC Press.
3 Grasmair, M. (2015). *Basics of Calculus of Variations*. Norwegian University of Science and Technology https://wiki.math.ntnu.no/_media/tma4180/2015v/calcvar.pdf (accessed August 2018).
4 Byron, F. and Fuller, R. (1970). *Mathematics of Classical and Quantum Physics*. Dover.
5 Calixto, W.P., Alvarenga, B., da Mota, J. et al. (2010). Electromagnetic problems solving by conformal mapping: a mathematical operator for optimization. *Mathematical Problems in Engineering* 2010: 742039. https://doi.org/10.1155/2010/742039.

6 Sullivan, D.M. (2000). *Electromagnetic Simulation Using the FDTD Method*. IEEE Press.

7 Edwards, T. (1992). *Foundations for Microstrip Circuit Design*, 2e. Wiley.

8 Reddy, J.N. (2003). *Introduction to the Finite Element Method*, 2e. McGraw-Hill.

9 Chew, W., Jin, J.-M., Lu, C.-C. et al. (1997). Fast solution methods in electromagnetics. *IEEE Transactions on Antennas and Propagation* 45 (3): 533–543.

10 Mittra, R. (2008). Whither computational electromagnetics? A practitioner's look at the crystal ball. In: *IEEE Computation in Electromagnetics Conference* (7–10 April 2008). Brighton, UK: IEEE, Inc.

11 Zhu, Y. and Cangellaris, A.C. (2006). *Multigrid Finite Element Methods for Electromagnetic Field Solving*. Wiley.

12 Paul, C.R. and Nasar, S.A. (1987). *Introduction to Electromagnetic Fields*, 2e. Wiley.

13 Balanis, C.A. (1989). *Advanced Engineering Electromagnetics*. Wiley.

14 Marshall, S.V., DuBroff, R.E., and Skitek, G.G. (1996). *Electromagnetic Concepts and Applications*, 4e. Prentice Hall.

15 Hayt, W.H. (1981). *Engineering Electromagnetics*, 4e. McGraw-Hill.

16 Ramo, S., Whinnery, J.R., and Van Duzer, T. (1994). *Fields and Waves in Communication Electronics*, 3e. Wiley.

17 Faculty of Mathematics (2015). *Parallel Plate Capacitor: Laplace's Equation*. University of Cambridge. https://www.maths.cam.ac.uk/sites/www.maths.cam.ac.uk/files/pre2014/undergrad/catam/IB/2pt2.pdf (accessed August 2018).

18 Persson, P.-O. and Strang, G. (2004). A simple mesh generator in MATLAB. *SIAM Review* 46 (2): 329–345.

19 Edwards, K. and Werner, C. (2002). *A Simple Guide to Generating a Finite Element Mesh of Linear Triangular Elements Using Battri*. University of North Carolina.

20 E. Tuncer (2015) *Conformal Mapping*, Chapter 4. http://weewave.mer.utexas.edu/MED_files/Former_Students/thesis_dssrtns/Tuncer_E_diss/CHAP4.PDF (accessed August 2018).

21 Olver, P.J. (2013). *Complex Analysis and Conformal Mapping*. University of Minnesota. http://www-users.math.umn.edu/~olver/ln_/cml.pdf (accessed August 2018).

22 Sernelius, B.E. (2014). *Conformal Mapping*. Linkoeping University. https://www.ifm.liu.se/courses/TFYY67/Lect5.pdf (accessed August 2018).

23 Nagel, J.R. (2009). *Solving the Generalized Poisson Equation Using the Finite Difference Method*. University of Utah. https://pubweb.eng.utah.edu/~cfurse/ece6340/LECTURE/FDFD/Numerical%20Poisson.pdf (accessed August 2018).

24 Riera, G., Carrasco, H., and Preiss, R. (2008). The Schwarz–Christoffel conformal mapping for "polygons" of infinitely many sides. *International Journal of Mathematics and Mathematical Sciences* 2008: 1–20.

25 Muller, H. (2013). *The Conformal Mapping of a Circle onto a Regular Polygon, with an Application to Image Distortion.* http://herbert-mueller.info/uploads/3/5/2/3/35235984/circletopolygon.pdf (accessed August 2018).

26 Teppati, V., Goano, M., and Ferrero, A. (2002). Conformal mapping design tools for coaxial couplers with complex cross sections. *IEEE Transactions on Microwave Theory and Techniques* 50: 2339–2345.

27 Abbott, J. (2011). Modeling of capacitor behavior of coplanar striplines and coplanar waveguides using simple functions. RIT thesis. Rochester, NY: Rochester Institute of Technology.

28 Li, Z. (2017). Finite difference methods for two-point boundary problems. *Computational Mathematics*, Chapter 3. http://www4.ncsu.edu/~zhilin/TEACHING/MA402/chapter3.pdf (accessed August 2018).

29 Frey, P. (2009). The finite difference method, Chapter 6. https://www.ljll.math.upmc.fr/frey/cours/UdC/ma691/ma691_ch6.pdf (accessed August 2018).

30 Hayakawa, K. (1988). Convergence of finite difference scheme and analytic data. *Research Institute of Mathematical Sciences* 24: 759–764.

31 Pozrikidis, C. (1998). *Numerical Computation in Science and Engineering.* Oxford University Press.

Part III

Power Integrity Analytics

8

Frequency-Domain Analysis

8.1 Introduction to FDA

As every PI engineer knows, frequency-domain investigations are fundamental to power integrity and are typically the starting point for many analyses. The scope of the subject in simple terms encompasses characterizing the power distribution network (PDN) through a given bandwidth and then determining where the resonances exist that may cause issues in the time domain when excited by a stimulus. This is why the study usually comes before a time-domain examination; by characterizing the link between the source and the load, an engineer has an idea of the frequency components that can cause potential issues in the power distribution path.

However, systems today have become much more complex than in the past, and thus, a simple impedance characterization is usually insufficient toward uncovering all of the issues that can affect the noise in a PDN; for example, drift in the PDN over time along with capacitance degradation or improper guardband characterization for specific droop behavior can be focal points for the PI engineer's study. Moreover, it is not always the case that a resonance in a PDN corresponds exactly with what one would expect the droop behavior to be in the time domain. This is often true when performing parametric analyses where one or more device components are changed in order to examine the effects on the droop or resonances for a given stimuli. These and others are areas which usually require further discussion where often the PDN must be broken down into individual sections or components rather than examining as a whole before one becomes convinced that there is good understanding of the overall behavior.

It is usually the case that the placement and type of capacitance used in a PDN construction are best understood first with an alternating current (AC) analysis rather than through studying the behavior in the time domain to start with. At a detailed level, using a field solver to find the parasitics (as discussed in some detail in Chapter 7) should result in a reasonably good extraction of the elements to a sufficient granularity to help the PDN designer *piece together* the

Power Integrity for Electrical and Computer Engineers, First Edition. J. Ted Dibene II and David Hockanson.
© 2020 John Wiley & Sons, Inc. Published 2020 by John Wiley & Sons, Inc.

structure for the forthcoming analysis. However, a frequency-domain analysis (FDA) is really a *what–if* analysis to the PDN designer. This is because an examination in the frequency domain gives an individual a way to determine where the main parasitics are in the PDN and what type of decoupling is required and quite often where to place them. It also is the start of the layout constraints in the system, which aids the power management integrated circuits (PMICs) and system designers in knowing what is required to maintain the efficacy of the power delivery system.

As we discussed in earlier chapters, there *does* exist for every PDN problem a methodology where an engineer can strive for a somewhat *ideal* PDN. Nonetheless, physical, system, and fundamental physics (usually of the practical makeup of the component structures) put constraints on the problem that prevent one from achieving that ideal. Nevertheless, knowing what to look for to achieve a more optimal design can be quite beneficial. This is where a number of these *what–if* analyses come into play with respect to the FDA. The designer can normally examine the type of PDN that would be most beneficial and then perform the analysis to see how it might be improved.

The other portion of the analysis often performed in the frequency domain is power loss. This is an often overlooked aspect of the PDN since so many system designs in the past have showed the power loss to be negligible compared with other losses in the system and is therefore often ignored. However, today, most power distribution designs are heavily constrained, and getting every joule out of a system (particularly battery-powered designs) is often critical to the success or failure of a product. As we have seen from our analyses in previous chapters, including Chapter 3, power is not constrained to simply the losses in the metal path between the source and the load. Capacitors have losses at frequencies that can be difficult to find or, for that matter, may not even be considered in the context of an examination. A FDA geared toward looking at the power loss can be extremely revealing (either negatively or positively) depending upon the excitation frequencies within the system. Along with tying the frequency content of the load to the PDN frequency-dependent effects, one can begin to determine where the most optimal point would be for operation in a system.

Thus, in this chapter, we will examine basic optimal (and, therefore by necessity, suboptimal) PDN designs for both power integrity and power consumption in general and give techniques that allow one to analyze with mathematical tools and Simulation Program with Integrated Circuit Emphasis (SPICE) to gain an understanding of the performance of a PDN. We will start with a discussion of what constitutes an *ideal* PDN and then look at a more realistic design. In the discussion to follow, we will start to pull this PDN apart and then use techniques to tune this structure for the parameters that are of interest. We will later use Monte Carlo techniques to find anomalies in the PDN while examining frequency-dependent behaviors in the decoupling to gain a deeper understanding of those effects. While we delve into the main subsections, we will introduce techniques similar to asymptotic analysis and then discuss lumped element subsection modeling. As in earlier parts of the chapter, the objective will be to expand upon the connection parasitics when inserting components into printed-circuit board (PCB) planes. Although the internal structures of how the silicon metal layers are created for power distribution is an important aspect to our learning, many of the details will be left for Chapter 10 at the end of the text.

As we broaden our understanding of the frequency-dependent effects of the PDN, we will begin analyzing power consumption and also reexamine both the excitation frequencies at the load and what we would expect those frequencies to look like from the power source.

In the following sections, we will begin to build upon what has been learned in the previous chapters and what will additionally be used in the upcoming chapters. Thus, the extraction of parasitic circuit elements will be assumed for the reader, meaning that we have already applied some of the techniques that we used earlier to generate these values for interconnections and will not be reintroducing these concepts. In some cases, the result will be to opt for using some of the formulae or have a basic knowledge of the components that make up the PDN to begin with.

Most of the analyses will be focused then around AC analysis in SPICE though the use of Math programs and analytical techniques to do some of the number crunching will be applied. Eventually, there will be a discussion on lumped element vs. distributed modeling methods and where one seems to apply better over the other.

Finally, the chapter will end with a full lumped element model along with some of the boundary conditions that are appropriate.

A final note about the tools used in this chapter. For engineers who work at corporations or are doing analysis in the context of a college class, there are many computer-based programs that extract parasitics and then integrate the results into SPICE to simulate the circuit behavior from the PCB and the silicon device. It is noted here that the lack of discussion on these toolsets in no way dismisses their importance (as mentioned in Chapter 7). Our goal, however, is to ensure that the engineer first and foremost is well equipped to understand the fundamentals behind these tools so that they might better utilize them. Thus, the authors here will continue to emphasize the basics since becoming proficient in any tool usage should be left to the reader's preference.

8.2 The PDN Structure, Physically and Electrically

The PDN structure for most subsystems within a platform is now considered one of the key critical aspects to the efficacy of the power delivery. As we have seen in so many cases, it is intricate to our ability to deliver power from the

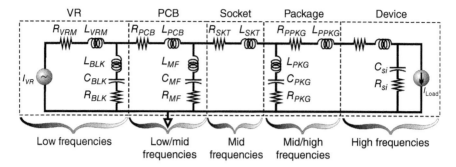

Figure 8.1 Power distribution network example from the source to the load.

source to the load. Physically, it is made up of metal layers within a PCB, package, and silicon and the adjoining decoupling that is distributed along the way. Thus, though it appears at first glance to be a rather simple conglomerate of passive devices, its effect on power is much more complex.

Figure 8.1 shows an example of how a power distribution network might be composed electrically section by section.[1] Depending upon which organization within a company has this responsibility, the output filtering from the VR may or may not be considered part of the electrical structure. However, to be complete, we will assume it is (for the moment) part of the PDN and thus will start from the output of the power converter switches as our source. From this point, one can build the network through examining both the interconnections and placement, type, and size of devices along the way.

For a starting analysis referring to Figure 8.1, this is what the PDN may look like within the context of a lumped element model. It is a simple way to understand the electrical behavior and is often where most PI engineers start the FDA process.

The *voltage regulator* is modeled with a simple RLC network where the voltage source may or may not be included in the model. In a SPICE simulation where one is looking back at the impedance of the network from the load, the voltage source itself is often ignored and the inductor in the lumped element network is placed to ground. The reason is that in an AC analysis, there is no direct current (DC) level signal coming from the load, and the voltage converter offset has no bearing on the results. However, there is frequency-dependent content generated from the voltage regulator, as seen from Chapter 2, which comes into play, and its effects on the load behavior should be considered at some point in the analysis.

---

1 This representation is consistent with most computer-based systems with high-performance processors. However, there are always exceptions based on packaging and decoupling that can vary this slightly.

Though only one lumped element per section is shown in the figure, this is usually never the case. There are multiple *sub-elements* which comprise the *PCB subsection* (as with the other subsections as well), and when examining the circuit board, it is usually considered the central part of the analysis of the PDN. This particular subsection, the PCB interconnect, will be elaborated on in more detail when analyzing these circuits later on in the chapter.

The next subsection shown in the figure is the *socket*. Though this can be important for such systems as servers and other platforms where the silicon device is replaceable, in many of today's systems, such as mobile platforms, this component does not exist since the device is soldered down. Thus, we will ignore this subsection in the context of this chapter.

The *package* is the other crucial part of the PDN that must be examined in detail. It often comprises a number of different capacitor types, for specific high- to mid-frequency decoupling, as well as both vertical and lateral interconnects to the silicon, and is of critical interest to the PI engineer.

The last subsection is the *silicon* device itself. This section warrants its own chapter due to the complexities of the silicon as a load, where routing power into such a device requires much more forethought. In this chapter, we will simplify this structure for now but will delve into it in further detail in Chapters 9 and 10.

It should be noted that the return path here was assumed ground for now. The return path is normally asymmetrical with the power path, and in the above model, the ground path was reflected back into the power path. This was done for simplification of the SPICE modeling to start. As models become more sophisticated, particularly with distributed modeling, the return path is usually split out accordingly.

In the next few sections, we discuss setting a target for our PDN structure with a simplified model and then look at each section of the electrical structure in Figure 8.1 and how it behaves in the frequency domain.

8.2.1 The Damped Transmission Line Approximation

Our first step is to examine both a somewhat *ideal* PDN and then look at how our overall lumped model fares in comparison. This exercise helps us to break down the sections of the model where the electrical behavior deviates from this ideal goal. Figure 8.2 shows an illustration of a damped transmission line PDN distribution as a function of frequency using the work from Chapter 5 on transmission lines. One can see immediately that the impedance of the structure is reduced as the frequency approaches 1 GHz. For all intents and purposes, this frequency is beyond the limits of the bandwidth of interest. This is because from the filtering in the PDN, frequencies above 500 MHz are usually attenuated and the effectiveness of the capacitance of the PDN is highly reduced. It is true that many loads can generate such high-frequency content, particularly processors.

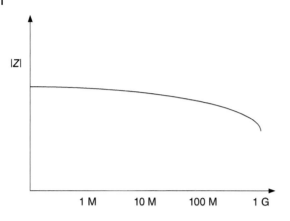

Figure 8.2 Impedance profile of damped transmission line.

However, due to the impedance of the network components near the load, these spectral components are usually not seen by the PDN – particularly out into the PCB. The more important thing to note is that what we are achieving here with a damped transmission line is a somewhat *flat* frequency response except at the highest frequencies where we want the impedance to be very low so that the energy is transferred efficiently. This is because the capacitance here is supposed to yield a relatively low impedance, and thus, the AC voltage across it should also be small.

Since the structure is still considered a *transmission line*, each element is assumed to be infinitesimally small, resulting in virtually no change from one section to the next. However, the shape of the impedance plot reveals that the structure is actually changing slightly as a function of frequency, which means that the PDN is not behaving precisely like an ideal transmission line. When representing losses in a *lossy* transmission line, it was seen earlier that in all practical nonideal lossy transmission line structures, the impedance *increased* with frequency as one would expect. However, the goal of our PDN is to *reduce* the impedance as the frequencies increase while simultaneously physically moving the components that cause the impedance to increase back toward the source and away from the load. Thus, electrically, this makes the PDN behave as a dampening filter whereby the lower spectral components at the load are less affected by the physical components near the source.

The result is as the frequency decreases, the noise of the system dampens slightly, resulting in higher impedance at the lower frequencies. Though this behavior is not completely ideal for power loss, it is effective in reducing much of the noise generated by the load in the system. We will study this in more detail in the time domain in Chapter 9.

It should be noted that the electrical path length for the PDN is also (normally) much shorter than what would necessitate a transmission-line model. To check this, we can compute the electrical path length of our simple transmission line model PDN. If the characteristics of the transmission path are known, one can estimate the time to traverse one side of the path to the other by determining the phase velocity. We do this by assuming a generally *lossless* transmission line to start. If the resistive elements are small, we recall the characteristic impedance of the transmission path to be,

$$Z_0 \cong \sqrt{\frac{L}{C}} \tag{8.1}$$

This was shown in Chapter 5 where the impedance is only dependent upon the inductance and capacitance of the structure for a lossless system.

The real components in the transmission path, however, will vary in value depending upon where they are physically located and for the type of decoupling required in the system. Each section is broken down into very small elements as shown in Figure 8.3, and the variation in the impedance is considered small. This is our assumption for the damped transmission line structure as well. The only difference is we have added a cascading set of RL networks that will vary infinitesimally which will have the effect of lowering the impedance at the higher frequencies. (Note that they are in parallel with each other as are the RLC components.) Each independent element (*L*, *R*, and *C*) is essentially the same from section to section. In a real transmission line structure, the assumption is that there are *n* elements that make up the distribution where the number *n* goes to infinity.

Since the objective is to only change the series impedance slightly from one end of the PDN to the other and from the low to high frequencies, we may simplify the model in Figure 8.3. This will help in the analysis process of determining the impedance of the transmission line structure. The first assumption is to change the series impedance elements to show only a small delta from the

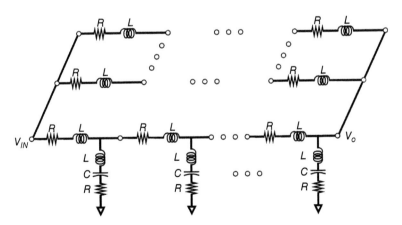

Figure 8.3 Damped transmission line model.

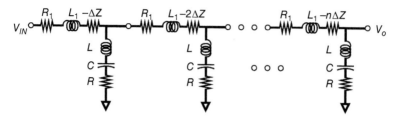

Figure 8.4 Simplified damped transmission line.

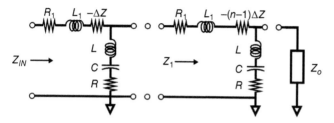

Figure 8.5 Damped transmission line network.

input to the output. The second change will be to increase this delta from left to right. This is shown in Figure 8.4.

The characteristic impedance may be found by looking into the section of the structure at the end and working back toward the source. Assuming the impedance of the last section is Z_0, the impedance of the damped transmission line structure appears as shown in Figure 8.5.

The impedance of the network in Figure 8.5 is now found from the following relation:

$$Z_1 = [R_1 + j\omega L_1 - (n-1)\Delta Z] + \frac{Z_0 \left[R + j\omega L + \frac{1}{j\omega C} \right]}{Z_0 + \left[R + j\omega L + \frac{1}{j\omega C} \right]} \tag{8.2}$$

However, Z_0 is equivalent to the following relation:

$$Z_0 = [R_1 + j\omega L_1 - (n)\Delta Z] + \left[R + j\omega L + \frac{1}{j\omega C} \right] = Z_1 + \Delta Z \tag{8.3}$$

or

$$Z_1 = Z_0 - \Delta Z \tag{8.4}$$

Substituting and rearranging terms in Eq. (8.4) gives us the right-hand side of the equation.

$$Z_0^2 - Z_0[R_1 + j\omega L_1 - (n-1)\Delta Z] - \Delta Z Z_0 - \Delta Z \left[R + j\omega L + \frac{1}{j\omega C} \right] \tag{8.5}$$

and the left-hand side as

$$[R_1 + j\omega L_1 - (n-1)\Delta Z]\left[R + j\omega L + \frac{1}{j\omega C}\right] \tag{8.6}$$

Now each circuit element shown in Figure 8.5 is physically made up of an infinitesimally small length such that it is electrically a per-unit-length parameter. This includes the delta impedance ΔZ. If we multiply through by this *length*, we can simplify both the right-hand and left-hand sides of the equation. Doing this and simplifying yields

$$Z_0{}^2 = \frac{R_1 + j\omega L_1 - (n-2)\Delta Z}{j\omega C} \tag{8.7}$$

Notice that the ΔZ section is still part of the result. Taking the square root of both sides yields the final impedance

$$Z_0 = \sqrt{\frac{R_1 + j\omega L_1 - (n-2)\Delta Z}{j\omega C}} \tag{8.8}$$

Basically, the characteristic impedance of the damped transmission line is a function of the series loss plus an additional very small series loss and the capacitance of the shunt element. Note that the $L-R$ portion of the shunt section does not play a role when defining the impedance. The series small ΔZ impedance is also an $L-R$ circuit, which means it is also frequency dependent. We may simply define this as an incremental portion of the series R and L circuit such that

$$Z_0 \cong \sqrt{\frac{R_1 - (n-2)\Delta R_1 + j\omega L_1 - j\omega(n-2)\Delta L_1}{j\omega C}} \tag{8.9}$$

where we have substituted for the delta impedance term

$$\Delta Z = j\omega\Delta L_1 + \Delta R_1 \tag{8.10}$$

For very high frequencies and near the load, we may assume that the series resistance terms will be negligible. Thus, we may simplify Eq. (8.9) to approximate the impedance at this point yielding

$$Z_0|_{HF} \cong \sqrt{\frac{L_1 - (n-2)\Delta L_1}{C}} \tag{8.11}$$

If n terms of the delta inductance is equivalent to say half of L_1, and $n \gg 1$, then the impedance becomes

$$Z_0|_{HF} \cong \sqrt{\frac{L_1}{2C}} \tag{8.12}$$

This now gives us some direction for how the damped transmission approximation will behave in our model. The number of segments and relative size of the delta elements are dependent upon the type of model one plans to construct, and thus the choices are up to the user.

In a standard transmission line, it was found that the velocity of propagation is related to the characteristic impedance of the line by the following relation:

$$v_0 = \frac{1}{\sqrt{LC}} = \frac{1}{C\sqrt{\frac{L}{C}}} = \frac{1}{Z_0 C} \tag{8.13}$$

The wavelength may then be determined from the velocity and therefore the frequency of the signal or

$$\lambda = \frac{v_0}{f} \tag{8.14}$$

Let us return now to our nonideal transmission line–based PDN. What is the electrical path length of this structure? Or more importantly, how long does it take to propagate a signal from one part of the PDN to another? If we assume that the impedance of our model is the one derived above, then we can determine the characteristics of the transmission line by looking at a simple subsection and then computing the wavelength. We can then compare that against an ideal transmission line using the actual physical length and determine if it indeed is acting like a transmission line. The next example illustrates this.

Example 8.1   Given the electrical characteristics of a PDN that is comprised of very small sub-elements (transmission line approximation), determine the time it takes to propagate a signal from one side to the other. Use the high-frequency approximation for the impedance. Also determine the quarter wavelength of the structure (approximation for a true transmission line given the physical and electrical values in Table 8.1). Compare this with an ideal transmission line model.

Solution:
The construction is basically a power plane that is 3 cm long with additional capacitance added to lower the impedance. We assume that this additional

Table 8.1   Values for transmission line computation of Example 8.1.

Metric	Value
Transmission line length	3 cm
Width of power plane	1 cm
Thickness of power plane	0.001 in.
Additional capacitance per cm	1 nF

capacitance (and the plane level capacitance) is distributed. We need to determine the impedance of the structure, but first we must approximate the inductance and capacitance per unit length. Note that for this approximation, a computation of the resistance was not necessary. However, from an earlier chapter, we could have easily estimated this as well.

The thickness of the dielectric is 1 mil (0.001 in.) which is 0.0254 mm. Given the width as 1 and 3 cm long, the capacitance would be (approximately)

$$C \cong \frac{A\varepsilon_r\varepsilon_0}{d} = \frac{(3 \times 10^{-2})(1 \times 10^{-2}) \times 4.2 \times 8.854 \times 10^{-12}}{0.0254 \times 10^{-3}} = 0.42 \text{ nF}$$

(8.15)

and for the inductance

$$L \cong \frac{lt\mu_0}{w} = \frac{(3 \times 10^{-2})(0.0254 \times 10^{-3})(4\pi \times 10^{-7})}{(1 \times 10^{-2})} = 95.8 \text{ pH}$$

(8.16)

Converting these two values to per-unit-length parameters yields

$$\frac{C}{m} \cong 1.39 \frac{\text{pF}}{\text{m}}, \quad \frac{L}{m} \cong 0.32 \frac{\text{pH}}{\text{m}}$$

(8.17)

Adding in the additional capacitance, we get the total distributed capacitance per meter as

$$\frac{C}{m} \cong 1.39 \frac{\text{pF}}{\text{m}} + 1.0 \frac{\text{nF}}{\text{m}}, \quad \cong 1 \frac{\text{nF}}{\text{m}}$$

(8.18)

We can now compute the impedance of the structure

$$Z_0 \cong \sqrt{\frac{L}{2C}} = \sqrt{\frac{0.32\text{pH}}{2 \times 1 \text{ nF}}} \cong 10.6 \text{ m}\Omega$$

(8.19)

The velocity of propagation is then

$$v_0 \cong 2.23e6 \text{ m/s}$$

(8.20)

This is quite slow. This is expected given that the capacitance per unit length is relatively large, which means the rise time will be long. The time then to propagate a signal across the 3 cm is

$$t \cong \frac{0.03 \text{ m}}{\left(2.23e6 \frac{\text{m}}{\text{s}}\right)} = 13.4 \text{ ns}$$

(8.21)

The next question we must answer is what is the frequency for a quarter wavelength of the structure? We know the velocity and the length now we can simply plug in

$$f = \frac{4v_0}{\lambda} = \frac{4\left(2.23e6 \frac{\text{m}}{\text{s}}\right)}{0.03} \cong 300 \text{ MHz}$$

(8.22)

Once again, this should not be a surprise given the size of the capacitance. The interesting thing to note is that if the capacitance is distributed across this small strip, then we may assume that signals in the range of 300 MHz or higher can resonate on this PDN.

The damped transmission line approximation is useful for examining aspects of the PDN where the PI engineer is looking into how to achieve the properties of their real design and where the deficiencies lay. However, there are significant differences between a transmission line approximation and a PDN represented as a lumped element model. As in the previous example, we examined the time to propagate across a 3 cm section. If the PDN is now represented as a lumped element model, what would be the actual time of flight?

Because each section is different and because it does not behave exactly like a transmission line, we need to estimate this time of flight in terms of the lumped elements. First, we break down each subsection and estimate the delay as a function now on a lumped element basis. This is akin to putting a step function on one side and then measuring how long it takes to reach the other. We can do this by assuming we have a *current* step function and then look at the time to propagate (a little more than one time constant) to reach the other side. We do this in the example below for the first section in the nonideal structure. The procedure to estimate the other sections will be similar.

Example 8.2 Estimate the time to propagate a step function for the first two lumped elements in Figure 8.6 given the following values for the components in the structure.

$$L = 1\,\mu H, \quad C = 100\,\mu F, \quad R = 1\,m\Omega \tag{8.23}$$

Solution:
Let us assume there are no initial conditions and use a simple solution to our problem given an RLC circuit where the voltage is across the capacitor as shown. Let us also assume for now that the load current is very small (negligible) which means that we can assume that we only want the voltage across the capacitor. The output voltage is then found from the relation in the s domain.

$$V_0(s) = \frac{V_i}{L\left[s^2 + s\frac{R}{L} + \frac{1}{LC}\right]} \tag{8.24}$$

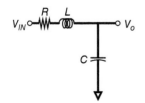

Figure 8.6 Simple RLC circuit.

Assuming $V_i(s)$ is a unit step function. We can complete the square to put Eq. (8.24) into a form that we may easily transform

$$V_0(s) = \frac{V_i\left[\frac{1}{LC} - \left(\frac{R}{2L}\right)^2\right]^{\frac{1}{2}}}{(LC)^{\frac{1}{2}}\left[\frac{1}{LC} - \left(\frac{R}{2L}\right)^2\right]^{\frac{1}{2}}\left[\left(s + \frac{R}{2L}\right)^2 + \left[\frac{1}{LC} - \left(\frac{R}{2L}\right)^2\right]\right]} \tag{8.25}$$

This is now in the form of a Laplace transform pair that we can manipulate

$$F(s) = \frac{b}{[(s-a)^2 + b^2]} \tag{8.26}$$

where

$$b = \left[\frac{1}{LC} - \left(\frac{R}{2L}\right)^2\right]^{\frac{1}{2}}, \quad a = -\frac{R}{2L} \tag{8.27}$$

Therefore, the solution in the time domain is now

$$v_0(t) = \frac{v_{i0}}{(LC)^{\frac{1}{2}}b}e^{at}\sin(bt) \tag{8.28}$$

We may plug in for the RLC values and plot the function. If the step occurs at $t = 0$, our goal would be to find when the voltage first reaches 90% of its maximum initially. This time delay will give us the value we are looking for. The plot is shown in Figure 8.7.

We can solve directly for t, but this is a transcendental equation which may take some mathematical manipulation. A more reasonable approach is

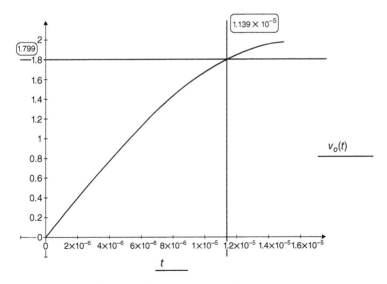

Figure 8.7 Plot of output voltage across capacitor.

to assume that the decay on the exponential function is small relative to the change in time and then compute the rise time on the sine function assuming that e^{at} decays very little over this range. Solving for t at 90% of the magnitude yields

$$
t|_{0.9\text{max}} = \frac{\sin^{-1}(0.9)}{b} \cong 11.4\,\mu s \tag{8.29}
$$

This corresponds to about 99.4% decay of the exponential function. Thus, we can assume that it will take ~11 μs to reach 90% of maximum value after a step function. This is also assuming no voltage on the capacitor and no current in the inductor initially. The interesting thing about this is that unlike a propagation time for a signal into a transmission line, a PDN truly behaves as a lumped structure for the most part when considering the bulk components that store energy: in this case the VR inductor and main storage capacitor. Note that we did not even examine the electrical path length, just the VR portion of the sub-circuit. The difference in delay between the 3 cm long transmission line and the sub-circuit was approximately three orders of magnitude in the delay or 1000× longer for the lumped element than the damped transmission line from input to output. Thus, when considering the behavior of a PDN in a certain frequency band, it is important to consider how it behaves as a set of lumped elements first and then look at the system as a whole later.

We now examine more closely the components which make up the PDN with respect to its frequency behavior.

The previous example not only gave us an interesting insight into the VR portion of the PDN but also let us know that the transfer of energy from these lumped elements is very slow. It is also a warning to us when we use a damped transmission line model for estimating certain time-domain functions, we should be aware those estimates can be inaccurate.

The reason for the large discrepancy is because the component values are rather large, and in the frequency domain, their impedance is also large at the higher frequencies; it is especially true for the power components and for the capacitance that is closer to the source than the load. Thus, it is also now clear, as we already knew, that other filtering will be necessary when the load switches at clock frequencies that are much higher than the bandwidth of the VR filter. This is because to ensure that the energy is transferred efficiently to the load, the impedance must be small at the spectral frequencies of the currents, which are of interest. This does not mean we abandon our damped transmission line model; only we use it with discretion as a guideline for trying to achieve a better more realistic based PDN.

In Section 8.2.2, we begin to break down the PDN section by section to gain insights into their behavior toward building a better performing PDN.

8.2.2 The Subcomponents of the PDN

Let us break down the PDN into its components and examine the structures using SPICE to understand how each subsection behaves electrically. Starting with the subsection of the VR, we can add in all of the main parasitics of this block and do a frequency sweep to look at its filtering behavior. It is clear that this would not serve as a complete PDN in a real system, but it helps to start to see how this portion behaves on its own.

Using the values for the RLC network that were given in the example, we may start to examine the frequency dependence of the structure. The parasitics from a circuit perspective are added in Figure 8.8. The inductor and capacitor parasitics are also included for clarity. The PDN portion would be on the right side of the diagram looking into the load.

When simulating such a structure in SPICE, we simply inject an AC load at the node points A and B and then sweep the frequency. The results are shown in Figure 8.9 for the impedance plots with and without the parasitics. The lighter trace is without the parasitics (both impedance and phase are shown). Note that, as expected, there is a definite resonant behavior to the structure that is due to the larger energy storage components of the voltage regulator. This is true in both cases. However, this is just the start of the examination process.

As can be seen in the plot in Figure 8.9, there is a definite difference between the two, especially at the higher frequencies. This is expected since these components were not intended to operate at these higher frequencies when the nonideal components are added. Moreover, as the reader probably guessed, the impedance network here is *well* filtered from the load behavior in a real

Figure 8.8 VR portion with parasitics added.

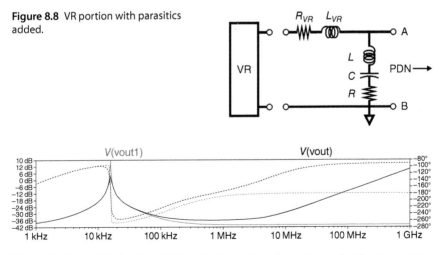

Figure 8.9 Plots of VR circuits in PDN – with (darker impedance line) and without parasitics.

PDN, which would prevent the bulk components from seeing most of the high-frequency signals from the load. It is instructive though to see why it is critical to add these extra elements in the path near the load and decouple the VR filtering from it. In reality, the bulk components are not designed to handle the higher frequencies, and thus, lowering the impedance with the proper decoupling capacitors is necessary.

Thus far, though, we still do not know from an overall PDN behavior how this data figures into our understanding of what an *ideal* PDN should look like. However, if we go back to the model for a moment of the damped transmission line in Figure 8.3 and examine it as a subsection of the PDN, we can see that the impedance is *flat* compared to the nonideal sub-circuit in Figure 8.8. The question is, can the sub-circuit portion be improved so that it begins to behave like the damped transmission line approximation? In this case, the components which make up the VR filter are chosen for a very specific reason. In particular, the inductor and capacitor were designed (as we saw in Chapter 2) to filter the PWM signal and store energy and remove the AC components of the signal as necessary. Moreover, the feedback of the system was *compensated* to ensure it was stable, and thus changing the values could have an impact on this stability. For the bulk components, it is usually difficult to get the power supply designer to change the values for the reasons just cited (though sometimes it is at least worth asking!). However, what about the parasitics? Some of them involve lay-out constraints, and others are intrinsic to the construction of the capacitor and inductor. The question is, can the electrical behavior be improved in any way? Particularly the resistive elements in both the capacitor and the interconnect. There is sometimes a trade-off between what is chosen for the main VR filtering for the power supply from the power designer's perspective and what may be needed as far as the PDN is concerned.

In this case, a large 100 μF bulk capacitor was chosen for filtering reasons. The ESR of the device is large, which results in dampening some of the components of ripple as well as in more loss. One way to improve the loss part of the equation is to use a group of smaller capacitors and lower the *effective* ESR of the filter capacitance. Distributing the capacitors over a region can also improve the electrical behavior. If we replace the 100 μF capacitor with two 22 μF capacitors and one 47 μF capacitor and then position them on the PCB such that their parasitics are minimized, we can determine if this helps the situation. This is shown in Figure 8.10.

We can see by the frequency sweep in Figure 8.11 that the resonance is at the same place, but the ESR is lower for the multi-capacitor solution (**vout3**). This has not only improved power loss at the frequency of 2 MHz (which is often close to the ripple frequency) but also helped with some of the mid-frequency harmonics, which may be generated by the load. Though we have likely com-promised on area and possibly cost to do this, often the trade-off can be worth-while. Moreover, as we shall see when we examine the next subsection of the

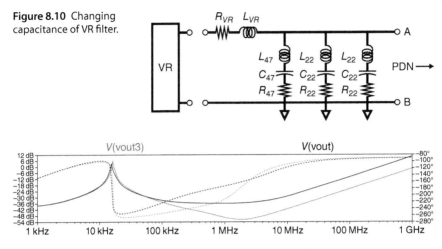

Figure 8.10 Changing capacitance of VR filter.

Figure 8.11 Frequency sweep with changing capacitance parasitics.

circuit model, this can help to reduce the need to add more high-frequency capacitance.

Another important note is the resonance at the lower frequency near 20 kHz. This is due to the LC tank circuit. If we do a computation, we can easily see this. From Chapter 2, the resonance may be found from the following equation:

$$f = \frac{1}{2\pi\sqrt{LC}} = \frac{1}{2\pi\sqrt{(1e-6)(100e-6)}} \cong 16\,\text{kHz} \qquad (8.30)$$

This is not a surprising result given the above plot. However, recall from Chapter 2 that with a real feedback system, the loop response of the system will basically remove the resonance of the system and will keep the impedance essentially flat in this range and below. Depending upon the loop bandwidth, this could be as high as 300 kHz. Thus, we can see clearly that at any frequency below this point, we can neglect the resonant effects due to the response of the voltage regulator. As voltage regulator responses increase, PI engineers can take advantage of this to simplify their PDN designs.

Let us turn to the next subsection of the circuit model as illustrated in the lumped-element model in Figure 8.1. Here we have a group of capacitors the purpose of which is to reduce the noise generated from the load after the high frequency on-die and on-package decoupling. We start with a group of eight capacitors of different sizes and electrical characteristics placed around a portion of the silicon load on the circuit board. Also added are some of the parasitics of the interconnections of the traces on the PCB. The lumped element equivalent circuit is shown in Figure 8.12.

Extracting the electrical behavior in a lumped element circuit for the capacitors only, we see clearly from the impedance profile shown in Figure 8.13 that it

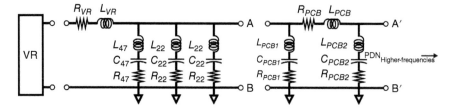

Figure 8.12 Addition of PCB capacitors to model.

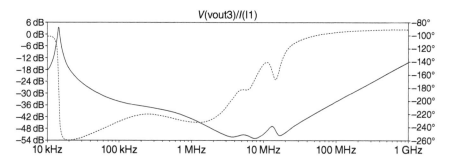

Figure 8.13 PCB capacitance and capacitor parasitics added to PDN.

behaves reasonably well in the mid-frequency range of the plot up to 10 MHz. This is a good start to our analysis.

However, the plot reveals that there is also a steady increase to the impedance in the higher frequency region of the plot. This increase begins in the 20 MHz regime. The unfortunate reality is that there will be significant harmonics from the load at these higher frequencies as the load currents are modulated by the activity in the device. We know this because of the fact that the clock source is likely in the 1 GHz frequency or higher and because the data patterns that are generated can be multiples of this fundamental frequency [1] as discussed earlier.

With respect to Figure 8.13, there are additional parasitics that are missing in the plot. These are the series interconnects that make up the connections between the capacitors on the PCB and the other portions of the PDN and load. These are shown unlabeled in Figure 8.12. In actuality, the capacitors on the PCB are often treated as a single lumped capacitance in SPICE when they are grouped together within proximity to the load; that is, the impedance between the load and these components is the same and the impedance between these capacitors and the next portion of the PDN, in this case the VR filter, is also the same. However, what if they were *not* grouped close together and have finite impedances between them?

Figure 8.14 shows the frequency sweep with one of the two instantiations of a potential layout variation for these components. The first one is what we

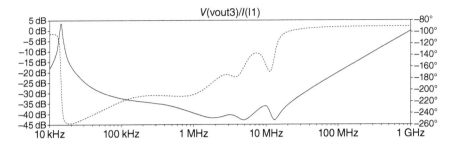

Figure 8.14 Capacitances with interconnect parasitics added.

have just examined and is shown in Figure 8.13. This was the impedance plot without the series parasitics. The second is a variation where the devices are *distributed* somewhat physically, and the series components have been added in Figure 8.14. A close examination shows that the addition of these parasitics has clearly made a difference.

Once again using SPICE, we performed a frequency sweep to look at the magnitude of the impedance. Though the parasitics of the interconnects appeared to be small, there is a noticeable increase at 10 MHz where the resonance has both *shifted* downward and the amplitude increased. The magnitude of the impedance may be found by doing a simple computation. The value from the graph is −36 dB which translates to ~15.8 mΩ for the impedance at this point. We compare this with the previous impedance of −45 dB or ~5.6 mΩ which was nearly 3× smaller. This can result in a substantial difference in voltage droop in the time domain assuming the stimulus is synchronized with this resonance (which is highly probable in high-frequency switching loads).

In the previous analysis, we ignored the phase information though we plotted it for each of the impedance profiles generated. For the previous plot in Figure 8.14 let us take a closer look at the phase information. As can be seen in the figure, there is a slight difference between the two (why?). The capacitor values are identical in both cases. Because some of the devices are physically and electrically further away from the load, we can clearly see that their effect on transferring current into the load will be different. However, the result will not necessarily be worse than the first one. This is because the purpose of the PDN is multifold, meaning the PDN serves to both help transfer energy (charge) from one region to the next and minimize noise from causing issues in the system. This is not only true at the load but at the source and other sub-blocks within the silicon as well where coupling can cause circuits to malfunction.

Though a higher impedance resonance is typically not desired in a PDN, the key to successful noise mitigation and energy transfer is both in the distribution of the storage elements (and interconnects between) and in the stimuli of the system. This is why certain excitation frequencies can cause perturbations on an apparently *good* PDN, resulting in larger voltage deviations than in a less optimal PDN design.

The next subsection model of the circuit model to add is the package elements. Often, these are comprised of low inductance interconnections (from the capacitors to the silicon load) and very high-quality high-frequency capacitance. Depending upon the type of load (processor or other), the number and type of capacitance may vary. In the lumped model below, we are adding a number of high-frequency capacitors around a load which is assumed highly dynamic. For now, we may assume that some of the capacitors are equidistant from the next set and can place the $L-R$ parasitics (interconnections) between them. Additionally, we want our capacitance to have very low inductance to mitigate much of the high-frequency content that is generated at the load. This section is added between A′–B′ and A″–B″ (Figure 8.15).

Adding to this the previous SPICE model, we can start to see the effects of the additional high-frequency capacitance as the PDN is constructed out toward the load (Figure 8.16).

The first thing that should be noted is that the resonance near 10 MHz has been essentially eliminated from the plot in Figure 8.14. This is due to

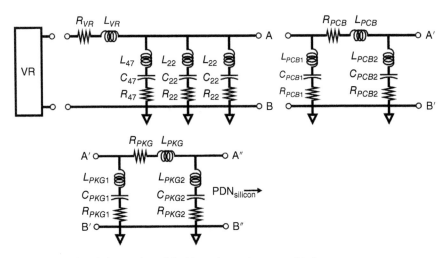

Figure 8.15 Simple lumped model with package elements added.

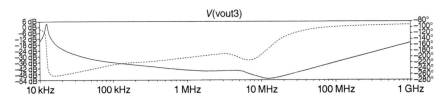

Figure 8.16 Frequency sweep with VR, PCB, and package elements added.

the high-frequency capacitors on the package that are designed to lower the impedance in this range. However, we still see the rise of the impedance as it increases out to 1 GHz. This is because the silicon level decoupling and parasitics have not been added yet. This portion of the model we will address in a moment, which will again have assumptions around the lumped component approximation. When we get to Chapter 10, we will begin to examine this in much more detail, which will help readers gain an understanding on how to model this along with the load and portions of the structure, which will matter to us.

The final subsection in Figure 8.1 is the silicon decoupling and interconnection. In most cases, the interconnect inductance on silicon is usually very low, while the resistive portion can be larger than on the package. The capacitance, however, is the more interesting aspect of the PDN on-die. With respect to a lumped element representation, the decoupling capacitance has an extremely low inductance such that it is often ignored. The series resistance, however, can be large enough that it is part of the final model.

Figure 8.17 shows the addition of the final silicon level model to our lumped element PDN representation. The entire distribution now has been built up from the VR filter to the load. We can start to see how this structure behaves electrically with respect to our ideal PDN model and then determine where improvements may be required. The first goal is to now simulate the entire structure and determine the impedance.

Figure 8.18 shows a simulation with the full PDN. As expected, we see a resonance near the 100 MHz regime. This is due to the quality of the on-die capacitance as well as the inductance between the package capacitance and the on-silicon decoupling. Once again, we can check this with a simple computation to determine how close the resonance is. If we include the shunt inductance

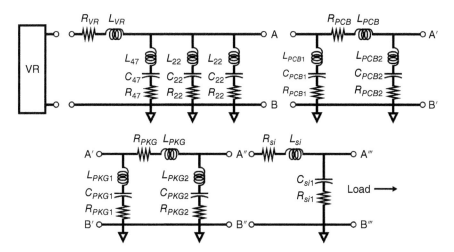

Figure 8.17 Full PDN lumped model with silicon components.

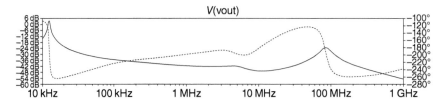

Figure 8.18 SPICE simulation of full PDN with silicon parasitics.

of the package capacitance (due to its size, it can dominate the computation), we see that the resonance is in the 80 MHz range which is very close to what the plot reveals.

The more important point is that this resonance dominates the impedance plot being the largest magnitude resonance of the PDN (ignoring the low-frequency components). For most PDN structures, this is expected. Unfortunately, as we shall see in Chapter 10, this resonance can play an important role in the power integrity near the load and thus is often a critical analysis point for many engineers. The goal, of course, is to minimize this peak resonance since the excitation frequencies can often line up with this resonance and cause spurious high-frequency noise near the silicon.

Before moving onto the sources of noise in PDN structures, it is appropriate to examine some techniques for analyzing lumped element PDNs. Often, breaking down the structures into subsections, as is done here, has its value in performing first-principles analyses to better understand that particular behavior. In addition, setting a goal with a damped transmission line structure and then analyzing the areas of deficiency help in achieving the desired electrical performance. Section 8.3 discusses both methods.

8.3 Analytical Methods

From the previous section(s), we have seen how each element behaves electrically using simple lumped element models and SPICE as our modeling tool. From a modeling technique for simple, and even complex structures, SPICE is often the preferred program for looking at the behavior for frequency-dependent analysis. However, it is also advantageous to have some analytical techniques at the power integrity engineer's disposal, especially when doing basic checks on a structure and to gain insights into the behavior of one subsection or another. One such technique is to use Bode diagram analysis based on a mathematical model to determine the electrical behavior of a subsection or even an entire PDN.

There are two ways to create a Bode plot that is applicable to our objectives here. The first way is to construct an *asymptotic* diagram and graph each subsection independently. This method has been around for many years and was utilized long before computer programs were available to the engineer. The second method is to construct the diagram piece by piece using a math program to show where the resonances occur in each subsection but forego the asymptotic method, opting for looking at the actual behavior directly. Since math programs and SPICE have made the job so much easier, we will go for the former approach.

It is, however, instructive to understand the objective of the asymptotic analysis before we generate the Bode plots for the impedance of the PDN subsections. The purpose of an asymptotic analysis is to give designers insights into portions of the PDN that are often not apparent when examining the entire PDN as a whole with a SPICE analysis. An asymptotic Bode diagram is simply a way of looking at different resonant points and slopes within the PDN to determine what components are dominating and what others are not. With the power of mathematical programs, breaking down the math section by section and showing the actual impedance effects is just as expeditious and more accurate. This simple method is illustrated below.

We start with examining the impedance of a structure looking back from the load as was done previously. Here, once again, the VR impedance is the starting point. The generic impedance of the VR section is shown in Figure 8.19.

As before, the objective is to determine the AC impedance by looking back toward the VR. This is simply the parallel combination of Z_1 and Z_2, the series and shunt impedances of the VR portion including the parasitics.

$$|Z_{21}(j\omega)| = \left| \frac{Z_1(j\omega)Z_2(j\omega)}{Z_1(j\omega) + Z_2(j\omega)} \right| \tag{8.31}$$

One can drop the functional frequency dependence for brevity since it is implied and use the following nomenclature:

$$|Z_{21}| = \left| \frac{Z_1 Z_2}{Z_1 + Z_2} \right| \tag{8.32}$$

This is the simplest model of the impedance that we will look at when analyzing each subsection. From here, the goal will be to add each portion and build up the PDN electrical impedance as we move toward the load. Each impedance element has a set of components or parasitics that should be considered when

Figure 8.19 Simplified VR impedance looking back from PDN.

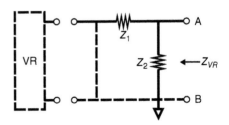

examining the behavior electrically. For example, Z_1 has the series parasitics of the interconnect and the inductor circuit elements added to it.

$$Z_1 = j\omega L_1 + R_1 \tag{8.33}$$

where L_1 and R_1 include both the parasitic inductance and resistance connecting to the Z_2.

The impedance for Z_2 is that of the shunt storage element with its parasitics added as well. Thus, mathematically, it is simply

$$Z_2 = j\omega L_2 + R_2 + \frac{1}{j\omega C_2} \tag{8.34}$$

The form of the equation that is desired for analyses with a Bode plot is

$$|Z| = C_n \left| \frac{\left(1 + \frac{j\omega}{\omega_{K1}}\right)\left(1 + \frac{j\omega}{\omega_{K2}}\right) \cdots \left(1 + \frac{j\omega}{\omega_{Kn}}\right)}{\left(1 + \frac{j\omega}{\omega_{Kn+1}}\right)\left(1 + \frac{j\omega}{\omega_{Kn+2}}\right) \cdots \left(1 + \frac{j\omega}{\omega_{Km+n}}\right)} \right| \tag{8.35}$$

where ω_{ki} are the break frequencies for the numerator and denominator. If we were to use asymptotic analysis techniques, we would put the equation in the form of Eq. (8.35). To construct the asymptotic Bode diagram, we would take the log of both sides of the equation with a goal to put everything in terms of decibels for plotting

$$\log(|Z|) = \log(C_n) + \log\left(1 + \frac{j\omega}{\omega_{K1}}\right) + \log\left(1 + \frac{j\omega}{\omega_{K2}}\right) + \cdots$$
$$+ \log\left(1 + \frac{j\omega}{\omega_{Kn}}\right) - \log\left(1 + \frac{j\omega}{\omega_{Kn+1}}\right)$$
$$- \log\left(1 + \frac{j\omega}{\omega_{Kn+2}}\right) - \cdots - \log\left(1 + \frac{j\omega}{\omega_{Kn+m}}\right) \tag{8.36}$$

We may examine the second term for a moment in Eq. (8.36). For an asymptotic analysis, the first thing to note is at very low frequencies, $\omega \ll \omega_{k1}$, the term goes zero or,

$$20 \log\left(1 + \frac{j\omega}{\omega_{K1}}\right)\bigg|_{LF} \cong 20 \log(1) = 0 \tag{8.37}$$

Thus, at the break frequency, the log of the term is close to but not quite at zero.

$$20 \log(1 + 1)|_{BF} \cong 20 \log(1) = 0 \tag{8.38}$$

This is helpful particularly if one were to construct a Bode plot of each subsection on a piece of paper and then graph the entire impedance by adding each individual graph together. In our case, we may leave the "1" in place under the log function.

The second thing to note is when the angular frequency is by a factor of 10 greater than the break frequency, the magnitude ramps 20× the log of the function. For example, when the frequency is approximately 10× greater than the break frequency, the value becomes

$$20 \, \log(1 + 10)|_{10 \times BF} \cong 20 \, \log(10) = 20 \tag{8.39}$$

and for a 100× increase in the break frequency, the result is

$$20 \, \log(1 + 100)|_{10 \times BF} \cong 20 \, \log(100) = 40 \tag{8.40}$$

In terms of decibels, it is apparent that there is a 20 dB increase here for every 10× increase in frequency in the magnitude. Thus, the previous term increases to 40 dB/dec. The key will be to determine these slopes when plotting the equations that make up the PDN impedance.

One can show how the plot is created by using values for the VR impedance in our previous example. Plugging into Eq. (8.32) for the VR impedance, the result becomes

$$|Z_{21}| = \left| \frac{(j\omega L_1 + R_1) \left(j\omega L_2 + R_2 + \frac{1}{j\omega C_2} \right)}{\left(j\omega L_1 + R_1 + j\omega L_2 + R_2 + \frac{1}{j\omega C_2} \right)} \right| \tag{8.41}$$

This equation needs to be placed into the form of Eq. (8.36) to manipulate it for plotting. In this case, however, there are two terms: one in the numerator and one in the denominator which have *complex conjugate* pairs. To handle this, the form of the equation is altered slightly. This is the same as when we plotted the gain and phase for the stability of the VR in the controller section back in Chapter 2. The form we are looking for is

$$H(j\omega) = \left[1 + 2\zeta \frac{j\omega}{\omega_0} + \left(\frac{j\omega}{\omega_0} \right)^2 \right] \tag{8.42}$$

where ω_0 is the resonant frequency. The other form of the equation shows explicitly the complex conjugates

$$H(j\omega) = \left[\left(1 + \frac{j\omega}{\alpha + j\beta} \right) \left(1 + \frac{j\omega}{\alpha - j\beta} \right) \right] \tag{8.43}$$

Multiplying out results in the quadratic equation

$$H(j\omega) = \left[\left(1 + 2\alpha \frac{j\omega}{\alpha^2 + \beta^2} + \frac{(j\omega)^2}{\alpha^2 + \beta^2} \right) \right] \tag{8.44}$$

which shows that

$$\omega_0 = \alpha^2 + \beta^2 ; \zeta = \alpha \tag{8.45}$$

This is the generic form where plugging in for both the numerator and denominator allows for generating the Bode diagram. To construct the Bode plot requires simply solving for the coefficients in the equation. The next example illustrates the method using the formulae above.

Example 8.3 Given the electrical parameters for the impedance of the VR portion of the PDN in Table 8.2, construct the Bode plot using the formulae in this section using a math program. Break down each section and plot individually on a graph to show how the final impedance is constructed. Compute the main resonance frequency.

Solution:
We first simplify and plug into the formula in Eq. (8.36) so that we can plot the impedance in terms of decibels.

$$20 \log|Z_{21}| = 20 \log \left| R_1 \frac{\left(1 + j\omega\frac{L_1}{R_1}\right)[1 + j\omega R_2 C_2 + (j\omega)^2 L_2 C_2]}{[1 + j\omega(R_1 + R_2)C_2 + (j\omega)^2(L_1 + L_2)C_2]} \right| \quad (8.46)$$

Each subsection may be broken down into simpler shunt and series terms in the magnitude equation separately by treating each equation under the parentheses as a transfer function

$$20 \log|Z_{21}| = 20 \log \left| C_n \frac{H_1(j\omega)H_2(j\omega)}{H_3(j\omega)} \right| \quad (8.47)$$

or

$$20 \log|Z_{21}| = 20 \log(C_n) + 20 \log(H_1(j\omega))$$
$$+ 20 \log(H_2(j\omega)) - 20 \log(H_3(j\omega)) \quad (8.48)$$

where the transfer functions are

$$H_1(j\omega) = \left(1 + j\omega\frac{L_1}{R_1}\right) \quad (8.49)$$

Table 8.2 Values for Bode plot of Example 8.3.

Metric	Value
L_1 (series)	$1\,\mu H$
C_2 (shunt)	$100\,\mu F$
R_1 (series)	$2\,m\Omega$
R_2 (shunt)	$5\,m\Omega$
L_2 (shunt)	$500\,pH$

and

$$H_2(j\omega) = [1 + j\omega R_2 C_2 + (j\omega)^2 L_2 C_2] \tag{8.50}$$

and,

$$H_3(j\omega) = [1 + j\omega(R_1 + R_2)C_2 + (j\omega)^2(L_1 + L_2)C_2] \tag{8.51}$$

The results are plotted in Figure 8.20. First, the equation Z_{VR} is plotted in the form without any adjustments as in Eq. (8.32). The second part of the analysis, as shown, illustrates portions of the entire function broken up for the various subsections (transfer functions) of the impedance profile all in decibels.

Note that both $Z_{\log}$ and Z_{vr} line up essentially on top of each other as we would expect. The only difference between them was that Z_{vr} was plotted without simplifying the function as in Eq. (8.32), whereas the other was broken up into elements as in Eq. (8.36). To check that the broken up function was computed correctly (though the plot for $Z_{\log}$ shows that it is indeed correct), we solve for the resonant frequency ω_0. From Eq. (8.30), the angular frequency is

$$\omega_0 = \frac{1}{\sqrt{(L_1 + L_2)C_2}} \cong 104 \, \text{kHz} \tag{8.52}$$

Note that this lines up exactly with the resonant frequency in Figure 8.20. As another check, we plotted the impedance earlier using SPICE, but the x-axis was in *frequency* rather than angular frequency. Computing the frequency now we get

$$f_0 = \frac{\omega_0}{2\pi} \cong 16 \, \text{kHz} \tag{8.53}$$

Plotting now $Z_{\log}$ in terms of *frequency* instead of angular frequency as in Figure 8.20, we see the following results:

Comparing this with Figure 8.9 shows that the resonances line up perfectly as expected. Also, note that we can see a second resonance more clearly between 1 and 2 MHz. This is found by computing the resonance (minima) in the numerator as

$$f_0 = \frac{\omega_0}{2\pi} = \frac{1}{2\pi\sqrt{L_2 C_2}} \cong 1.46 \, \text{MHz} \tag{8.54}$$

Again, the result lines up with predictions, and thus our check is complete.

Before we move onto the other subsections, we should examine the results for a moment that we observed in Figure 8.21. The first resonance that we computed was at 16 kHz and had an impedance that was relatively high compared to where we wished the PDN to be (more on that will be discussed later). Thus, at first glance, one might be tempted to want to look at the filter of the VR and determine what could be improved. However, recall that the voltage regulator response (as mentioned) has a bandwidth that will likely be higher than

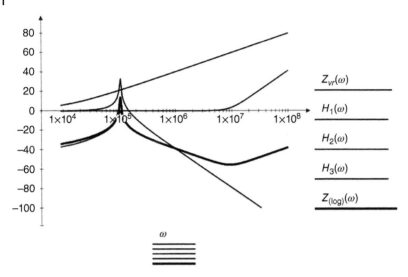

Figure 8.20 Components of impedance plot in decibels.

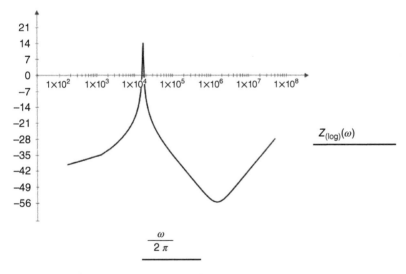

Figure 8.21 $Z \log(\omega)$ plotted in terms of frequency.

100 kHz. This means that the actual impedance in this frequency range will be lower than what the frequency analysis predicts because the loop response will limit the voltage perturbations, resulting in an effective lower impedance below the bandwidth of the converter. Thus, if we were to include the VR model in this analysis, we would find that the impedance looks back into the VR at this point below 100 kHz, which would be substantially lower than what is shown.

This is a very important consideration to the PI engineer since as the regulator's response increases, the requirement to manage the PDN below this bandwidth is lessoned, making the job of the engineer easier.

The other more important thing to note is where the second resonance resides and what sets the minima at this point. This frequency was determined to be at 1.46 MHz: This is very likely beyond the response of the regulator, and from this point on, the impedance increases at 20 dB/dec. This may be seen by examining the plot in Figure 8.20. H_2 and H_3 start to cancel each other out above 1.46 MHz (or about 10 MHz angular frequency). However, H_1 increases linearly at 20 dB/dec. Thus, this is due to the inductor and its parasitics. This is not surprising since the inductor is intended to operate effectively at a lower frequency and is not designed (typically) for this frequency response. This gives one an insight into what is required for the next stage in the sub-circuit design; that is, a shunt element that starts to affect the impedance near this resonant point to counter the increase in impedance due to the VR filter effects. Therefore, when the next subsection of the circuit model is added, we have now developed an insight to the required PDN design changes that we wish to achieve.

We can extend the computation for determining mathematically what happens to the impedance of the structure when we add a subsection to our base model looking back into the voltage regulator. One of the more elegant and easier ways to do this is to simply add it in as we move toward the load. This means that every time we add either a *shunt* or a *series* element, we recompute the new impedance and examine the effects. This is done very simply using our math program. This way one can see how each parasitic subsection impacts the effectiveness of the PDN.

For example, looking back at Figure 8.12 for a moment, if we add one of the series elements for a portion of the PCB parasitics, we get the following simple impedance equation:

$$Z_{VRP1sr} = Z_{VR} + Z_{P1sr} = \frac{Z_1 Z_2}{Z_1 + Z_2} + Z_{P1sr} \tag{8.55}$$

Adding a shunt element takes the previous impedance in parallel with the new shunt element

$$Z_{VRP1ss} = Z_{VRP1sr} \| Z_{P1sh} = \frac{Z_{VRP1sr} \left\{ \frac{Z_1 Z_2}{Z_1 + Z_2} + Z_{P1sr} \right\}}{Z_{VRP1sr} + \left\{ \frac{Z_1 Z_2}{Z_1 + Z_2} + Z_{P1sr} \right\}} \tag{8.56}$$

Having just plotted the impedance for the VR section, we can now examine what effect there is when we add the series element of the PCB subsection. This is shown below in the plot in Figure 8.22.

As illustrated in the figure, the addition of the series package parasitics shifts the impedance slightly. This is because it is a resistance and inductance that is

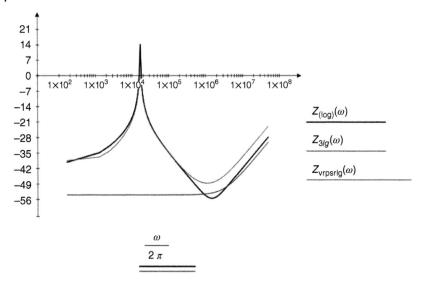

Figure 8.22 Impedance plots with series elements for PCB added.

added to the existing impedance directly. The inductance effects start to come into play in the higher frequency bands while shifting the resonance upward and slightly to the left. Though the impedance of the series elements are small, there is a marketable effect, especially in the higher frequency band. Often this indicates that the parasitics in the layout of the PCB need to be changed. This can mean lowering the inductance and/or resistance between the next set of capacitors on the PCB. It should also be noted that one cannot simply *add* the components in terms of the log equations. The computation of the series/parallel impedance elements must be performed first; then, the magnitude inside of the log function must be taken as:

$$20 \log(|Z_{VRP1sr}|) = 20 \log(|Z_{VR} + Z_{P1sr}|) = 20 \log \left(\left| \frac{Z_1 Z_2}{Z_1 + Z_2} + Z_{P1sr} \right| \right)$$

$$(8.57)$$

In general, one can build the impedance structure by simply adding series and shunt terms looking back into the source. To gain insights into how each subsection affects the PDN, using a math program to construct and graph these sections is a very elegant methodology and such programs may be used effectively for such tasks. By mathematically adding each subsection, the PI engineer is essentially constructing a *nested* fraction. As an example, look at the impedance network shown in Figure 8.23.

Figure 8.23 Simple impedance network.

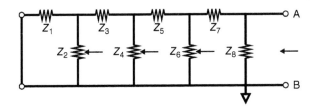

The impedance may be constructed by looking back into the network from port A–B. Grounding Z_1, the equivalent impedance looking back into Z_2 is simply

$$Z_{21} = \frac{Z_1 Z_2}{Z_1 + Z_2} \tag{8.58}$$

The impedance of the next subsection of the circuit model is

$$Z_{31} = Z_3 + \frac{Z_1 Z_2}{Z_1 + Z_2} \tag{8.59}$$

This is the impedance of the section looking back from Z_3 all the way to Z_1. In this case, it is the addition of the series impedance of Z_3 and Z_{21}. Thus, the next equivalent impedance is

$$Z_{41} = \frac{Z_4 \left[Z_3 + \frac{Z_1 Z_2}{Z_1 + Z_2} \right]}{Z_4 + \left[Z_3 + \frac{Z_1 Z_2}{Z_1 + Z_2} \right]} \tag{8.60}$$

This is the impedance looking back from Z_4 all the way to Z_1. Constructing the next series and shunts elements results in

$$Z_{61} = \frac{Z_6 \left(Z_5 + \frac{Z_4 \left[Z_3 + \frac{Z_1 Z_2}{Z_1 + Z_2} \right]}{Z_4 + \left[Z_3 + \frac{Z_1 Z_2}{Z_1 + Z_2} \right]} \right)}{Z_6 + \left(Z_5 + \frac{Z_4 \left[Z_3 + \frac{Z_1 Z_2}{Z_1 + Z_2} \right]}{Z_4 + \left[Z_3 + \frac{Z_1 Z_2}{Z_1 + Z_2} \right]} \right)} \tag{8.61}$$

and so on. Note that the original parallel impedance for Z_1 and Z_2 is nested within the parallel impedance with Z_3 and Z_4 and so on. Using a math program, it is simple to build up the entire impedance network by replacing the previous impedance by its equivalent. Substituting in for Eq. (8.61) results in

$$Z_{61} = \frac{Z_6 (Z_5 + Z_{43})}{Z_6 + (Z_5 + Z_{43})} \tag{8.62}$$

We may now add the next section of the circuit model and plot essentially Z_{41}. This is shown in the plot in Figure 8.24. Essentially what we have done is

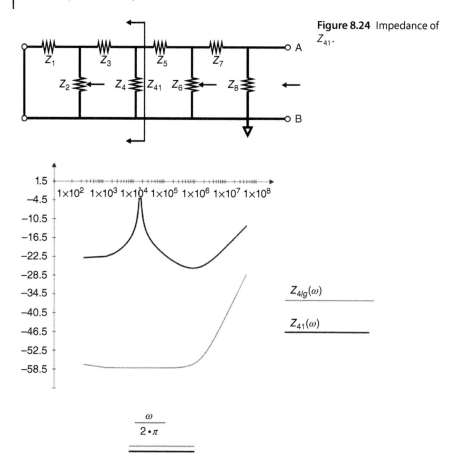

Figure 8.24 Impedance of Z_{41}.

Figure 8.25 Plot of Z_{41}.

plot the series and shunt elements for the PCB subsection that is adjacent to the VR subsection. This is the part shown in Figure 8.24.

Mathematically, we can add this and show the electrical effects in our impedance plot. This is shown in Figure 8.25.

Note that the parallel combination of the PCB capacitance of Z_4 with its parasitics reduced the overall impedance of the PDN up to this point. What is even more interesting is that the capacitor itself did not start to rise in impedance until about 1 MHz. This shows that this added capacitance is lowering the *effective* impedance in the lower to mid-frequency range that we have designed it for.

We may add the addition of the PCB capacitors Z_6 and the series elements of Z_5 now to see how the higher frequency effects have been mitigated. The addition of Z_6 should have a higher SRF or series resonant frequency as discussed earlier in the text. We can see this by plotting the impedance of Z_6 by itself

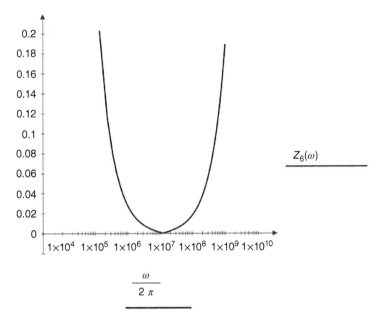

Figure 8.26 Z_6 impedance profile showing SRF of capacitor.

below. The plot shows the actual impedance rather than the value in decibels to illustrate the impedance at the SRF (Figure 8.26).

The impedance profile shows where the capacitor (and its parasitics) will have the greatest impact in lowering the effective impedance. This is around 10 MHz. The plot of Z_{61} (in decibels) in Figure 8.27 illustrates the effect that adding this capacitor has on the PDN in this frequency range.

Note that the impedance at 10 MHz has been lowered in the mid-frequency band as predicted. However, the impedance is increasing steadily at 20 dB/dec from that point on. Plotting now in terms of impedance, one can see clearly that this increase could present an issue at the higher frequency bands. Note that the impedance is already 20 mΩ at 100 MHz in Figure 8.28.

The next step is to add in the last impedance terms to see how they can reduce the effects of this increase in impedance. It is simple now to add the series interconnect parasitics and then taking this in parallel with the on-silicon capacitance near the load.

The plot in Figure 8.29 shows the impedance through the range of interest. It is clear that there is a large resonance around 100 MHz. This is consistent with our earlier analysis as well. The on-silicon capacitance was able to minimize the peak of this resonance by working against the parasitic inductances on the package and PCB. However, it has not completely removed it as is evident by the plot.

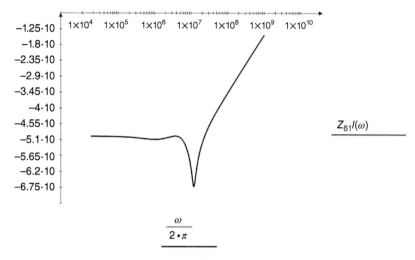

Figure 8.27 Impedance plot (decibels) of Z_{61}.

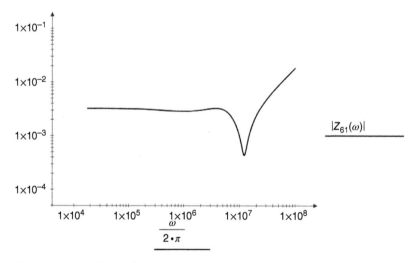

Figure 8.28 Impedance of Z_{61}.

The more interesting item about the result is where the resonance peak occurs. At first glance, one would have expected the resonance to be mainly due to the on-silicon capacitance and the package/silicon inductance. However, this path is very low, and thus a quick computation reveals that this is not the only effect on the resonance that we are seeing. In fact, it is clear that there are other components that are affecting this resonance. This is an important

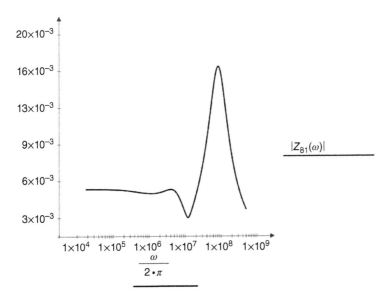

Figure 8.29 Plot of impedance for Z_{81}.

observation since it tells us that reducing the series inductance will not change the resonant point significantly. We can check this by reducing the series inductance by a factor of 10 to compare with.

Figure 8.30 shows this change. Note that the impedance is barely affected by the reduction in inductance between the load and the package capacitors and parasitics in this case. It should be pointed out here that this is true only in this example. In many cases, the inductance between the package and the load can be significant, and thus, having the layout engineers do their best to reduce this is always good practice. Nevertheless, this example does illustrate the importance of examining all of the effects of the system that can impact the peak resonances in a PDN design.

Where the excitation energy comes from is also a critical aspect of understanding FDA. It is not enough to simply design a PDN and then allow the excitations of the load drive the voltage deviations, which can affect the efficacy of the data and power. Having a clear understanding of the components in a PDN, where to place them, and where the resonances are (which may cause large voltage deviations) is certainly a critical part of the power integrity engineer's skill set. In Section 8.4, we examine how excitation occurs at the PDN load and source. This is done with respect to FDA but is obviously an important part to time-domain and silicon-level power integrity. Thus, we will investigate this further in the following chapters in addition to this one.

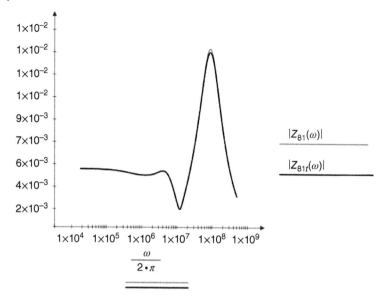

Figure 8.30 Impedance plots of Z_{81} with changing inductance in Z_7 by factor of 10.

8.4 Excitation in PDN Systems

Because the load in most silicon devices generates essentially random currents, engineers tend to look at the frequency dependence of the PDN independent of the source that excites it and assume that any signal below the clock frequency can put energy into the system. Though fundamentally this is true for high-frequency clock synchronous systems, the fact remains that the excitation from the load does indeed generate specific spectral components as well, which can either be mitigated or amplified by the PDN itself. Moreover, these harmonics are often very repeatable given the activity around the workloads that are exercised in the active logic. For this reason alone, it is instructive to examine the harmonics of the load and determine the potential behavior, in particular, worst-case scenarios that may exacerbate the noise and droop generated in the system.

High-frequency data processing activity, particularly those associated with processor cores and groups of cores, is quite complex. A general processing core operates synchronously to a clock source, but determining exactly which groups of logic inside the core are active, and to what level, at any one time can be a difficult task. In most cases, design groups focus on the performance aspects of these units and less on the power or current consumption, at least to the detail that they focus on the performance. Moreover, correlating the activity of the processing units to the actual current seen near these loads on the silicon is just as difficult to predict. Additionally, showing the load behavior of a processing unit can reveal certain design aspects of the processor which most organizations wish to avoid. This is why many companies generate representative load profiles for their power converter suppliers and other groups rather than show actual simulated or detailed measurements of the currents correlating to processing activity.

This usually makes the job of the power integrity engineer more difficult since predicting the response of the PDN to load behavior without having an in-depth understanding of the load is challenging to say the least. Fortunately, with respect to the power distribution, there are specific current profiles that may excite a PDN more than others. Moreover, knowing the basic mode of operation and activity in the load, in general, can be very helpful in revealing what can happen to the voltage behavior (droop) in a system PDN.

In this section and chapter, we intend to describe some basic fundamental behavior of the load without getting into the specifics of the processing units activity. As a matter of fact, power integrity engineers do not really require a deep understanding of the activity of a processing unit, only how the current that is generated by this activity can manifest itself so that engineers may mitigate the noise on the PDN appropriately. This subject will be delved into in more detail in Chapter 10 when on-silicon power integrity is discussed. As a note, it is not required for readers to spend time reading through Chapter 10 before we address into this topic. The following discussion is intended to be focused around a specific load profile excitation to aid the engineer in determining how to mitigate noise generated at various frequencies. In that vein, we start with a given load profile and then subsequently analyze it in the frequency domain.

A load profile in the time domain has explicit frequency-dependent characteristics as expected. The time dependence of the load profile is investigated in Chapter 9 in some detail as well. However, it is good to show a typical load profile first before generating the frequency spectrum.

A load profile, by definition, is exactly that; it is a profile of the current near the logic and is intended to represent worst-case conditions. Its function is to illustrate the shape of the current from that load such that an engineer may diagnose the harshest response in one form or another. Figure 8.31 shows an example load profile of a possible processing unit that a power integrity engineer might typically see from a developer. Usually, this is a starting point for their analysis.

As illustrated earlier, normally, the load comprises a series of pulses (which often vary) that have a set of rise and fall times associated with them where the harmonic content is intended to excite some of the higher frequency resonances in the system. Along with that, a step response is usually generated at some point in the profile, which is intended to cause a longer-term droop event where the voltage regulators response comes into play. Normally the rise and fall times and amplitudes are defined to a point to allow the power integrity

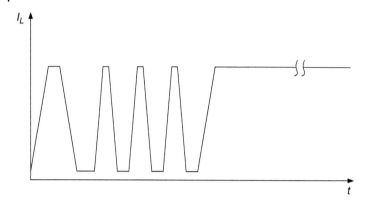

Figure 8.31 Simple load profile.

engineer to gain an insight into the voltage power bus behavior. Note also in this figure there is a slightly longer pulse that has a somewhat slower rise and fall time but has the same amplitude as the other pulses. This is usually put into the profile to excite other aspects of the PDN. Also, note the current does not start at zero. This is because to create the largest amplitude of current in the system (or change in current) requires the processing unit to be in the highest *p-state* and that both the leakage current and minimal activity in the core create some finite current offset in the system. Since this is typically an AC response issue, many engineers ignore the DC offset. However, when analyzing power loss in the system, it is a key metric that cannot be ignored.

Engineers understandably usually ask key questions about the load profile when they are first given one. The first question is how was it generated? This question we will address in the Chapter 9 in more detail, however. As mentioned previously, the load profile is directly related to the activity of the processing units. This correlation is key to how the load profile is generated of course.

The next question is around whether the load profile actually represents the load behavior correctly and if such a profile is even possible given the activity of the digital logic. In most cases, the answer to both is normally "yes". The load *can* (in general) generate a profile similar to the one shown. This is highly dependent upon what the actual rise/fall times and amplitudes are of course, which are not shown in the figure since these numbers are highly dependent upon the device creating them. Additionally, developers of the profile add in *margin* due to the difficulty in predicting the worst-case profile a load can generate. The reason for this is quite simple; software in most industries is changing continually and it is not clear how some set of low-level commands in the processor in some specific order can generate a higher current demand than the previously defined worst-case workload that is measured at the time an initial load profile is generated or simulated. This is always true, even in systems that have gone into production and are on the market for some time. There is always the possibility of an application and system condition that can create some current profile that was not presently seen or predicted by the design team. This is fundamentally why developers of the current loads often error on the conservative side when creating such patterns.

The third question that arises is with respect to the series of pulses, quantity, and order. For example, one may wonder why there isn't an infinite set of short current pulses (like the three in the figure) that causes continual excitation of the PDN at a given frequency. The reason the number of pulses is finite (and usually limited from a few to may be 8 or 10 at the most) is because normally the machine (or machines) generating such pulses typically cannot generate high current pulses for an endless period of time. This is because to generate a large amplitude, along with a high-frequency set of pulses, usually requires many logic units to be active and to synchronize their activity within a certain time window. Once that activity is done, there are latencies in the system requiring the machine to get more data and instructions to restart the process in order to keep such activity going. During this time, the machine will have a much lower activity and amplitude, and the duration of the pulses will change substantially (toward that of lower amplitude and longer time period). This is true in all complex processors. Latencies in the form of memory fetching as well as scheduling is a common source of this delay. It is still true that more than one core can be loading down a given power bus which would exacerbate this problem. However, in that case, the load profile would assume the cores were somewhat synchronized. The result would be that any profile with less than all of the cores active and synchronized could not generate a worst-case profile.

Once the time-dependent profile is generated, the power integrity engineer can examine it in the frequency domain to see the harmonic content. This is useful since alignment to the impedance profile of the PDN can often show if there should be concern with respect to the higher impedance resonances in the PDN. We can take a given profile in the time domain and generate an FFT of it to see where the spectral energy resides. In Chapter 6, we developed a simple Fourier series methodology which helped us convert a time-dependent signal into the frequency domain. In the next example, we use SPICE to generate the spectral components of a given load profile.

Example 8.4 A developer generates a load profile as shown in Figure 8.32 along with the characteristics of the profile in the adjoining table. Generate the spectral profile in the frequency domain using SPICE for the given time-domain load profile and speculate on the characteristics of the PDN from the data (Table 8.3).

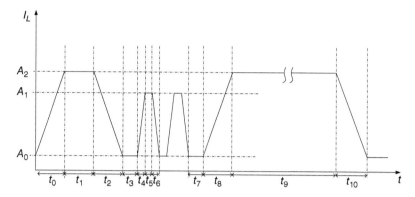

Figure 8.32 Load profile for example.

Table 8.3 Values for transmission line computation of Example 8.4.

Metric	Value
A_0	0.5 A
A_1	2 A
A_2	3 A
T_0	40 ns
T_1	50 ns
T_2	40 ns
T_3	10 ns
T_4	10 ns
T_5	10 ns
T_6	10 ns
T_7	20 ns
T_8	40 ns
T_9	10 μs
T_{10}	40 ns

Solution:
Before generating the waveform in a program, it is instructive to look at the profile first and make a few observations. The first one is that there are three distinct pulses in the profile. The first and third have the largest amplitudes. However, they also have slower rise times than the two in the middle, which are identical. The last one is obviously representative of a step response, and it is clear that the spectral energy there should be within the bandwidth of the

voltage regulator. However, the rise time is fast enough that most of the energy will need to come from the storage elements in the frequency range of the harmonics that are created by such a pulse.

There are a few ways to solve this problem. The easiest of course is to generate a PWL file in SPICE and then take the FFT of the waveform. This is what we will do below. It is simple to generate the SPICE file for the load profile and then simply plot the current waveform. We can measure the current from the source and then look at the spectral components of the waveform. There are a multitude of options for SPICE, and we choose the following to zoom in on some of the salient behaviors of the harmonic content:

- time range: 1 ms
- number of data points: 262 144
- window function: Gaussian
- Gaussian parameter, σ: 0.9

The plot of the waveform over the range of 12 μs is shown in Figure 8.33.

The FFT is shown in Figure 8.34. First, we need to note the strong low-frequency harmonic in the 250 kHz range along with its subharmonics. If the waveform truly resembled that of the pulse function, we would expect strong harmonic content from the second harmonic on up. Note that if this were repetitive, then the PDN would need to decouple this. The good news is the low-frequency VR filter is designed to handle this behavior. As the harmonics get higher in frequency, we note that there is a strong harmonic in the 10 MHz regime as well as some in the lower frequency bands. As the frequency content and spectral energy increase in this range, the filter system

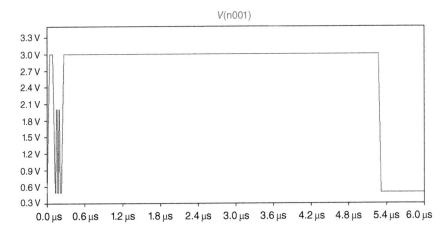

Figure 8.33 Pulse function over 6 μs.

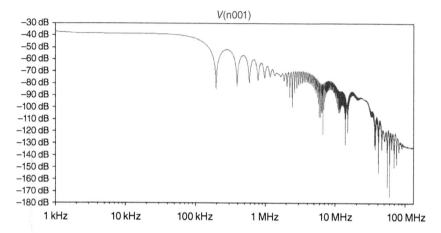

Figure 8.34 FFT of pulse waveform.

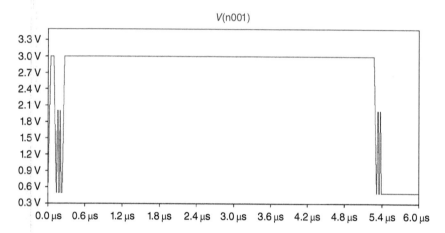

Figure 8.35 Addition of two pulses at end of step response.

will need to handle this energy and reduce the effects in the time domain. The higher frequency components are of interest here as well.

The two short pulses do not show a large spectral content relative to the rest of the waveforms. However, if these pulses were repeated, what would we expect to see? This is illustrated in the waveforms below where we have added two more pulses at the end of the step response (Figure 8.35).

Note where the harmonic content is for the waveform in the figure and where the frequencies of interest are now? We can speculate on where the impedance needs to be to minimize the droop events if a current profile like this was generated at the load of a real device. From our previous analytical work, it would be clear that harmonic energy in the 100 MHz regime could still be problematic

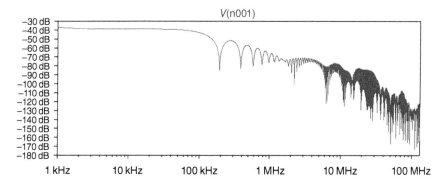

Figure 8.36 FFT with two additional pulses at end of step response.

due to the resonances in that area. Thus, good high-frequency decoupling near the load would also be important here (Figure 8.36).

It is possible to plot this current profile analytically as well. This would reveal a more distinct set of harmonics. However, the goal here was to see how the spectral energy was distributed across the frequency window of interest. We have seen that while some of the energy may be controlled by the VR and its filter (and loop response), the higher order harmonics would still need to be contained.

To examine the frequency content further around the noise source, we can use techniques discussed in Chapter 6. Here, we generate the spectral amplitudes of the signal and assume that it is periodic in nature.

Let us assume that we have a square wave current source at the load that resembles the waveform shown in Figure 8.37.

We know from Chapter 6 that we can represent the spectral amplitude by the following Fourier coefficient.

$$A(n) = \frac{A_0}{n\pi}\left(1 - \cos\left(\frac{2\pi n\tau}{T}\right)\right) \tag{8.63}$$

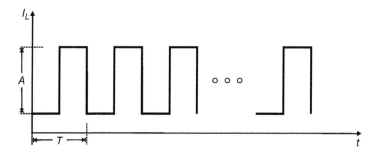

Figure 8.37 Simple square wave load profile.

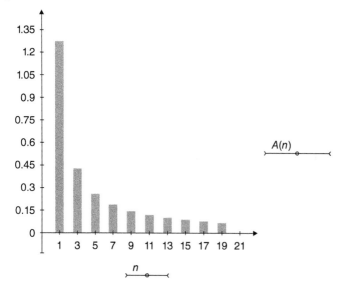

Figure 8.38 Spectral amplitudes of square wave load profile.

This is only true for a half-wave symmetric wave or a 50% duty cycle square wave. We may plug in for the constants to get the spectral amplitudes of the square wave over the first 10 harmonics to look at the spectral energy.

The profile in Figure 8.38 shows where the spectral amplitude of the load current is coming from for a 50% duty cycle function. As expected, the fundamental has the largest amplitude. Another way to look at this is in the frequency domain with actual values as in Figure 8.39.

Note that we only have the *odd* harmonics in this instance. We can change the duty cycle, however, to see the effect on the amplitude of the harmonic content. If the duty cycle is now 60% instead of 50%, we first note that the waveform itself has changed (Figure 8.40).

Here, one can see that *even* harmonics are now present. The signal from the Fourier spectrum is plotted below (the actual signal is a perfect square wave with an infinite Fourier spectrum). The more important thing to note is that with a fundamental of 1 MHz, we see that there are significant even harmonics with energy much further out in time. If the measurement point is close to the load and there is very little dampening in the system, then this tells us we can have significant energy out to reasonably high frequencies.

However, we should note that this is assuming a square wave function with a finite harmonic spectrum. In fact, because of this, the resulting time-domain signal will have a finite edge rate. Nonetheless, we can gain more insight if we focus on decomposing a signal that has finite rise and fall times associated with it. This was the goal of the previous example. Here, we will take one of the

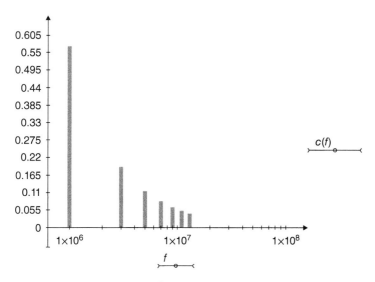

Figure 8.39 Spectral content of 50% duty cycle current wave.

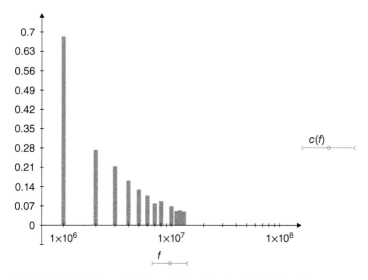

Figure 8.40 Square wave with $n = 13$ harmonics, 60% duty cycle.

high-frequency pulses from that example and assume a finite rise and fall time (Figure 8.41).

We generate two plots for the envelope function. The first is with an edge rate of ¼ the period, which is fairly slow. The frequency is 40 MHz. The envelope below clearly shows that the higher order harmonics are much lower than

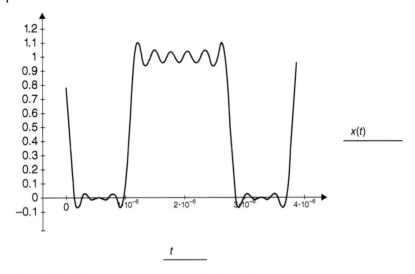

Figure 8.41 60% duty cycle signal compiled from a Fourier series.

the fundamental. In fact, the graph needed to be scaled to even show them (Figure 8.42).

The next plot shows the same scale when the edge rate is increased to $T/40$ (Figure 8.43).

The figure illustrates what can happen when the edge rate of a reasonably high-frequency periodic function has a fast edge rate generated at the load. The harmonic content and energy can require the PI engineer to add decoupling that specifically attenuates the energy of the higher order harmonics simply because they can effectively be generated by the load.

When a PI engineer is examining the noise at the load, typically they look at the highest clock frequency that the processing unit can run at and then try to imagine that the harmonic content in this frequency range is possible. In fact, the amount of energy at the operating frequency is normally small compared to other lower harmonics of the fundamental clock. The reason is because with every clock pulse, a limited amount of switching occurs in the data path, and there is a probability of some of that logic sinking current while others are sourcing it. Because of this, the average current at the clock frequency is typically not that high. However, over time, the current can ramp up at a rate that lines up with the peak resonance of the highest harmonic of the PDN. This is where the edge rate of the current pulse can be significant. It is not the clock frequency that is of concern, but the lower order harmonics that lead to a current pulse with a significantly fast edge rate. This is one reason why it is important that PI engineers question the current plots that generate noise in the range

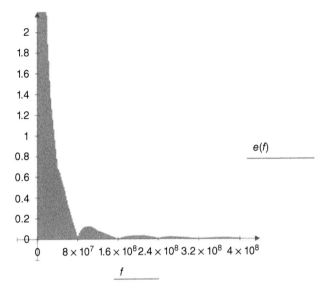

Figure 8.42 Envelope function with $f = 40\,\text{MHz}$ and edge rate of $T/4$.

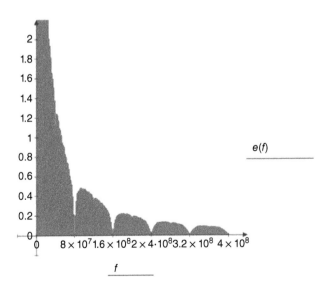

Figure 8.43 Envelope function with $f = 40\,\text{MHz}$ and $tr = T/40$.

of the highest harmonic in the impedance profile when they are given a load profile. If the load profile does not comprehend this, then it is possible that the power integrity engineer may not necessarily capture the worst-case droop in the system near the load.

In Section 8.5, we examine some basic techniques to optimize the PDN in the frequency domain.

8.5 PDN Optimization

One of the more important aspects and learnings from FDA is to determine how to optimize the PDN before moving into analysis in the time domain. The term *optimize* though should not be overstressed, however. As discussed earlier concerning the damped transmission line, it is usually impractical to try to achieve a perfectly damped transmission line effect with a PDN. Moreover, without knowing the exact excitation frequencies as discussed in the last section, it is nearly impossible to keep from having some voltage droop within the system PDN design, particularly with high-performance loads. Thus, the next best thing is to take the existing PDN and determine where the frequency-dependent behavior can be improved. Then, once this is done, the PI engineer can examine the new design through time-domain modeling to further tune the structure. The final goal is to minimize the voltage droops throughout the excitation load and frequency range of the load.

In reality, the impedance as a function of frequency is just that: a function. Thus, it has both minima and maxima at particular resonant points in the structure. For system loads that are clock synchronous, which is the predominant form of excitation from the digital loads that are dealt with in the industry, there are fundamental harmonics, which are of practical concerns in which the PI engineer should focus on. From our previous discussion on excitation sources and harmonics, it should be noted that there is indeed an infinite number of potential patterns that can excite a PDN. However, from a practical perspective, it is not uncommon to focus on the ones in the system that appear to cause more noise than others do. This is fundamentally related to the operation of the processing unit, in particular, harmonics from the main clock frequency and activity as it relates to current excitation. Breaking these patterns down into harmonics that excite the PDN resonances is the first step. This should be done in concert with the PDN resonances that exist in the structure the PI engineer is given. The reason is because if the key excitation frequencies align with certain resonances, it gives the designer a practical starting point to address the noise.

Finding the key maxima and minima can be done by simply plotting the impedance of the PDN and observing where the resonances exist and then through an iterative process, add in passive networks to minimize the effects. Another way is through differentiating the impedance profile and examining the behavior analytically. The goal here is to make the function go to zero throughout the band of interest. This can be cumbersome when there are many elements in the path unless a math program is used to perform the work for us. The key here is once the impedance plot is realized, it may be differentiated, plotted, and then examined quite easily. Once the function is plotted, the key is to examine the derivative slope changes in the structure to determine where the changes occur most dramatically before and after the resonances. Of course, the second derivative can also be used analytically to determine which are maxima and which are minima, but the plot of the first derivative is normally much more informative and is readily available with minimal manipulation. Once these plots are found, the next step is to add in new elements to zero out the derivative of the impedance plot such that the slope tends to approach the axis for most of the frequency band. We may do this analytically first for the simpler portions of the PDN to illustrate the methodology.

In Section 8.3, we introduced some analytical techniques for analyzing the PDN behavior section by section. The impedance of the VR filter was found to be

$$|Z_{21}| = \left| R_1 \frac{\left(1 + j\omega\frac{L_1}{R_1}\right)[1 + j\omega R_2 C_2 + (j\omega)^2 L_2 C_2]}{[1 + j\omega(R_1 + R_2)C_2 + (j\omega)^2(L_1 + L_2)C_2]} \right| \tag{8.64}$$

From our previous work, we know that placing the filter next to the load directly without adding lower impedance components (at higher frequencies) to the PDN will result in large high-frequency droops. Let us assume for the moment, however, that all we have is the VR filter and that we are starting the process of building our PDN from there. This is often where the PDN designer often begins anyway. There will be a resistance and inductance that will be present to represent the interconnect from the VR filter to the load, but for now we will ignore this. Our first objective is to determine where the VR filter falls short of being able to do the job throughout the frequency band (if it is adequate, our job would be done and we could simply move on to the next project).

We can place the equation into a form that is more easily manipulated. Notice that taking the magnitude of a complex function is simply taking the square of the real and complex components under the square root of the function or

$$|Z_{21}| = \left| \frac{A + jB}{C + jD} \right| = \sqrt{\frac{A^2 + B^2}{C^2 + D^2}} \tag{8.65}$$

Performing the math in Eq. (8.64) yields

$$|Z_{21}| = R_1 \left| \frac{(1 - \omega^2 K_1) + j\omega(K_2 - \omega^2 K_3)}{[1 - \omega^2 K_4 + j\omega K_5]} \right| \tag{8.66}$$

where the constants K_x are defined as

$$K_1 = L_2 C_2 + L_1 C_2 \frac{R_2}{R_1}, \quad K_2 = R_1 C_2 + \frac{L_1}{R_1}, \quad K_3 = L_2 C_2 \frac{L_1}{R_1} \tag{8.67}$$

and

$$K_4 = (L_1 + L_2)C_2, \quad K_5 = (R_1 + R_2)C_2 \tag{8.68}$$

Equation (8.68) can be expanded out as in Eq. (8.66) where we substitute for the constants now

$$|Z_{21}| = R_1 \sqrt{\frac{(1 - \omega^2 K_1)^2 + \omega^2 (K_2 - \omega^2 K_3)^2}{(1 - \omega^2 K_4)^2 + (\omega K_5)^2}} \tag{8.69}$$

The next step is to take the derivative with respect to ω. Using the quotient rule to differentiate the function in Eq. (8.69)

$$\frac{df}{dg} = \frac{gf' - fg'}{g^2} \tag{8.70}$$

where we have made the following substitutions:

$$f = R_1 \sqrt{(1 - \omega^2 K_1)^2 + \omega^2 (K_2 - \omega^2 K_3)^2} \tag{8.71}$$

and

$$g = \sqrt{(1 - \omega^2 K_4)^2 + (\omega K_5)^2} \tag{8.72}$$

Taking the derivative of function in Eq. (8.69) for the numerator yields,

$$f' = \frac{R_1}{2\sqrt{(1 - \omega^2 K_1)^2 + \omega^2 (K_2 - \omega^2 K_3)^2}} \times [-4\omega K_1 (1 - \omega^2 K_1) + 2\omega (K_2 - \omega^2 K_3)^2 - 4\omega^3 K_3 (K_2 - \omega^2 K_3)] \tag{8.73}$$

and for the denominator

$$g' = \frac{1}{2\sqrt{(1 - \omega^2 K_4)^2 + (\omega K_5)^2}} [-4\omega K_4 (1 - \omega^2 K_4) + 2\omega K_5] \tag{8.74}$$

The square of the function g is simply

$$g^2 = [(1 - \omega^2 K_4)^2 + (\omega K_5)^2] \tag{8.75}$$

We can simplify the combined equation or we can let our math program do the work and then plot the function. As stated before, if we set the function to zero, we could solve for the minima and maxima for values where ω crosses the axis. However, simply plotting the function will reveal the information that we are interested in.

To start, let us look at the original impedance plot again. The magnitude of the impedance is shown once more in Figure 8.44. There is a very sharp increase

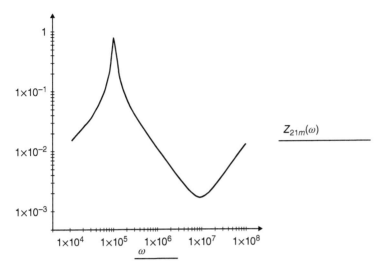

Figure 8.44 Plot of impedance Z_{21}.

around the angular frequency of 100 kHz. The question is does this resonant cause issues in the overall design of our PDN and can something be done to improve it? As we determined earlier, the response of the voltage regulator is clearly able to reduce this peak and control the impedance through the loop bandwidth, so we do not have to worry about this resonance. The second resonance is a minimum, which shows that there is an increase starting at about 10 MHz and increasing at 20 dB/dec. This is to be expected since it is dominated by the inductance of the circuit. Certainly, the minima here is a good thing, and there does not appear to be any issue with this resonance. However, the rise in impedance is rather sharp, and at some point, an excitation above a certain frequency could be an issue. One of the ways to realize this is to assume that we have an AC current at a given frequency and then determine what the droop would be at that frequency. If there is an excitation at 100 MHz of 10 A, the result would be

$$IZ \cong 10 \times 10^{-2} = 100 \, \text{mV} \tag{8.76}$$

This is obviously an approximation since the voltage droop is dependent upon the initial state of the system, and we have made no assumptions about this. However, it does give us an idea that if we saw an excitation at this frequency, the droop could be quite significant. Let us now assume that we want to change this. By examining the slope in this region, we can find out what slope we would need and where to apply it. It is reasonable to assume that at some point, a capacitor that countermanded this slope and that made the slope in this range go to zero would be beneficial toward meeting our goal. Let us assume that we wish to reduce this from 100 to 30 mV. Thus, this would mean changing the impedance here from 10 down to $3\,m\Omega$, which is slightly higher than the minima of the function at $10\,MHz$.

It is clear from the plot of the impedance that it increases steadily from the minima which we assume will not be acceptable. One way to address this is to add in a capacitance that has a negative slope that corrects for this. Mathematically, we know that we are trying to put an impedance in parallel that results in an impedance that is flat for a given frequency range. Mathematically, this looks like the following:

$$Z_N = \frac{Z_{21}Z_C}{Z_{21} + Z_C} \cong 3\,m\Omega \tag{8.77}$$

Our goal is to find the function Z_c that yields a *flat* impedance. Solving for the *compensation* impedance yields

$$Z_C = \frac{Z_{21}Z_N}{Z_{21} - Z_N} \tag{8.78}$$

This is now a new function that comprises the impedance of the parallel function and the new impedance, which is simply a flat resistance (for now) that we wish to achieve.

Figure 8.45 is a plot of the derivative of the function. As can be seen in the plot, it is readily easy to see where the function crosses the horizontal axis in one of the cases. This is for the lower frequency resonance. Though the peak of the derivative is small, it rises quickly. Moreover, we can also see where the function monotonically increases before and after these minima and maxima. By plotting this and the original function, it is evident now where improvements would be needed to flatten out the frequency response of the function.

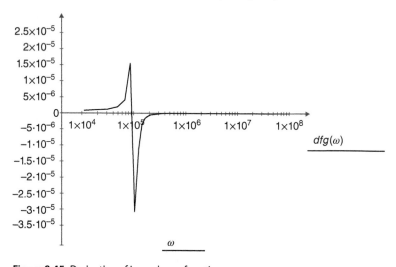

Figure 8.45 Derivative of impedance function.

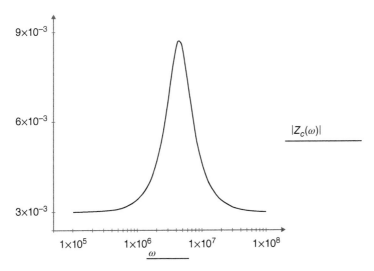

Figure 8.46 Plot of compensation function $Z_c(\omega)$.

The goal is to now add in a capacitance that addresses this increase. As in an earlier part of the chapter when we broke down the sub-circuits section by section, we now add in a capacitor. We now plot our new compensation function to see the results (Figure 8.46).

As expected, the impedance is essentially flat throughout the band of interest. Our job now is to construct a similar graph using capacitor networks to align with the shape of the function $Z_c(\omega)$. If we have an infinite set of capacitors and room on our circuit board, we can certainly construct something very close to our new compensated function. By shifting the SRF of the capacitor slightly upwards in frequency and maintaining the impedance across this band, one can realize the compensation function.

$$\frac{1}{Z_C} = \frac{1}{Z_{n1}} + \frac{1}{Z_{n2}} + \frac{1}{Z_{n3}} + \cdots + \frac{1}{Z_{nk}} = \sum_{i=1}^{k} \frac{1}{Z_{ni}} \tag{8.79}$$

The result is the sum and then the inversion of the function. As k gets large, the function starts to approach the exact value for Z_c. Each value for Z_n has a shift in its resonance. Thus, there is a slight shift upward for each Z_n or

$$\omega_{0(i+1)} = \omega_{0(i)} + \Delta\omega_0 = \frac{1}{\sqrt{L_i C_i}} + \Delta\left(\frac{1}{\sqrt{(LC)}}\right) \tag{8.80}$$

Thus, we want a small increment of the resonant frequency of each parallel circuit as we move upward in frequency. The practical side of this is clear; however, there is a fundamental limit to the SRF of the capacitor network as well as the parasitics internal to the circuit. Moreover, engineers cannot place an infinite

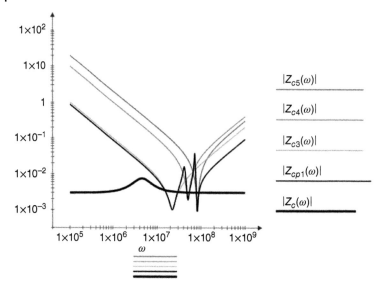

Figure 8.47 Added networks for compensation.

number of capacitors within a region of a device. It is therefore good practice to create a series of parallel circuits and then tune the overall network once the physical and practical limits have been achieved.

This methodology may be more effective than simple trial and error, which very often engineers begin with when trying to tune and optimize a PDN within a certain bandwidth. The final PDN looks like Figure 8.47 when we add in the three capacitive networks from Figure 8.46.

As it is, there are practical limitations to this. Instead, we will add three capacitor networks each with different resonant points as we move from the slope of our new function out to the higher frequencies. The three capacitive networks are captured in Figure 8.47 along with the parallel combination and our compensated ideal network.

It is of note that our goal here was to follow the compensation as close as possible, and though the new network has multiple resonances in it, it is clear that we have taken steps toward improving the response in our PDN by adding this new compensated network. The final result is shown in Figure 8.48. Though there are multiple resonances now above the lower resonance, the total impedance has been lowered and the frequency response increased.

Since we are interested in the negative going slope on the right side of the graph, we now add a series of capacitive networks with a slope similar to that in the graph in Figure 8.48.

Section 8.5.1 delves into examining the effects over time of a PDN and what changes can occur in the components in the frequency response.

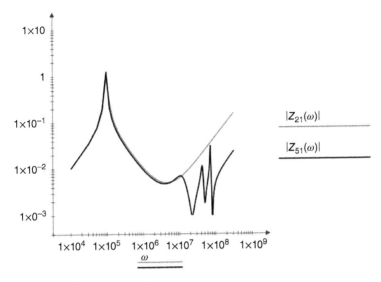

Figure 8.48 Compensation function added to Z_{21}.

8.5.1 Monte Carlo Analysis

In the end, the power integrity engineer is concerned about the efficacy of the voltage droop in the system in the time domain due to a large current excitation as we have seen. This will be examined in more detail in Chapter 9. However, in the frequency domain, because the PDN impedance does not remain static over time, it is of interest to the PI engineer to understand the relationship between the frequency response and the time-dependent effects. Normally, engineers focus only on the start of life of the PDN effects and tend to forget or neglect the decay of components over time and/or the effects of temperature. Even the circuit board parasitics change as a function of temperature, and though the electrical behavior may be appropriate for a given design when using start of life data from the component vendors, after time, capacitors degrade and the PDN impedance shifts around to some extent. As examined in earlier chapters, the capacitance of a multilayer ceramic chip (MLCC) may change by 50% over a few years of use, and thermal changes may cause the capacitor to drift nearly as much. It is critical then to examine this behavior numerically to see how the PDN is affected overall. In particular, it is critical to see how the resonances shift and increase to make sure that in the end the power delivery path still meets the required specifications.

One very power tool at the disposal of PI engineers is the ability to run Monte Carlo investigations on the PDN to determine how tolerances and component variations due to temperature and time affect the efficacy of the PDN. Normally, this is done with SPICE where each component is varied over its range and a number of plots are generated. When these profiles are generated, normally two effects are observed: first, the resonance frequencies shift around and second the amplitudes of these resonances change.

To the casual observer, when examining these changes in a log-dependent graph which is often the case when analyzing PDN impedances, the changes can sometimes appear small and inconsequential. However, even subtle changes can have significant effects with respect to electrical behavior.

One such effect is on the frequency response of the voltage regulator. If the capacitance of the main filter and a number of the higher frequency decoupling drift in a certain direction, a shift in a lower frequency resonant to a higher frequency may result in that resonance moving out of the loop bandwidth of the voltage regulator. If the PDN designer is counting on the VR to remove certain frequency droop events, a shift in resonance may countermand the ability of the regulator to address this.

Even more critical is when a shift in a resonance causes an excitation at a subharmonic frequency that the load happens to generate on a regular basis where the system guardband is then affected. The result could not only lead the design team to shift the guardband of the system higher, which would cause the load to burn more power, but might also move the device operation windows for the manufacturing of the product. This could affect yields in such a way that the impact is not only electrical, but there could be a direct negative impact on the financial side, resulting in fewer devices passing the reliability standards.

Thus, having a good understanding of the potential variations in components over time before the design goes into production is prudent and with complex electronics today a necessity.

In reality, however, Monte Carlo analysis has more than one goal. The first is to ensure that the given design has enough margins over the life of the product that even when the components vary over time and temperature the PDN will still satisfactorily meet the specifications. The second aspect to this is to understand which components may make the system more susceptible to electrical effects. It is true that MLCCs and other decoupling capacitor technologies will not vary at the same rate or have the move in the same direction as a function of temperature. This understanding is crucial toward determining when a failure can occur over the life of a product.

Additionally, when it comes to time-domain voltage droop, the worst-case effects do not necessarily occur when the capacitance of all the networks has reached their lowest tolerance points; meaning that even if all the capacitors have drifted to their lowest values, the resulting electrical behavior in the PDN may not be the poorest. This is because there are sub-resonances that can occur between devices that can result in larger droop events when a combination of changes to the PDN occurs over a range of tolerance effects.

So what is Monte Carlo analysis? It is a statistical analysis based on the probability an element or system can vary over a given set of parameters, resulting in a distribution of possible outcomes. For circuits, it is the variation of a component or set of components statistically over time and other environmental and design factors that can result in a distribution of variation in performance. In the case of a PDN, the parasitics and capacitor components themselves can vary over time, temperature, and electrical excitation. At first thought, it is easy to convince oneself that since Monte Carlo algorithms are inherently based on random variations, it will not necessarily capture the variations in capacitors (for example) of a PDN mainly because MLCC's tend to degrade in one direction over time. Though this is generally true, the fact remains that the value of capacitance can change in both directions as a function of electrical excitation (AC voltage, for example) as well as temperature. Because of this, it is good to assign a tolerance to each capacitance and then use the random variation of Monte Carlo within this tolerance to determine the variation of the PDN over this range. Moreover, as we have seen in earlier chapters, the initial capacitance of a device is also tolerance based, which can result the value of the device starting at a statistically random value based on this tolerance. Thus, it behooves the engineer to determine this distribution in order to become familiar with the change in the PDN and its electrical properties. This can also become important when power loss in the PDN is examined in Section 8.6.

To simplify the analysis, it is best to look at the variation of one particular component within the PDN and then vary the rest of them statistically to look at the total effect. We start with the circuit from earlier in the chapter as a baseline and examine the impedance of the function over frequency (Figure 8.49).

For now, we will assume that only the inductor varies as a function of its tolerance. Most power inductors have a $\pm 20\%$ variation over time and temperature, so we will use this value to determine the change in the impedance plot. Running through a large number of iterations allows us to examine the total shmoo of the impedance profile. This is shown in the plot (Figure 8.50).

We can see clearly the effect on the impedance profile when we vary just one of the components, in this case, the inductor. As expected, it has the greatest effect down in the lower frequency regime of the plot. However, as we already know, this effect will be *nulled* out by the loop response of the voltage regulator. Thus, in essence, the variation in the inductor should have little effect.[2]

Now we vary the rest of the components to see the overall effect on the impedance profile. We recall from Chapter 4 that the capacitance of an MLCC can vary over its life significantly and lose half of its value. Including temperature drift, we can safely assume that for some devices, 60–70% change is possible. Putting in these tolerances to our Monte Carlo analysis gives us the new impedance profile.

---

2 Except in the loop bandwidth which we know from Chapter 2, indeed the Bode plot will change. Most of the time, VR engineers take these variations into account in their designs.

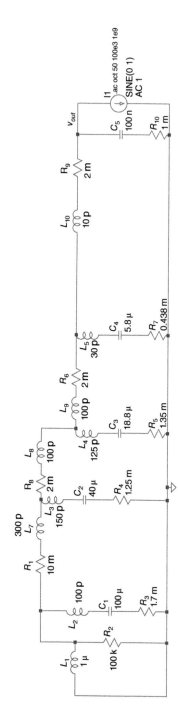

Figure 8.49 Basic circuit of PDN, AC analysis.

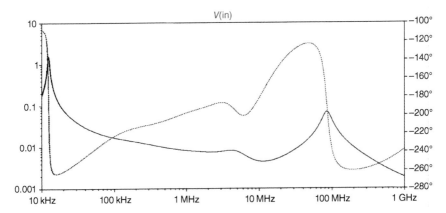

Figure 8.50 Single component Monte Carlo variation.

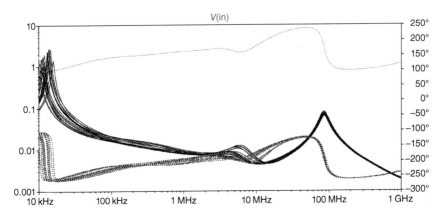

Figure 8.51 New impedance profile with all components varying.

The new profile in Figure 8.51 illustrates some important aspects. First, not only has the impedance varied significantly in the low-frequency regime but in the mid-frequency range the main resonance has moved, and in some cases, it has increased significantly. This is because the variation in the capacitance of some of the devices was large enough to impact a sensitive area in the frequency response. It is now evident that though an initial or start of life examination of the impedance may indeed yield a satisfactory result, after some time, it is probable that the impedance profile could shift enough such that the effects on the power delivery could be quite detrimental to the electrical behavior of the system. As we shall see in Chapter 9, this type of shift can increase droop with certain workloads that can result in the system not meeting its overall performance.

8.6 Power Loss in PDN Systems

After looking at the distribution of the PDN from tolerance variations, it should be clear that this would also affect power loss in the network. When developing a PDN, most engineers are much more interested in the electrical performance of the network rather than how much power it consumes. This is understandable since mitigating large voltage deviations is the main purpose of the PDN to begin with. Moreover, the DC portion of the network is largely determined by the placement of the source and load relative to each other. Thus, from this standpoint, it seems reasonable to assume at first glance that the PI engineer can do little to affect the power loss in the network. Moreover, in the virtually every case, the power loss in the system is dominated by the load behavior and the efficiency of the converter. This means that to an even larger extent, the PDN power loss would appear negligible in comparison.

However, as systems become more complex and power consumption at virtually every level of the system – particularly in mobile platforms – is a main focus for obvious reasons, even the PDN itself needs to be scrutinized for power consumption. Besides the variations due to the DC portion of the PDN, the capacitors themselves also consume power. The question is, within a PDN structure, where is the real power dissipated.

It is true that when current is charging a capacitor or going into an inductor, that power is transferred from one part of the circuit to the other. This is called reactive power. In an ideal inductor or capacitor, the total power dissipated in the device will be zero. However, in a nonideal capacitor or inductor, there are resistive elements, and depending upon the frequency of excitation throughout a network, more or less power may be burned. If we look at the power consumed in a network, there is an apparent power and a real power described by Eq. (8.81).

$$P = Re(P) + jIm(P) \tag{8.81}$$

where the actual instantaneous power burned in the system is the real part or

$$P_R = Re(P) = Re(V \cdot I) \tag{8.82}$$

When an AC analysis is performed on the impedance of the network, the current source is normalized so that we can examine the results directly. In this case, we can see in the figure both the instantaneous power and the real power consumed in the network (Figure 8.52).

Note that from the figure, the apparent power has a phase relationship, but the real power does not. However, if one were to look at the results closely, they would notice that the real power consumed in the PDN is not dramatically different from the apparent power. This is because while the DC portion

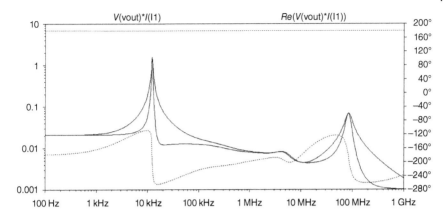

Figure 8.52 Instantaneous vs. real power consumed.

of the curve (below 1 kHz) clearly shows the loss in the network from the series components, the parallel components come into play in the higher frequency regimes. In fact, the resistive portions seem to dominate near the highest resonance. This is due to the fact that with respect to the other capacitors in the network, the energy is mainly reactive, and thus very little is lost. However, near the high-frequency resonance, the energy is clearly being burned at that particular frequency. If a series of current pulses were to run at this resonant frequency for some time, we would expect the losses to be significant. This is one reason to understand how the impedance of the network behaves relative to the activity of the load. Though, statistically, the load characteristics (as we have seen) are indeterministic, the periodicity of certain common workloads is more predictable. If an engineer knows that certain frequencies are more common than others for current activity near the load, it is sometimes prudent to look for decoupling that will move the impedance of the network away from these common resonant points to aid in the performance of the system.

There are a couple ways to see how power can be consumed in certain areas of the PDN when the load current is periodic with a certain frequency or set of frequencies. One way to investigate this is to sense the currents through a number of the decoupling networks and look at their relationship to the power dissipated overall. Another way is to choose a sinusoidal load and then eliminate a number of the network components and see how the power changes. We will use the first approach here in the next example.

Example 8.5 Determine where the power loss in the PDN is in Figure 8.17 through looking at the currents in the key decoupling networks and plotting them to see how the loss has changed.

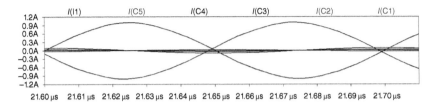

Figure 8.53 Currents in each of capacitor networks.

Solution:
This method is simple. We look at the currents in each network with a sinusoidal 10 MHz current source and determine where the currents are coming from in the network.

Figure 8.53 shows the results.

It is clear that the currents are mainly coming from the second network (through C4) rather than the other networks. This would mean that for this frequency of excitation, the power loss would be (likely) dominated by the ESR of this network. Fortunately, for this case, the ESR is ~0.5 mΩ, which is reasonably low for a capacitor or group of capacitor networks. For a 1 A current, we can safely say this loss would be less than 1 mW. Compared to an 85% efficient VR, the loss would be

$$P_L = P_O \left(\frac{1}{\eta} - 1 \right) = 1 \left(\frac{1}{0.85} - 1 \right) = 176 \, \text{mW} \tag{8.83}$$

Thus, for this case, the VR would clearly dominate the losses in the system. However, what if the AC signal was operating closer to the resonance of the PDN in the higher frequency regime? This is left as an exercise at the end of the chapter.

8.7 Summary

FDA is one of the key techniques at the disposal of the power integrity engineer. Often engineers start with FDA before moving onto time-domain analysis in order to understand where the resonances are and where decoupling is needed the most. The PDN as we have seen is made up of a number of components including (often) the VR passive components. Engineers can choose to examine the networks in a distributive fashion; that is, the components of the network are distributed about a number of subcomponents. Or, a lumped element analysis may be more than sufficient toward understanding the efficacy of the PDN. In this chapter, we focused on lumped element analysis.

It was also seen that power integrity engineers could strive toward a somewhat ideal PDN by examining damped transmission line models. The damped transmission line has essentially a flat impedance profile with some dampening to mitigate the higher frequency droops near the load. Though impractical to achieve completely, it does give a starting point to shoot for where the engineer can see where their current design needs more work.

The PDN is indeed a conglomerate of many subcomponents from the VR output stage all the way to the load itself. Each subcomponent has its purpose; as the frequency content increases closer to the load, the PDN components (decoupling) must also have better performance to mitigate the droop events by having acceptably low impedance relative to the requirements of the system.

As discussed in this chapter, there are a number of analytical methods that are available to the PI engineer for analyzing the PDN; asymptotic analysis and Bode diagrams are quite useful especially with math programs which enable engineers to break up the subsections and analyze their behavior together or independently.

After analyzing the PDN, prior to running time-domain analysis, engineers often need to try to optimize their designs. By examining the PDN as a function, one can not only determine the slopes and peak resonances but also use techniques to reengineer portions of their PDN to achieve a more optimal design.

In addition, it is important that engineers consider not only optimization but also where the PDN can change over time, temperature, and excitation. Monte Carlo analysis is very helpful in determining this key behavior. Power loss is also now an important aspect to PDN design where excitation at the right frequency and over a longer period of time may result in higher loss in the system than desired. It is best to perform a power loss analysis as a check in the system to ensure that certain frequencies of excitation cause larger than expected losses during operation.

Problems

8.1 Determine Eq. (8.8) when the change in impedance is $\frac{1}{4} R_{DC}$. What is the value of the impedance after 10 lumped subsections?

8.2 Reanalyze Example 8.2 using the values in Table 8.4.
How does the impedance compare with that in the example? Is this expected? Why?

8.3 Determine Z_{41} as in Eq. (8.60). Replace the impedance values with the R's, L's, and C's. Plot the function analytically with a math program using the same values as in the section. Determine if it aligns with the same plots as in the section.

8.4 Determine the numerator and denominator in Eq. (8.4) and perform an asymptotic Bode plot. How close does it resemble the Bode plot obtained from Problem 8.3?

Table 8.4 Values for Bode plot of Problem 8.2.

Metric	Value
L_1 (series)	0.1 μH
C_2 (shunt)	10 μF
R_1 (series)	5 mΩ
R_2 (shunt)	10 mΩ
L_2 (shunt)	2 nH

8.5   Take the derivative of the magnitude of Z_{31} from Section 8.5. Plot the function up to 100 MHz. What are the maxima and minima to the function?

8.6   Perform a Monte Carlo analysis on the circuit of Figure 8.49. Adjust the tolerances for all components to 5% and plot the functions. Adjust all of the tolerances to 35%. How far did the resonances shift? What was the largest peak resonance in the impedance plot?

8.7   Determine the power loss in the impedance plot of Problem 8.7 when the excitation is a sine wave of 1 A at 1 MHz.

8.8   Perform the same analysis as in Problem 8.7 but at 10 and 50 MHz. What was the difference in power loss between them? Which circuits were responsible for most of the power loss?

Bibliography

1 Zhong, B. (2016). *Lossy Transmission Line Modeling and Simulation Using Special Functions*. University of Arizona.

2 Fizesan, R. D. Pitica (2010). Simulation for power integrity to design a PCB for an optimum cost. *SIITME*. IEEE.

3 Novak, I. (1999). Accuracy considerations of power-ground plane models. *EPEP*, IEEE.

4 Novak, I. L. Smith, and T. Roy (2000). Low impedance power planes with self damping. *EPEP*. IEEE.

5 Pathhak, A. S. Mandal, R. Nagpal, and R. Malik (2009). Modelling and analysis of power-ground plane for high speed VLSI system. *IEEE India Conference*, IEEE.

6 Swaminathan, M. D. Chung, S. Grivet-Talocia, et al. (2010). Designing and modeling for power integrity. *IEEE EMC Conference*, IEEE.

7 Hockanson, D.H. and DiBene, J.T. II. (2007). Power delivery for high performance microprocessor packages – Part I. *IEEE EMC Symposium Tutorial Presentation*, IEEE.

8 Sinkar, A.A. (2012). Workload-aware voltage regulator optimization for power efficient multi-core processors. *IEEE Design, Automation & Test in Europe Conference*, pp. 1134–1137.

9 Novak, I., Smith, L., and Roy, T. (2000). Low impedance power planes with self damping. *IEEE Conference on EPEP*. IEEE.

10 Wang, J.S.-H. and Dai, W.W.-M. (1994). Optimal design of self-damped lossy transmission lines for multi-chip modules. *IEEE International Conference on Computer Design VLSI in Computers and Processors*. IEEE.

11 Ren, Y. K. Yao, and F. Lee (2004). Analysis of the power delivery path from the 12-V VR to the microprocessor. *IEEE Transactions on Power Electronics*. IEEE.

12 Selli, G. J. Drewniak, R. Dubroff et al. (2005). Complex power distribution network investigation using SPICE based extraction from first principle formulations. *EPEP*. IEEE.

13 Rosen, G., Coghill, A., and Tunca, N. (1992). Prediction of connectors long term performance from accelerated thermal aging tests. *Electrical Contacts – 1992 Proceedings of the 38th IEEE Holm Conference on Electrical Contacts*. pp. 257–263.

14 Murphy, A.T. and Young, F.J. (1991). High frequency performance of capacitors. *ECTC*. IEEE.

15 Rostamzadeh, C., Canavero, F., Kashefi, F., and Darbandi, M. (2012). *Effectiveness of Multilayer Ceramic Capacitors*. InCompliance.

16 Brown, R.W. (2007). Distributed circuit modeling of multilayer capacitor parameters related to the metal film layer. *IEEE Transactions on Components and Packaging Technologies* 30: 764–773.

17 Shi, H., Sha, F., Drewniak, J. et al. (1997). An experimental procedure for characterizing interconnects to the DC power bus on a multilayer printed circuit board. *IEEE Transactions on EMC* 39: 279–285.

18 Van Valkenburg, M.E. (1974). *Network Analysis*, 3e. Prentice Hall.

19 Mezhiba, A.V. and Friedman, E.G. (2003). Electrical characteristics of multi-layer power distribution grids. *ISCAS*. IEEE.

20 Reddy, E.K. (2016). *The Performance Trends in Computer System Architecture*. American Association for Science and Technology Communications.

21 Marshall, S.V., DuBroff, R.E., and Skitek, G.G. (1996). *Electromagnetic Concepts and Applications*, 4e. Prentice Hall.

22 Ramo, S., Whinnery, J.R., and Van Duzer, T. (1994). *Fields and Waves in Communication Electronics*, 3e. Wiley.

9

Time-Domain Analysis

9.1 Introduction to TDA

The previous chapter focused on analyzing the power distribution network (PDN) in the frequency domain. This topic normally comes before time-domain analysis simply because it helps to set the stage for what most power integrity engineers usually focus on when they are asked to determine the efficacy of the power distribution path. In the end, the instantaneous state of the voltage relative to the silicon load is what is important in a computer system design. If the voltage deviates out of its specified range, not only may the performance of the device be impacted, but the integrity of the data may be compromised as well. Thus, there is a reason why, in the end, performing a time-domain analysis on the full PDN is considered one of the most crucial activities in the design process for the PDN designer.

In most system designs, however, there is more than one aspect to the voltage deviations in the PDN (or normally termed the *voltage droop*) that are examined. Moreover, the location where the voltage droop is analyzed or measured is just as critical. When the sense point is very close to the load, high-frequency voltage excursions are usually present, and thus, understanding the decoupling near and on the silicon die is usually the focus (along with the parasitics between the load and the decoupling). As the circuit moves further back toward the voltage regulator, the frequency content is naturally lower, and thus the droop behavior is different. This was clear in the previous chapter when examining different portions of the PDN and their purpose in maintaining the efficacy of the power delivery path.

When a designer develops the PDN for a given voltage rail and load, there are a number of considerations that must be taken into account to ensure that it operates correctly across the dynamic behavior of that load. As discussed in Chapter 3, the load line and guardband of that rail must be considered as well as what voltage droop is tolerable within the requirements for the given operational state. For most high-performance silicon loads, there is more than one voltage droop limit that needs to be considered when the load is active.

Power Integrity for Electrical and Computer Engineers, First Edition. J. Ted Dibene II and David Hockanson.

Thus, not only must the designer develop the PDN to handle the worst-case conditions, the PDN must also be constructed so that it works under less stringent conditions as well. Because, the frequency behavior of the load can change, depending upon the workload and frequency of operation, it is imperative that these conditions be considered by the power integrity engineer. This was the purpose behind understanding the excitation in the frequency domain in the previous chapter. When the load is excited with a different clock frequency, the current waveforms that are generated create different harmonics that can actually exacerbate the voltage droop, and thus, even though the worst-case operational states of the processor or groups of processors usually lead to the largest voltage droops, they are not necessarily the most difficult to manage.

This is one reason why when the processor state changes from a higher to a lower state (e.g. the frequency drops), the guardband for that state also changes. This means that the maximum allowable voltage deviation is smaller than that in the previous state. This is certainly expected since by design, the load cannot change as dramatically compared to the previous state. However, the next lowest state also has two aspects to it, which are different from the previous state: the frequency of the clock signal and the nominal voltage associate with it. Both are smaller than that in the previous *p-state*, and thus, there is less guardband for the system designer to manage the voltage changes through. This is just one of the complexities that a power integrity engineer must be cognizant of when analyzing and developing a PDN for a high-performance silicon load.

The purpose of this chapter will be to delve into the aspects of the load a bit more and to analyze the droop behavior of the PDN for a number of the processor states that a high-performance silicon load may have. To begin with, we will develop standard time-domain analysis in Simulation Program with Integrated Circuit Emphasis (SPICE) and look at the basic voltage droop of a system from close to the load all the way out toward the converter.

9.1.1 Data and Power Integrity

Though it is often implied but is less clear for most system analyses, there is a direct relationship between the efficacy of the power delivery path and the data integrity in a system. Data integrity is defined as the correctness or robustness of a group or set of data with respect to its propagation through a combination of logic, which results in the lack of error or errors from that execution. Corruption to data can take on many aspects and sources independent of the state of the voltage rail at any given instant in time. However, there is still a correlation between the power delivery and the data as it propagates through a silicon device that is important for power integrity engineers to understand. Though this relationship is not limited to any specific device, it is particularly important when it comes to processing units.

When a functional block within a processing unit executes on a set of data, there is a relationship between the clock and the data that must be maintained

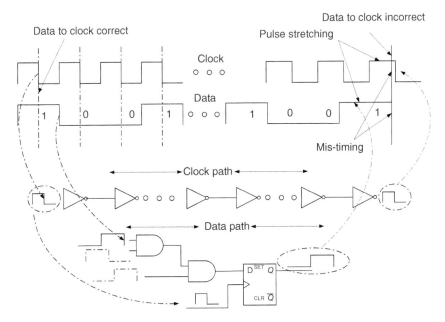

Figure 9.1 Example clock and data relationship with logic.

throughout. This relationship is governed by the timing between that clock and data which has specific rules associated with it [1]. An example of this is shown in Figure 9.1.

When a clock and data signal relationship is established correctly, as in the upper left side of Figure 9.1 with respect to the signals, the data signal is properly latched or transitioned in a logic cell prior to the clock edge coming through to force the signal to propagate into the next level logic. As the signal propagates through additional logic paths, a *skew* between the clock and data edge can occur, resulting in mistiming between them. For example, when data is latched into a flip-flop, it must be valid for a certain amount of time around the incoming clock-edge. The time that the data must be fixed before the clock edge arrives is called the *setup time*. This is akin to the data transitioning early (or the clock edge coming too late) in the propagation of data from one point to the next. *Hold time* is the time the data must be fixed after the clock edge has arrived in order to make it valid. If the data transitions before this fixed time, then a hold violation occurs.

Both forms of timing violations may cause the data to be corrupted. Even with error correction algorithms and circuitry, the performance of the system at the very least may be degraded. This is why it is so important to ensure that the data integrity be maintained throughout the execution process.

In the diagram on the right in Figure 9.1, one can see that the data and clock have both *stretched*, causing the relationship with the edges of both signals to be off. This *skew* can occur because delays have increased from one set of logic to the next. Delays of this sort may be driven by a number of factors including temperature and voltage. Voltage-induced delay is often a consequence of either the global voltage for that processing unit being low relative to its proper set point or possibly a local drop in voltage causing the same phenomenon.

Local voltage drops in regions of the die can occur when a functional block draws a large current for a short duration, causing the voltage to droop temporarily in that particular region. Depending upon the size and length of time that the droop occurs, the clock and data that is propagating in the region can slow down. In many cases, if the time is short, the number of stages (fan-out) from that point for both the data and clock paths can be small, resulting in a minimal delay or skew between the clock and the data. If this occurs and the clock to data relationship is retimed or maintained, no errors will occur as function of the timing. However, if the time is lengthened, then data corruption is possible. This is one reason why longer droop events, with relatively less voltage excursions, can be more problematic than shorter durations with larger voltage changes as we discussed in earlier chapters.

In any silicon design, it is important to comprehend the relationship between the power delivery and the data integrity. This is usually the job of the system teams responsible for power and guardband analysis. The objective is to ensure for any operational state for the device, that the voltage is maintained to ensure both data integrity and high performance.

To help understand how certain droop behavior can cause data corruption or degrade the system performance, it is practical to define the type of droop so that engineers can better focus their solutions on the problem at hand. Section 9.2 discusses droop types and where they come from.

9.2 Voltage Droop Definitions

There have been a number of definitions for droop in the industry based on the PDN behavior. Many organizations use a numbering scheme assigning the duration of the droop in the time domain to the type of voltage deviation that one might see. We will also adopt this nomenclature to help simplify the discussion and analysis. Typically, these droop shapes are assigned *first-*, *second-*, and *third-*order droop events [2]. This is shown in Figure 9.2.

The third-order droop is usually associated with the feedback of the VR and is the slowest of the voltage deviations since the bandwidth of the converter plays a significant role here. In contrast, the first-order droop is normally one that occurs within a very short period of time, typically in the tens of nanoseconds or less. This droop behavior is mainly due to the load changing very quickly, the dominant factors being the parasitics near and on the silicon. However, there are differences in the behavior of the processor concerning fast

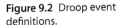
Figure 9.2 Droop event
definitions.

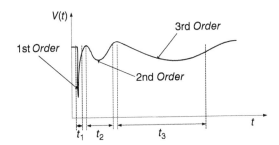

and slower droop events that need to be considered when designing a system. This is because deviations in the voltage when the changes are very fast usually have a local effect to them, whereas when the changes are slow, the effect is much more global. This will become more evident as we address these issues later on in the chapter.

The second-order droop is normally a consequence of a particular resonance in the PDN profile and is due to the parasitics and capacitance that are responsible for that particular frequency response. Because this is a *mid-frequency* event, the second-order droop is outside of the loop bandwidth of the converter but is also slower than the higher frequency droop events that are local to the silicon. In many cases, second-order droop events are ignored or neglected simply because their effect is less than the other two droop events. However, because processing units can create random current excursions, a series of second-order droop events in succession can cause the voltage to deviate in a downward progression for a while before the voltage regulator can respond. This is why the load profile, as we shall see, usually includes excitation pulses to exercise this type of droop event to ensure most causes of droop in the system are covered.

In the next few sections, we will discuss the differences in the point at which droop is measured and also what each event means with respect to the state of the digital block that is active. Since the focus here is on highly dynamic and high-performance loads, such as microprocessors and cores, the discussion will be around how these devices operate and what we would expect the voltage deviations to look like here.

9.3 Droop Behavior and Dynamic Loads

The definition of voltage droop for a given load is relative to the state of the device and set point of the voltage and guardband that has been assigned to that state. This means, especially for processing units, that there is more than one (typically) dynamic voltage range for a given load at any time. This also means that the acceptable voltage droop for that device is also dependent

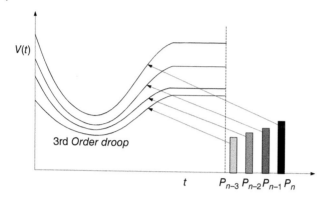

Figure 9.3 Droop and *p-state* association.

upon the state of the processing unit [4]. Figure 9.3 illustrates this effect with respect to the *p-state* of such a load. At any given functional setting for the frequency/voltage set point pair (*p-state*), there will be an acceptable range in which the voltage can dynamically change within those boundaries [8]. Normally, as shown in Figure 9.3 after the droop event, there can also be an associated *load line* with each *p-state*. This means the voltage will settle at a specified level based upon the VR load-line setting. For advanced systems with multiple cores, this is desired since in the end the load line helps to reduce the overall power consumed by the load by adjusting the V/I slope relative to the state of the processing unit [10].

Note that each *p-state* voltage starts at a point that is greater than the previous setting when the *p-state* increases. Moreover, the droop event will be greater because the load current and activity in the processor are, by definition, higher.

From a process perspective, when a device requests a voltage change from a lower voltage to a higher one, as in the very simplified process flow of Figure 9.4, the system clock for the processing unit is adjusted and the PLL (phase lock loop) is locked onto the correct frequency. Once this occurs and lock is complete and the processing unit is informed of the state of the clock, the processor can start processing data at the new frequency. When this happens, it is assumed that the VR is also adjusted to handle the current and the proper load changes prior to the start of activity. This includes adjustment of the load line to the correct V/I slope in the VR feedback system as described in Chapter 3.

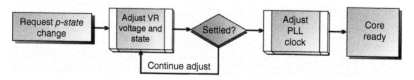

Figure 9.4 Simple *p-state* process flow.

With respect to the integrity of the power delivery at any given *p-state*, both the frequency content and amplitude are subject to the load changing within the bounds of the operation of that particular state. This means that even though the voltage droop may be less, all things being equal, for a lower *p-state*, the actual boundary conditions for that state can be more stringent.

As an example, we discussed in Chapter 3 both the load line and guardband effects within a given processor state and how the window of the guardband changed between *p-states*. In the case of a processor moving from a higher to a lower *p-state*, one would expect to have a *decrease* not only in the voltage/frequency pair but also in the guardband. This is because, by definition, the voltage droop should be less since the amplitude of the current waveform must be smaller. Nonetheless, with respect to the PDN design, it is not always the case that this means that the voltage droop will meet the boundary requirements. The reason is that the load frequency content may align both in phase and frequency with a resonance in the case of a lower *p-state* load activity and not align in a higher one. This is illustrated in Figure 9.5.

In the case of the higher frequency periodic load current, the load step harmonics (darker bars) do not align with the resonance of the PDN in the first case, which results in the droop being above the guardband limit (V_{pn_gb}). In this case, of the droop for the P_n state, the voltage meets the guardband limit

Figure 9.5 Voltage droop for two different *p-states*.

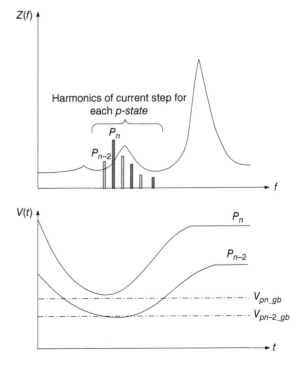

and the system stays within its boundaries. However, in the lower frequency periodic load step, the current waveform contains harmonics that *do* align with the resonance, and thus we see a larger voltage deviation, resulting in the voltage droop exceeding the range of the boundary conditions (V_{pn-2_gb}), in this case with the guardband.

Alignment is a *relative* term here. Because in the frequency domain, the mathematical representation between the voltage and current is a multiplication, in the time domain, it is a convolution and the representation of the droop is more complex. This was evident when we examined Laplace and Fourier signal analyses earlier.

Because the harmonic content of the load step matters with respect to the filtering capabilities of the PDN, it is crucial that a number of cases in the time domain be run. When analyzing the system droop behavior, the changes with the PDN should also be varied to ensure that the PDN meets the requirements over the entire system range.

Part of the job of a PI engineer is to ensure that over the worst-case conditions of the system, the efficacy of the power delivery is maintained. This is, of course, a challenge since it is impossible to run every particular case that a given load can produce. In fact, because often the activity of the processor is highly dependent upon the sequence of functions that are called upon at the execution level, over a reasonably large period of time, the potential sequences become astronomical in number. This is why the PI engineer usually works with one or more groups inside of an organization to build a set of test loads that emulate what the actual load can do.

Nonetheless, coming up with a clear set of load profiles is extremely challenging. The first reason is that if the test load is too conservative, the final design may add too much cost to the end solution and actually limit performance of the platform. However, if the design is not conservative enough, there is a chance that failures can occur in the field, which could be catastrophic to a company's bottom line. Thus, it is crucial that the correct design limitations be put in place to ensure that the system PDN meets the requirements overall.

Typically, there are three general tests that are used to check the efficacy of the power delivery for a given platform (Figure 9.6). The first is a simple step response. This usually comprises a current load step that is intended to exercise the PDN and the power source with the goal of determining the worst-case slow droop in the system. There are usually subsets of these whereby a step response with some percentage overhead (amplitude of current pulse) is used for each *p-state* to ensure that the PDN is exercised within each of the limits for a given operation of the processing unit.

The second item is a repetitive high-frequency set of pulses with the purpose of exercising certain resonances in the PDN. These are bounded by the activity that is possible in a particular processing unit, and normally, the end goal is to determine the lowest voltage point at the load to ensure it meets the system conditions.

Figure 9.6 Different load test examples.

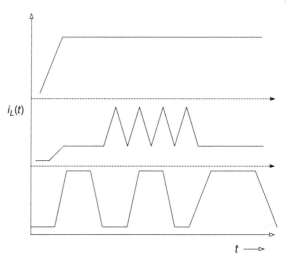

$i_L(t)$

$t \longrightarrow$

The third one is a specific targeted set of current loads that are intended to exercise a particular operational state of the load where it is observed that the given system may be susceptible. As we saw in the previous chapter, usually this is found by looking at the frequency behavior of the PDN and then doing a sweep of current pulses at different frequencies with the goal of finding the set that causes the lowest voltage deviations overall.

In the first case, the step response includes a model of the VR to show the true closed-loop response of the system so that the low-frequency or third-order droop is captured. In the second two cases, unless the voltage regulator response is within the bandwidth of the current pulse, normally the VR is not included because the energy usually comes exclusively from the capacitive networks in the PDN. To determine the efficacy of the PDN, however, all three cases, with multiple subcases, are usually run by the PI engineer for a given set of *p-states*.

9.3.1 Step Response

Let us examine a simple step response first. The purpose of the step response, as mentioned earlier, is to determine the third droop response of the VR. The important thing to note about the step response that needs to be comprehended is the amplitude and the slew rate. In many dynamic loads, such as processors, the step current can be very large. However, there are limits to how fast the load can slew under such conditions. As we discussed previously, to get the maximum current swing on a load, it can often take a longer time than to generate a smaller load change that can slew very quickly (sometimes in the tens of nanoseconds or less). The reason is because the processor must activate a number of units in succession to create such a large current change and that can take

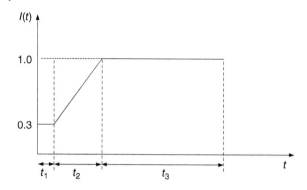

Figure 9.7 Step response characteristic.

many clock cycles. Thus, when looking at the step current, both parameters need to be defined by the system designer.

In the next example, we show a simple step response using the closed loop VR designed from Chapter 2. The step current will be varied in both slew rate and amplitude to show how the VR responds.

Example 9.1 First, use the closed loop model from Chapter 2 and the step response in Figure 9.7, to look at the voltage droop. Add in the parasitics for the first portion of the PDN as shown below and show the droop behavior. Place a 2 Ω resistor at the output so that the direct current (DC) offset is ~0.5 A (Table 9.1).

Second, add in another current step 125 μs later as shown in Figure 9.8 and examine the second droop in the system. Measure the effect on the voltage droop at the output of the VR filter and at the load. What was the effect? What would you expect to happen if you added another step 55 μs after the second step?

Solution:
The first step is to set up the problem with the closed loop model developed in Chapter 2. We also need to add in the PDN parasitics as shown in Figure 9.9.

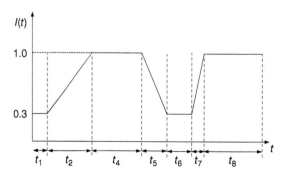

Figure 9.8 Second step response characteristics.

Table 9.1 Parameters for step response.

Metric	Value
t_1	1 ms
t_2	10 μs
t_3	1.5 ms
t_4	124 μs
t_5	10 μs
t_6	125 μs
t_7	10 μs
t_8	500 μs

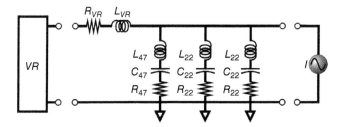

Figure 9.9 Figure for adding portion of PDN.

We need to let the closed loop model settle, which is why the latency between the start of the simulation and the step current is 1 ms. Thus, we see that we have a current that goes from 0.5 to 2.5 A (approximately). The bias current is expected. The absolute voltage is not important in this case, but it hovers around 800 mV. The result is shown in Figure 9.10.

Using the measurement in linear technology SPICE program (LTSPICE), we see the droop is about 145 mV from the original voltage setting to the minima. We also note that the response is ~50 μs here; that is, it takes about this long for the low-frequency droop to start to recover. Since our bandwidth for our converter was quite slow to begin with, this is understandable. Now, we add in the next step current.

With the addition of the second step current, we would expect that the VR has settled again. Thus, we would not expect the droop to be more than the first one. And in fact, it is not. However, we do note that there is an overshoot due to the unloading. If the voltage was near the maximum for the silicon, this might be a problem. Usually, the VR designer would examine the overshoot problem to ensure that this was acceptable from a reliability point of view (Figure 9.11).

There is another important aspect to this. Notice that the first two runs were measuring at the output of the VR. However, the last one we measured at the

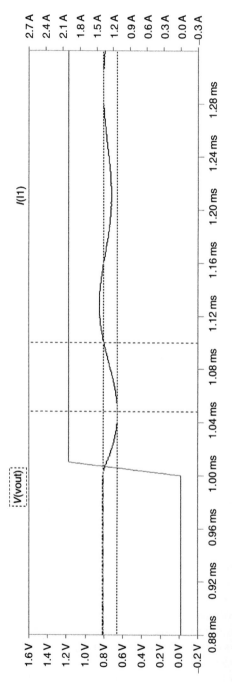

Figure 9.10 Droop with first step response.

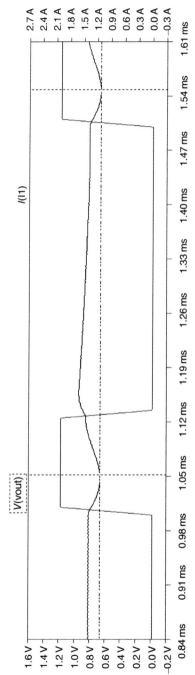

Figure 9.11 Voltage droop in example with second step current.

load. Note that the voltage is quite a bit lower. This is because of the parasitics in between the VR sense point and the load. In most high-frequency (HF) designs today, the sense point for the VR is at the load rather than at the output of filter. This is the reason why. Because of this extra voltage drop, the actual voltage sensed at the load could be lower than what the VR sees, and thus, the guard-band could be compromised (Figure 9.12).

Now for curiosity sake, we add in another step about 120 μs later just to see the extra step causes. Though it is subtle, the droop is actually increasing slightly. The first droop event at the output of the regulator was about at 660 mV, whereas the third one dropped to about 655 mV. We can also see that the effect at the load is slightly more pronounced as well (Figure 9.13).

These effects are important to examine when one is analyzing droop with an active voltage regulator. The reason is that in the case of a single or multiple step responses, the bandwidth of the VR can have an effect on the droop that can affect the efficacy of the PDN and the power delivery. Normally, this means that both the PDN and VR designer must work closely together to ensure that the overall system behavior is always met.

We saw in the previous example the effects of adding in multiple step responses into a closed loop system. This was definitely a low-frequency event that involved the bandwidth of the VR. The next thing to examine is to see what happens when the high-frequency portion of the PDN is looked at that does not involve (directly) the bandwidth of the VR. To see this, we need to go back to the full PDN lumped element network and then perform a high-frequency resonant analysis. Normally, the best way to start this is to look at the PDN frequency response and then look at where the high-frequency resonant behavior was.

9.3.2 High-Frequency Pulse Droop

From the previous chapter, we saw that for our PDN example, the high-frequency resonant peak for the impedance appeared at around 100 MHz. This is shown in Figure 9.14.

Thus, our starting point then would be to develop a load behavior of multiple step functions that had a fundamental frequency of about 100 MHz. This is, in part, only if we had little to no direction from the system teams in the creation of a high-frequency current profile that would exercise the PDN and power delivery properly.

Figure 9.15 shows a typical high-frequency pattern where T is the period of the pulse which is equivalent in our case to,

$$f = \frac{1}{T} \cong 100 \, \text{MHz} \tag{9.1}$$

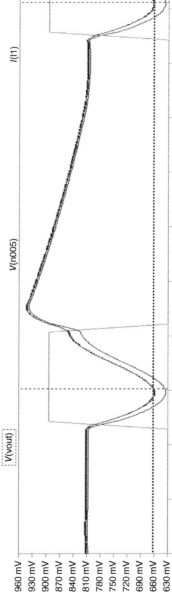

Figure 9.12 First and second voltage droop step measured at load and output of VR.

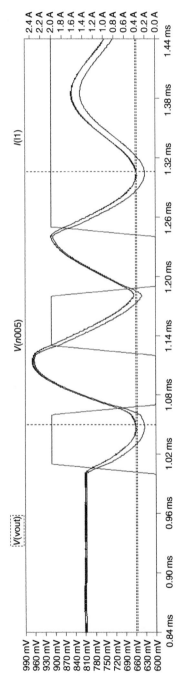

Figure 9.13 Addition of third step.

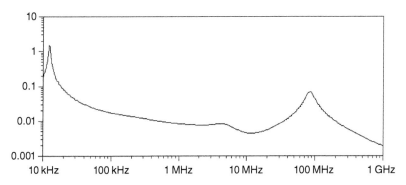

Figure 9.14 Impedance profile for PDN.

Figure 9.15 High-frequency
pulse pattern.

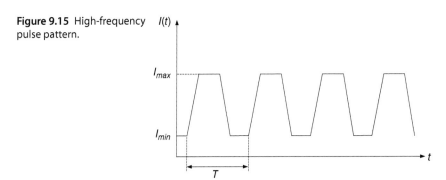

Thus, our goal now will be to put a model together and excite our PDN with a series of pulses that occur around this periodicity. We can do this with or without the closed loop model. However, for most VR designs (except of course a very high-frequency converter design), we would not expect the loop bandwidth to have any impact on the result. We will show this, however, for both cases: first, we will run a simulation on a simple open-loop lumped element model and then add in the closed loop model to see how the behavior changes. The reason is that the state of the voltage droop is quite important here. The state of the voltage at a given load point in time, when a set of high-frequency current pulses occur, is just as important as the event itself.

To start with, we can use the simple open-loop lumped model from the previous chapter to build our model. Because the excitation frequency is very high at the load and we want to capture the behavior there, we need to make sure that the full PDN is present.[1] Our starting point will be at a lower frequency current pulse to see what the behavior looks like at one decade lower.

---

1 There is no reason to not use the closed-loop model in this case unless one is trying to expedite the simulation.

Our expectation would be that the excitation for a lower frequency current pulse, for the same amplitude, should have a smaller overall droop than one that aligned with a particular resonant that was larger. The simulation for the lower frequency current pulse is shown in Figure 9.16.

Figure 9.16 illustrates three high-frequency droop events in succession at a frequency of 10 MHz. The result indicates that the droop actually *increases* after a certain point but starts to level off. This can be seen by looking at the two cursor markers from the first pulse to the last. This is obviously an *open-loop* system without any feedback even though the closed loop model had a VR model in it. It is clear that because the amplitude is less than the step response, if the event occurs when the voltage is at steady state, then the system should maintain its guardband. However, what if the pulses were activated at a point in time after a step response? That is, what if the high-frequency pulses occurred shortly after the unloading of the low-frequency step (Figure 9.17)?

We see in Figure 9.18, when we unload the system at the minima of the step response. Because the VR is just starting to respond, we would expect the voltage would steadily but slowly increase from that point onward. And in fact, this is what we actually see. Additionally, what we also observe is that the second pulse is at about the same point as that of the first pulse. This is because the rate of voltage increase is nearly matching the decrease pulse by pulse from the high-frequency pulses. Note also where the minima is here relative to Figure 9.17. Because of the large step function, the voltage droop minima is closer to the step function droop minima which is approximately 700 mV. Moreover, this droop is *higher* than the minima due to the droop, which we will see in Figure 9.18.

Because of the unloading of the step current, it is evident that the voltage has recovered nearly 100 mV from its minima before the HF pulses occur. This is an important insight since the slew rate of the slow step function has a large impact on where the voltage will be relative to the next set of high-frequency pulses. We know by definition that we cannot get an immediate high-frequency set of pulses since this would exceed the maximum current capability of the processing unit. The question then is what is possible at this point with respect to the high-frequency pulses that may be generated by the load?

Interestingly, this is not necessarily the worst-case droop point when this could occur. If we move the starting point of the high-frequency pulses just before this low point, the VR has yet to respond, and thus we would expect the trajectory would still be going downward with each successive pulse. However, this is not the case as Figure 9.19 shows.

The unloading effect has moved the energy around such that the localized voltage trajectory is now going upward again. In many ways, this is expected. The pulses are within an impedance/frequency range that the energy from the system is able to feed the capacitors at the load, and thus, the voltage trajectory is *increasing* here. However, the pulses generated here were at a point in

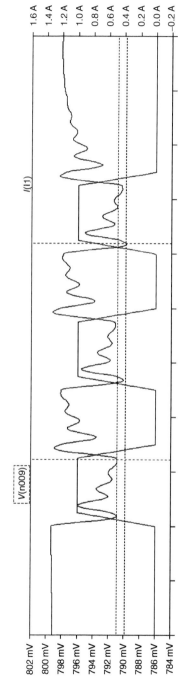

Figure 9.16 High-frequency droop behavior.

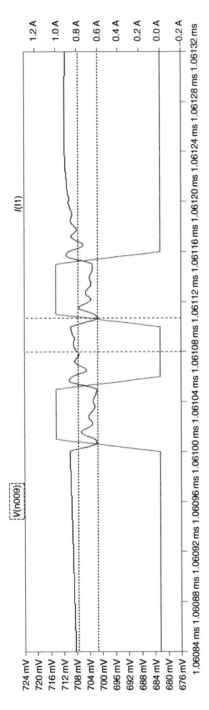

Figure 9.17 Successive pulses after unload of step function.

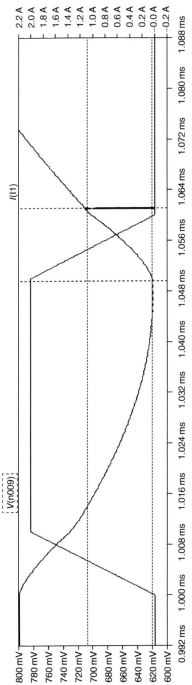

Figure 9.18 Droop minima with HF current step and slow step.

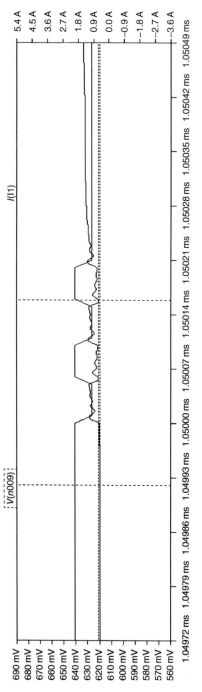

Figure 9.19 Droop event from successive pulses after step response.

the PDN impedance where it was relatively low. If we were to increase the frequency by 100 MHz, what would result then be?

Figure 9.20 shows what happens to the localized droop when the frequency is increased. In the previous case, the voltage droop was less as one increased the number of pulses. However, in this case, the voltage droop *increased* with successive pulses.

The reason has to do with where the energy is commutating in the capacitors in the PDN. It also has to do with the impedance of the PDN at that frequency. Note from Figure 9.14 that the impedance was much higher at the 100 MHz resonant than at the 10 MHz point. This explains, in part, why the droop was larger.

Because we are now looking at the droop at the load point and not back near the VR output, it is evident that the charge stored in the capacitors near the VR are trying to feed the charge into the higher frequency capacitors. However, there is a *time-lag* here between when that charge can be delivered, the high-frequency pulses which are being driven at the load, and the feedback of the VR. This delay is due to the impedances between the large slower capacitors (effectively higher impedance in this frequency range) and the smaller faster capacitors. Essentially, the capacitors near the load are being *charge-starved*, and there is no enough energy to maintain the proper voltage at the load.

Before the PI engineer panics over this situation, it is best to go back to the developer of the pulse train (normally from a processor systems team) and determine exactly what type of pulse train is possible. It is highly likely that a 100 MHz series of pulses is probable given the architecture of most processors today. The question though will be how many successive pulses are possible and what is the rate at which these can occur? Normally, to answer these questions requires some level of sophisticated modeling by one or more design teams, but often the question can be relevant – especially when the choice could be between failure to meet the voltage requirements and possibly having to add an excessive amount of capacitors, which could limit the ability to add other features to the platform.

In the next chapter, we will discuss how processors generate (in general) certain dynamic loads. For now, we have to make some assumptions as to the periodicity of the load, the quantity, and the rise and fall times. It is true that most processor cores can generate fast di/dt pulses in succession. The question though remains as to characteristics of these pulses. One way to determine the efficacy of a given power delivery system is to create a pulse pattern and then determine when or if the minima occurs with respect to the guardband of the system. Our next example illustrates the method by developing a set of successive pulses in a closed-loop model and then setting the guardband to a given point to see if that value was reached.

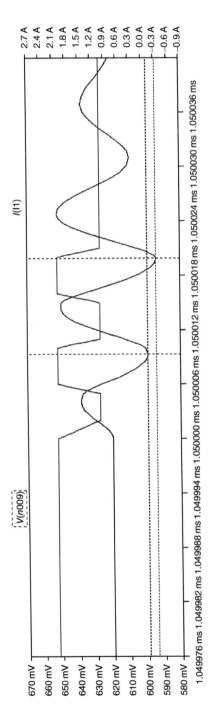

Figure 9.20 100 MHz droop from two pulses.

Table 9.2 Parameters for load profile and guardband.

Metric	Value
t_1	1 ms
t_2	10 μs
t_3	1.5 ms
t_4	124 μs
t_5	10 μs
t_6	125 μs
t_7	10 μs
t_8	500 μs
V_{gb_min}	580 mV

Example 9.2   An engineer is examining the high-frequency droop behavior of a dynamic multicore rail. The individual has been given a set of pulses that the system processing team believes is possible, but they do not know how many can occur within the architecture given a set of commands within a workload.

The engineer needs to find how many pulses occur before the minimum guardband is hit. Given the PDN from the previous example and the closed loop VR model, along with the minimum voltage set point for the state, run the load profile over a set of pulses (from two to nine) and determine if the minima reaches the guardband of the system. Use the data in Table 9.2 as criteria and the same pulse train that was illustrated in this section.

Does the droop go beyond the set guardband minimum in any of the cases? Explain your answer.

Solution:
Once again, we need to set up the problem correctly. We use the data in the table to characterize our pulse pattern profile and guardband voltage point at the load. Our starting point is essentially the same model that was set up previously except we are going to add pulses and determine when the voltage hits the minima. The first simulation below shows where the minima occurred after two pulses (Figure 9.21).

With the line drawn at the 580 mV mark, it is clear that with two pulses, there is still guardband in the PDN. Now we increase the number of pulses to four to see if it reaches the minima of the guardband.

The result is actually quite interesting. Our assumption was that the increase in pulses would result in an increase in the droop, but in fact, the droop minima actually stopped decreasing after two pulses and started to increase. The reason is not intuitive, but it has to do with the previous discussion about energy

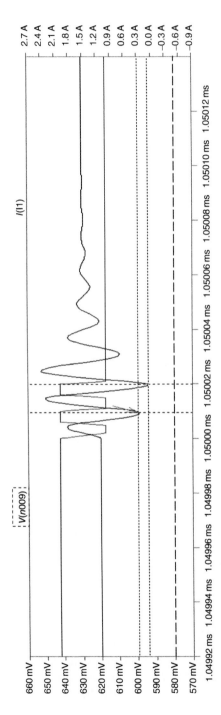

Figure 9.21 Minima after two pulses.

commutating in the capacitors. What has happened is that the delay in charge transfer from the slower capacitors to the faster ones has now been taken up by the time window within the rate of the successive pulses. Basically, the rate of charge increase into the higher frequency capacitors has increased over time, and thus, the voltage droop that is local is actually decreasing. This has as much to do with the quality of the capacitors in the PDN (at varying points in the PDN) as it does with the interconnect impedance between them (Figure 9.22).

For completeness, we add two additional pulses to see if indeed the droop has hit its minima. And, as expected, as shown in Figure 9.23, the droop is definitely decreasing.

We can now answer the question of whether or not the increase in pulses caused the droop to exceed the minimum voltage guardband; the answer is *no*. Another way to look at this is that the energy being taken away from the HF pulses is less than the energy that is supplied back by the system at the same time.

When examining the droop behavior in a PDN, it is normally a good idea to examine the voltage drop across enough conditions such that the engineer begins to understand the PDN characteristics. This way, if a problem is encountered, they know the correct direction to go with (quantity of capacitance, type, etc.) in order to mitigate the problem in the future.

9.3.3 Susceptible State Voltage Droop

As we have seen in the previous two sections, the PDN responds to current load changes at the load point quite differently depending upon the amplitude of the pulse(s), the frequency, the repetition rate, and the characteristics (rise, fall, pulse width, etc.). It is not always obvious what particular current load profile will exercise the PDN to its maximum extent, particularly when the operational state of the load is varying.

Most power integrity engineers start out examining the highest *p-state* of the load and spend their time modeling the load behavior with worst-case profiles to ensure they cover the conditions they believe could challenge the efficacy of the power delivery path. However, more often than not, because of constraints on the other *p-states*, it is not always clear that the highest *p-state* will be the state, which creates the largest problem in the PDN system. The reason is because the other states of the load must also be guardbanded, and the design team wants to get the most performance out of each state while minimizing the power consumed at the load; they will push the limits of the guardband of every state regardless of the load. Moreover, because many platforms have workloads that spend much more time in these lower *p-states*, often the systems design teams want to push the guardband to as low as possible voltage to minimize the energy consumed in the platform overall.

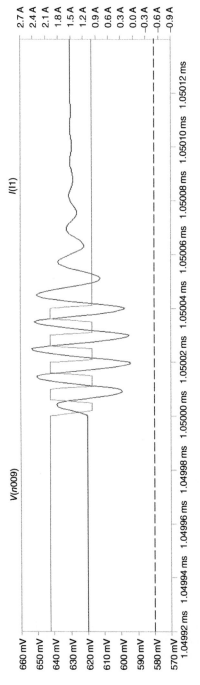

Figure 9.22 High-frequency droop with four successive pulses.

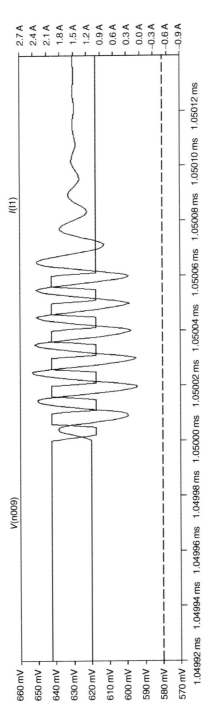

Figure 9.23 High-frequency droop with six successive pulses.

It is for these reasons and others that the power integrity engineer often finds himself or herself having to investigate in detail not only the highest voltage processor state but also the other states. This is an important activity for power integrity engineers in many ways because normally the load profile can be quite different between the different load states, resulting in frequencies that excite resonances in the PDN that do not normally result in voltage deviations that can cause issues in other states, including the large current generators.

It is true that the frequency behavior of the load is tied to the clock frequency of the processing or logic unit that is the load. However, so is the current profile, which becomes very important when investigating the droop behavior. Normally, it is not a consistent set of periodic pulses that causes the worst-case droop conditions, but more often than not, it is a *distinct* set of load pulses that occur in some specific order, frequency, and amplitude that usually force the voltage to deviate beyond the guardband limits. Additionally, because this profile is very indeterministic, it is very difficult to predict by the system teams. One way a PI engineer can address this is to put together a series of different pulses based on guidance from the load profile developer with the purpose of exciting the PDN toward its worst-case performance state, or voltage droop minimum.

The start of this investigation starts with an analysis of the guardband and load-line settings for the VR and the system. As we saw in Chapter 8, the programmed load line is normally different for each *p-state* simply because if there were only one load line, the system power loss would eventually be higher than required. This is because the system team would have to program the load line to the highest current activity, or *p-state*, which would result in a much more conservative voltage setting for the lower *p-states*, and thus, the system would burn more power than necessary.

This is shown in Figure 9.24 where one can see the difference in creating a load line that has the same slope as a higher *p-state* and then attaching that to the next lower *p-state*. The goal should always be to minimize the power loss in a platform. However, when a systems designer creates the full system guardband and load line for a particular *p-state*, what is often neglected is the effect that it has on the PDN.

When a load profile is created for a given *p-state*, it is difficult to predict where the excitation frequencies will end up. It is true that the number of possible harmonics may decrease due to the slower clock frequency, but that does not guarantee that the voltage droop will always go down linearly. The reason, as we saw in Section 9.3.2, is that there are resonances in the PDN, which can be excited by combinations of different pulses which have harmonics that align to the impedance of the PDN at a given state. We can see this through sweeping a set of pulses with 50% duty cycle and then examining their droop behavior to find out which ones caused the largest event.

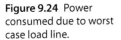

Figure 9.24 Power consumed due to worst case load line.

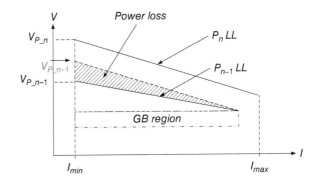

To determine these effects, we start with a step response again and then add in a series of pulses to determine the effects. For now, we will start two pulses and then shift the frequency and duty cycle of those pulses after the step response. The goal here is to see if we can cause the droop to exceed the lowest point from the larger step response.

The first attempt is in Figure 9.25. After the step response, we drive the load with two 2-amp steps at a frequency of 1 MHz. As shown in the plot, it is clear that there is enough time to recharge the capacitors near the load and thus respond to the charge depletion such that by the time the second pulse occurs the voltage has raised enough to recover effectively.

The next attempt is shown in Figure 9.26 where we increase the frequency to 10 MHz to see the effect. We keep the rise and fall times the same (20 ns) and then determine the effects from the two pulses. To show the similarities between them, we keep the voltage scale the same as in Figure 9.25. Note that the voltage droops are much closer together but have not approached the lowest point of the first step (Figure 9.26).

The next try is with a 50 MHz set of pulses. Again, we use the same edge rates and scale as Figures 9.25 and 9.26. It is clear that two things occurred when we shortened the pulses. First, the droop did not appear to go down as far as the previous set, which is likely due to the lack of energy being pulled out of the system with respect to the time constants of the capacitors. Because of this, we see that the droops did not go quite as low as the previous 10 MHz pulses (Figure 9.27).

For our next attempt, we now extend the length of the two pulses instead of keeping 50% duty cycle. However, we keep the same rise and fall times. We also shift the point of the first step function to the left slightly to see if the step function minima has had an impact. The result is certainly more interesting than our previous three attempts. Note that by extending the two pulses, we were able to determine the minima of each successive droop and then drive the load toward the point where the voltage droop of the first pulse now exceeds the point of the first step function (Figure 9.28).

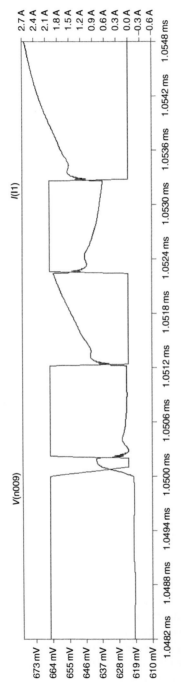

Figure 9.25 Step function and repetitive 1 MHz pulses.

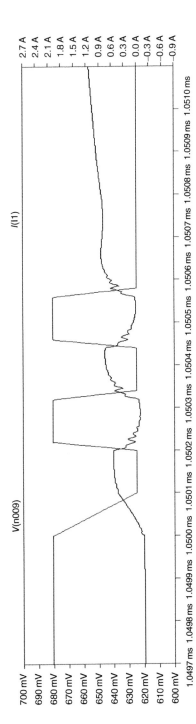

Figure 9.26 10 MHz pulses after step function.

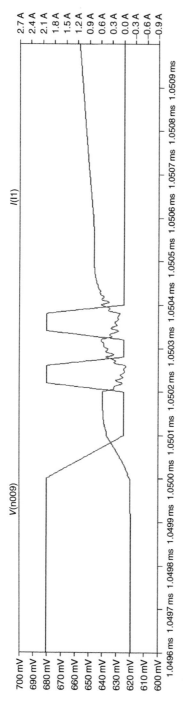

Figure 9.27 50 MHz pulses after step function.

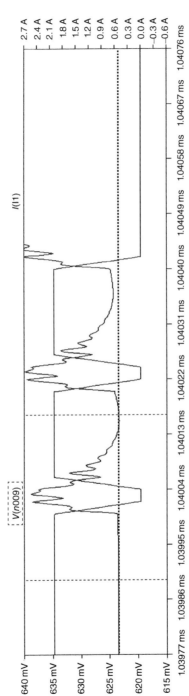

Figure 9.28 Extension of duty cycles from Figure 9.27.

To do this, we needed to catch the first droop on its trajectory downward. Note the voltage is slightly higher than in Figure 9.27 because of this. Thus, the specific combination of where the step function stops and the combination of extending the duty cycle of the pulses appears to take enough energy out of the system that the voltage droop from the first HF pulse is lower than the second pulse function. If it turns out that this minima is the lowest point that one can achieve given the data, then it is likely that for the mid-frequency pulse events, the minima here has been hit.

In Section 9.4, we discuss some simple analytical techniques with respect to looking for the minima mathematically.

9.4 Analytical Approach to Step Response

Analytical methods for the time domain are somewhat different from the frequency domain. It is still true that we look for minima in the waveforms, but we also examine the effects with respect to the guardband and load line, which we examined in an earlier chapter. In this section, we look at different methods to analyze the droop behavior to gain some insight into determining when the minima can occur in a system.

To begin, we go back to the load line and the guardband for an example system given a certain *p-state*. Often, not only do system designers provide load profiles, guardband and load lines, as references for the power integrity engineers, but they also (to a large extent) expect some data coming back from a PDN designer to determine if their settings are truly optimal. The main reason is usually that they want to see if their settings are appropriate for that *p-state* because if they were overly conservative (or not conservative enough), this could be a problem in the end for the product. Thus, it is often up to the PI engineer to do an analysis to see if the exercising of a particular *p-state* resulted in the optimal design point.

To start with, for a given *p-state*, the PI engineer is given both the load line and the guardband from the systems team along with the profile and then asked to perform a series of simulations to see where the droop ends up. Because modeling has an element of tolerance with regard to the results, there will always be some built-in guardband to account for the error. In addition, because the components in the PDN are not necessarily at the exact values used in the simulation (as illustrated in the Monte Carlo analyses), the voltage droop results will certainly have some error with respect to the lab measurements and the droop over the life of the product.

Figure 9.29 shows an example load line and guardband for a given *p-state* voltage where the minimum voltage droop is targeted. As we examined in an earlier chapter, there was an accumulation of the guardband due to a number of factors. Normally, these are fixed elements, and it is up to the systems team

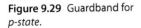
Figure 9.29 Guardband for
p-state.

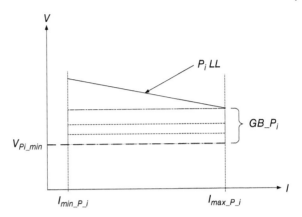

to address them. However, there are some, which involve the voltage regulator
and the PDN, which should be examined closely. For example, the filter for
the VR and its bandwidth can have a big effect on the step response minimum
droop point. Because this is often the most significant droop in a typical
high-performance system, understanding the minima and where in time it
occurs for a given system can be beneficial. This answers the question of *"why
did the minimum occur at that point"* rather than just examining a SPICE
output with the actual parasitics. Analytical insights to such events usually
yield other insights into the design of the PDN. This leads the PI engineer to
drive toward more creative solutions because not only through insights from
the SPICE modeling do they glean the behavior of the PDN, but with analytics
there is an additional benefit into specific behavior that can help down the
road when addressing a future PDN issue.

With regard to high-frequency resonant issues, the behavior of the load
becomes the focal point. Moreover, the type of capacitance near the load has a
strong effect on the HF resonance near there, which will affect the final droop
point as well. However, the HF droop events, as we have discussed, do not
involve the VR. Moreover, it is often simpler to just run a SPICE simulation
as in Section 9.3 to determine where the droop resides. This is not to say that
analytically modeling the load and its nearest passive elements of the PDN
is not beneficial. It is normally more expeditious to use standard simulation
methods to understand this behavior. This topic will be discussed in greater
detail in the next chapter.

For the low-frequency droop in the system, there is usually an expectation
that is involved with detailed analysis here. When these droop events are
examined, the question is, normally can the minimum voltage point for a given
p-state be set at a different point to help either save power or give more margin
to the design?

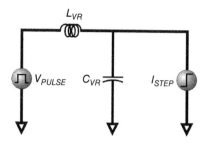

Figure 9.30 Simple circuit for analytical modeling.

For the low-frequency droop events, modeling mathematically the entire feedback of the system can be cumbersome to say the least. However, even with current mode controller-based systems, the duty cycle does not change substantially over a large number of switching cycles. Thus, it is reasonable to assume that the switching signal can be represented by a simple Heaviside function. As for the load, the main objective is to determine the minima, so a simple step response function is normally adequate. The next question is, what parts of the PDN and VR filter should be modeled? If the ripple effects are important, then it is possible that some parasitics may need to be added.

Figure 9.30 shows the minimum set of building blocks that need to be modeled. The objective is to find the minima for the droop using this simple model and understand mathematically what that means to the system overall, including the power loss. Before we get to the effect on the system, first we need to build our analytical model.

The representation of the model starts with solving a simple equation. We can use a Laplace analysis to solve the problem. The total current at any instant is equal to the load current; in this case

$$I_{ld} = I_1 - I_2 \tag{9.2}$$

where I_1 and I_2 are the currents shown in the circuit diagram (Figure 9.31).

Our goal is to solve for the output voltage V_0. Thus, plugging in for the currents in Eq. (9.2) yields

$$I_{ld} = \frac{V_S - V_0}{Z_L} - \frac{V_0}{Z_C} \tag{9.3}$$

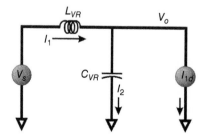

Figure 9.31 Details for model.

The key is to get the proper initial conditions on the voltage across the capacitor and the current through the inductor. The current through the inductor is assumed to be zero when the pulse starts. However, the capacitor has an initial voltage of one across it. Thus, we get the conditions for the inductor and the capacitor.

$$L \rightarrow V_L = sLI_L(s) - Li_L(0^-) = sLI_L(s) \tag{9.4}$$

$$C \rightarrow I_C = sCV_C(s) - Cv_c(0^-) = sCV_0(s) - C(1) \tag{9.5}$$

In addition, we need to know what the source looks like. This is a series of Heaviside functions where we generate a series of pulses. The important thing to note is that we want to have enough pulses to switch through the droop event. Thus, we want a series of pulses that allow us to determine when the droop starts to come back. A pulse function is shown in Figure 9.32.

The figure shows two successive pulses separated by a distance $b-a$. This separation is essentially the *commutating* cycle time. We can adjust this such that our frequency and time correspond to the switching frequency of the VR and for a simple voltage mode control use the pulse width (duty cycle, time $= 1$) for the energy going into the filter.

To start with, we assume that there are N pulses for the droop event that we need to cover.

$$V_S(s) = V_{S0} \sum_{i=0}^{N} \left(\frac{e^{-sbi}}{s} - \frac{e^{-s(bi+a)}}{s} \right) \tag{9.6}$$

This now gives us all of the functions we need to plug into our equation. The load current is also a step function, but in our case, it is simply a weighted unit step of magnitude I_{ld}. For now, we can simplify the problem and assume that the source V_S is just a simple step function. Since we know the other parameters,

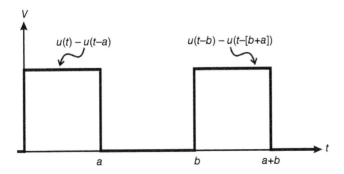

Figure 9.32 Two pulses represented by step functions.

we can plug in for I_1 and I_2 now.

$$\frac{I_{ld}}{s} = \frac{\frac{V_S}{s} - V_0}{sL} - C(sV_0 - 1) \tag{9.7}$$

Solving for V_0 we get

$$V_0(s) = \frac{-sI_{ld}L + V_S + s^2LC}{s(s^2LC + 1)} = \frac{-s\frac{I_{ld}}{C} + \frac{V_S}{LC} + s^2}{s\left(s^2 + \frac{1}{LC}\right)} \tag{9.8}$$

This is the function that we need to solve now using a partial fraction expansion. We examine this simpler case when the voltage is just a step response at the source. This is illustrated in the following example.

Example 9.3 Using Eq. (9.8), solve for the function for a simple step response. Use the values in Table 9.3 and plot the function to look at the droop at V_0.

Solution:
We can use the formula in Eq. (9.8), which does not include any exponential terms in the numerator. Thus, the Laplace transform is much simplified since V_S is now just the unit step. This takes on the form of the solution of the following:

$$V_0 = \frac{A_1}{s} + \frac{A_2}{s + \frac{j}{\sqrt{LC}}} + \frac{A_3}{s - \frac{j}{\sqrt{LC}}} \tag{9.9}$$

We multiply through on either side of Eq. (9.9) and then solve for the coefficients. Solving for the coefficients yields

$$A_1 = V_S, \quad A_2 = \frac{1}{2}(V_S - 1) - \frac{jI_{ld}}{2}\sqrt{\frac{L}{C}}, \quad A_3 = \frac{1}{2}(V_S - 1) + \frac{jI_{ld}}{2}\sqrt{\frac{L}{C}} \tag{9.10}$$

Table 9.3 Values for analytical step response in Example 9.3.

Definition	Parameter	Value
Inductor	L	$1\,\mu H$
Capacitor	C	$360\,\mu F$
Source voltage	V_S	$1\,V$
Load current	I_{ld}	$1\,A$

The solution is now achieved by simply taking the inverse of the Laplace transform in Eq. (9.9). Doing this, we arrive at our final solution.

$$v_0(t) = V_S u(t) + (V_S - 1)\cos(\omega_0 t) - I_{ld}\sqrt{\frac{L}{C}}\sin(\omega_0 t) \tag{9.11}$$

where we have designated the angular frequency as

$$\omega_0 = \frac{1}{\sqrt{LC}} \tag{9.12}$$

We can plug in values from the table and examine the shape of the step response. Note that the second term goes to zero since $V_S = 1$. Since the voltage starts at 1 V, we would expect that the droop to begin here as well. In addition, because V_S is a constant, the voltage should droop temporarily and then come back to a point at DC that is essentially the same. The reason is that there is no resistance (damping) in the path. If there was, we should see the settling point to be lower than the starting voltage.

As shown in Figure 9.33, we see the droop behavior clearly. The solution hits a minima around 937 mV and starts to turn back up. This is because the capacitor starts to recharge, and the voltage will begin to increase. We can change the filter characteristics to see how this behavior might change. For example, if we were to increase the inductance by a factor of 5, would the droop go down or up? Figure 9.34 gives us the answer.

The droop has not only increased, but it has taken longer to hit a minimum. This is one reason why the filter of the VR plays such an important role in the time-domain droop response of the system. It is even more critical when the

Figure 9.33 Droop derived from analytical methods.

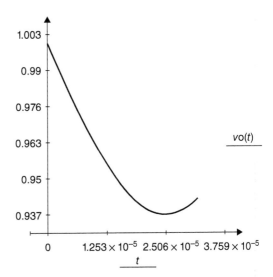

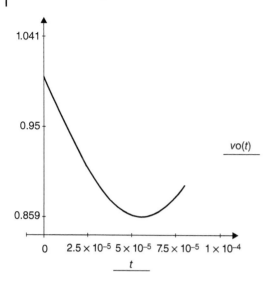

Figure 9.34 Droop with increase in inductance.

switching nature of the VR comes into play. This behavior is illustrated in a number of problems at the end of the chapter.

Looking at the behavior analytically sometimes gives insights into portions of the PDN behavior – in particular, the voltage droop, which is the critical time-dependent behavior of the PDN. In Section 9.5, we analyze the boundary budget for the system and look for where the system can be improved.

9.5 Boundary Budget System Discussion

After going through different load events that can cause voltage droops and understanding an analytical approach to the droop for the slowest event, the next objective is to examine some boundary analysis with respect to the PDN. This is similar to using Monte Carlo analysis to determine variances in the droop due to change in the components of the PDN over time, temperature, and other effects.

It is not always clear that when a PI engineer starts a droop analysis that the voltage droop event will always meet the requirements over the life of the product. As discussed previously, we also have seen that it is not always the case that the worst-case droop event is due to the largest current step at the highest *p-state*. Thus, it is important to have an idea in what could be the biggest problem for the given platform. This is as much a part of the analysis process as it is comprehending the boundaries that were given by the systems teams toward understanding what those limits are.

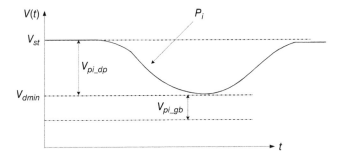

Figure 9.35 Typical boundary window for budget.

To start with, the PI engineer usually sets up the boundary conditions to make sure that they understand where the guardband is in the system. Doing a boundary budget analysis first is always helpful to make sure that the margins are understood before the PI engineer starts doing a statistical boundary budget study. Figure 9.35 shows a simple visual for the boundaries that one would start with and how they stackup.

The droop event for a particular *p-state* (p_i) will be assessed first from a starting point (V_{st}). This is normally determined from the maximum step that is possible for P_i. Normally, the guardband for the system allows for an accumulation of the statistical variances to ensure that all of the tolerances were accounted for and that the droop event would never exceed that minimum value. Thus, there is a *predicted* V_{dmin} that is intended to be the lowest point in the droop over a given operational range.

However, this is just the initial analysis. If we add in all of the statistical variances of the components of the PDN, we can estimate what that lowest value should be. This is illustrated in Figure 9.36.

As discussed in Chapter 4, it is clear that many of the guardband elements are statistical in nature. Thus, the first thing is to compute this as an estimate. The second thing is to check what the voltage regulator set point should be. Because the VR also varies over time, the engineer needs to account for this. This is an accumulation of errors from the amplifier(s) drift, internal resistor tolerances,

Figure 9.36 Droop event due to statistical variances in PDN.

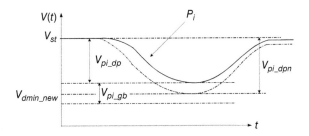

and filter errors. Because these add up and are statistical as well, they need to be added into the budget.

With respect to the voltage regulator, engineers often want to add in to their analysis – assuming that they have access to the model – the changes to the components in the VR. This is certainly possible but very difficult to do in general. Mainly because the VR componentry is very complicated overall and trying to capture every element in the VR that can drift over time is challenging at best. Thus, it is often more judicious to simply use a statistical accumulation and then add that into the guardband.

The goal here is to understand where the system designer may take advantage of excess margin in order to (i) save power in the system during an operating state or (ii) increase performance due to the extra margin. As we discussed earlier, if there is more margin in the droop event, then this can translate to [potentially] a shallower load line. This directly affects the power consumed in the system for a given operational mode or sets of modes. With respect to guardband, this can additionally contribute to a total reduction in the guardband given that the event is less than predicted.

The boundary budget starts with a particular operating and design point, and then the margins are added in the system. However, this is performed within a particular *mode* for the power supply. This is important since the power supply creates different noise and margin for different modes of operation. For example, for a higher current load (sustained), it is common to operate a multiphase buck design with most if not all of their phases on. When this occurs, and when the phase alignment between the phases is tuned to minimize ripple and noise, the guardband changes as a function of the number of phases and the amount of ripple that is generated. This is illustrated in Figure 9.37 as the VR is transitioning from a lower current to a higher current due to the load increasing.

Normally, the guardband is already accounted for and assumed to be at its worst case for all of the modes of operation (the ripple is usually the highest for the smallest number of phases operating) for the VR during a particular

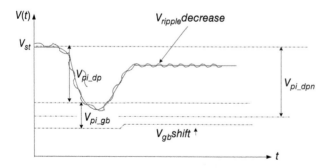

Figure 9.37 Illustration of guardband shift due to larger ripple.

p-state. Unfortunately, this results in additional power loss in the system. The main problem, of course, is that the mode of the VR cannot be predicted under all circumstances, and thus to ensure margin in the system, it is always prudent to adjust the guardband to the worst case.

If the VR is also operating within a particular load line, this can cause the guardband on the *load line* to change as well. Very seldom, if ever, will a VR change its load line during a particular operational state. This means that, for a given voltage set point/*p-state*, the load line will remain constant during the entire workload. In many cases, the load line may be fixed even across the *p-states*, even though this usually results in a suboptimal design. Thus, when the VR is active, the voltage will continue to adjust back and forth along the load-line curve as the current modulates.

However, as shown earlier, the VR will also *mode-switch* across these load changes automatically. For example, the VR can increase the number of phases to flatten out the efficiency curve and respond to the current demand as required. As mentioned, this can have an effect on the ripple in the system and usually results in a smaller ripple than in a previously lower current load. The question is can system teams take advantage of this additional margin?

The answer can sometimes be quite complicated because mode switching and load behavior are both complex events and interact with each other continually. The mode switching of a multiphase supply occurs within *windows*, and depending upon the load current, the voltage could be drooping at the time the mode-switching occurs. Thus, additional margin is often required to handle this additional switchover noise to allow for the VR to respond to the change in load current. As we discussed in Chapter 2, the feedback of the voltage regulator plays an important role here and to ensure stability the switchover event often requires a dwell time as the number of phases are increased or decreased as needed.

When designing a PDN and when considering all of the system noise and guardband, it is important to understand what can cause the worst-case conditions and thus design the margins into the system such that the final operational modes result in continued data integrity under all system conditions. In Section 9.6, we examine the power loss aspects of the PDN with regard to these system issues.

9.6 Power Loss Due to the PDN

Though for many platforms power loss is less of a concern, it is true for mobile-based systems that any power loss can contribute to less battery life and even performance. In the past, the impact to the system from the power loss in the PDN was ignored simply because its contribution appeared to be small. However, the requirement to make solutions smaller and thinner has

Table 9.4 Sources of power loss in PDN.

Source	PDN system cause
Dynamic silicon power	Load line and guardband
Silicon leakage	Load line and guardband
DC interconnect loss	Resistive network from source to load
AC power loss	Capacitors in PDN, load and VR switching ripple

had the effect of increasing the resistance between the power supply and the load. Though the effect seems small, this slight increase has started to contribute to the overall power loss in the system. Moreover, guardband and load-line slopes can have a significant effect on the power loss in the system due to the increase in the voltages at the source and load. When one adds in the effects due to ripple, the overall impact can be more than marginal.

As we discussed in Chapter 2, the efficiency of the converter is usually the first place engineers look at when power is being consumed in the power delivery system. Though efficiency in the VR is important, compared to the effects at the load (data integrity in particular), it is often less than 10% of the overall system impact to power loss. Interestingly, the largest impact to the power loss occurs at the load when active, since small changes in current and voltage result in substantial power changes. This is why an examination of the PDN and its impacts on power is an important aspect to power integrity.

Table 9.4 shows the potential areas where a PI engineer needs to focus on when examining power loss in the PDN. There are obviously many more sources relative to the silicon, but those are out of the purview of the power integrity engineer's area. The ones that are of most interest are those that affect the largest effect on silicon, that is, the load line and guardband [5].

We saw earlier in this chapter, and in Chapter 3, how changing the slope of the load line can have an impact on power. We can see the overall impact analytically by looking at what happens within a given *p-state* when changing the slope by a small amount. However, before we do this, we need to examine how power is consumed in the first place at the load.

9.6.1 Dynamic Silicon Power and Leakage

Reviewing our discussion on load line and silicon power loss again from Chapter 3, there are two sources of power that are typically examined when looking at the silicon load: dynamic and leakage. The dynamic power is related to the dynamic capacitance of the FUB (Functional Unit Block) and is represented in simple terms mathematically by the following equation:

$$P_D = AF \cdot C_{dyn} V^2 f + V I_{leak} \tag{9.13}$$

The second part of the loss is due to leakage of the silicon. This is considered a *nondynamic* event and occurs when a voltage is applied to the silicon but when no activity, or processing, is evident. Leakage occurs in advanced silicon devices and is highly dependent on the voltage applied to the device. Often leakage power is expressed as a power expression.

$$P_{leak} = KV^\alpha \tag{9.14}$$

where K is a constant to be determined and α is typically between 2.5 and 3.5 for advanced silicon processes.

The leakage current, in general, is a function of the transistor leakage, which can be estimated for an NMOS (negative [channel] metal-oxide semiconductor) transistor from the following:

$$I_{NL} = \mu_N C_{OX} \frac{W}{L} V_T^2 e^{\frac{V_{GS} - V_{THN}}{nV_T}} \left(1 - e^{-\frac{V_{DS}}{V_T}}\right) \tag{9.15}$$

This, however, is only an approximation for advanced processes. It does illustrate though that the leakage current can be complex which is why the use of Eq. (9.15) is normally used to examine the leakage globally in a silicon FUB.

For dynamic power, the activity factor (AF) is related to the workload in the processing unit. Normally, the activity factor does not exceed 60% since it is related to the number of logic cells that are active at any given instant. The activity factor is usually associated with a particular benchmark for that processing unit and system. For more active benchmarks, 40% is a reasonable activity factor. For less active benchmarks, 15–20% is usually the norm.

The dynamic capacitance is related to the aggregate capacitance for the FUB and represents (roughly) the capacitance for that unit that can switch actively when the unit is doing work. Often, an entire processor is assigned a dynamic capacitance based on the number of transistors or the amount of logic in the core. As one can see, the other factors are the frequency of switching and the voltage. Thus, even if the activity factor and C_{dyn} are constant, an increase from one *p-state* to the next will result in the square of the voltage and a linear increase in the frequency together. It is no wonder that the dynamic power in a processing unit can swing significantly from the lowest to the highest *p-state*.

How the system PDN and design affect both the dynamic and leakage power at the load is related to the load line and guardband of the system. This is intimately related to the design of the VR as much as anything is. However, if the interconnect of the PDN is large, then this increases the voltage drop from the source to the load which requires the voltage to be set higher at the output of the filter. Normally, this would not have an effect on the load since the VR could sense the voltage close to the silicon and thus the voltage at the load would not change. However, when the interconnect is further away from the load, normally this means the capacitive networks are as well. If the decoupling is further away, this means the response of the voltage regulator should also be

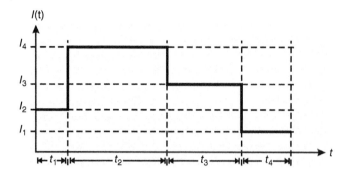

Figure 9.38 Nominal active current profile.

slower. By necessity, it can mean that the droop can be larger overall. To miti-gate this problem, systems teams normally increase the slope of the load line to accommodate it. As we have seen, this results in more power loss since if the load line maximum current points are the same between a shallower load line and a steeper one, invariably there will be more power loss in the system as one approaches the maximum current. The alternative to changing the load-line slope would be to increase the guardband in the system. However, this will increase the power loss across the entire operating range and could result in higher power loss overall.

The next example illustrates how system PDN load line and guardband can increase the power at the load for a given workload.

Example 9.4   A power integrity engineer is given a representative workload behavior (Figure 9.38) that is similar to the average of all of the workloads that a server core will see in an actual application. The data represents over a short period of time the *active* current spent during that workload for a particular *p-state* along with the voltages that are associated with it. The related times and modes are shown in Table 9.5. The other parameters for the power compu-tations are provided in Table 9.6. The system team wants to know if by trading off guardband and load line for two different system setups which will result in better power loss overall. Given the data in the table, compute the power con-sumed during the particular load windows and estimate the energy consumed overall for the two different conditions for load line and guardband.

Solution:
Part I: The first step is to compute the active power consumed (and energy) given the first load line and guardband. The workload operates at a high *p-state* and operational frequency (which is similar to many server workloads), and thus, when the processor is active, the power will be fairly high. We also note that the workload spends its time essentially in two modes. Thus, we need to

Table 9.5 Data for workload/current profile in Example 9.4.

Parameter	Value	Mode
I_2	$0.15 \times I_{max}$	P1-initialization state
I_4	$0.7 \times I_{max}$	Mode 1 – high active state
I_3	$0.4 \times I_{max}$	Mode 2 – mid-active state
I_1	I_{min}	Leakage nonactive state
t_1	0.2 ms	P1-initialization state
t_2	0.5 ms	Mode 1 – high active state
t_3	4 ms	Mode 2 – mid-active state
t_4	3 ms	Leakage nonactive state

Table 9.6 Parameters for power loss problem in Example 9.4.

Definition	Parameter	Value
Activity factor (init mode)	AF_init	0.05
Activity factor (high mode)	AF_h	0.4
Activity factor (low mode)	AF_l	0.2
Load-line slope	R_{LL}	0.044/0.051 Ω
Dynamic capacitance	C_{dyn}	10 nF
Frequency of operation	F	2.5 GHz
Guardband voltage	$V_{gb1/2}$	100/80 mV
V_{min1}/V_{min2}	$V_{min1/2}$	1.0/0.98 V
V_{max1}/V_{max2}	$V_{max1/2}$	1.2/1.21 V
Maximum current	I_{max}	5 A
Minimum current	I_{min}	0.5 A
Leakage power	P_{1_L}/P_{2_L}	0.1/0.4 W
Leakage voltage	V_{1_L}/V_{2_L}	0.6/1 V

compute the voltage and currents for these particular modes, along with the leakage. We also need to compute the leakage power when it is not running.

To simplify the analysis, we will assume that the transition from one mode (current load) to the next is very fast, resulting in minimal transition power loss. Our first step is to compute the correct voltages and currents for the load line and guardband. We have been given information on I_{max} and I_{min} and the associated starting points for the voltages for those two data points for the active current consumed. We can compute the voltages for the different points, and then we need to compute the power from the active modes.

We are given the load-line information and the end points. The load line is a negative slope. However, we do not know the y intercept of the line. This information will be important to compute because we need to know how to compute the voltage during every mode in the profile to get the actual power and energy. The equation for the voltage can be derived as

$$V_{mode} = -R_{LL}I_M + V_0 \tag{9.16}$$

where V_{mode} is the voltage, R_{LL} is the load-line resistance (slope), I_m is the modal current, and V_0 is the $I = 0$ intercept point. Note that V_{gb} and V_{min} have already been taken into account once V_0 is determined (see Eq. (9.17)). If we take the two data points that we are given, we can solve for the y intercept or where $I = 0$.

$$V_0 = \frac{1}{2}[(V_{min} + V_{max}) + R_{LL}(I_{max} + I_{min})] \tag{9.17}$$

Solving Eq. (9.17) gives us the last voltage that we need to compute the dynamic power. The next part of this analysis involves computing the leakage power. We have been given two points to help build our equation. We may estimate leakage power by using Eq. (9.14). First, we solve the coefficient α,

$$\alpha = \frac{\log\left(\frac{P_{L2}}{P_{L1}}\right)}{\log\left(\frac{V_{L2}}{V_{L1}}\right)} \cong 2.714 \tag{9.18}$$

This value is consistent with what we would expect from an advanced process node. We can then solve the constant K by plugging in.

$$K = \frac{P_{L1}}{V_{L1}^{\alpha}} \cong 0.4 \tag{9.19}$$

Using the equations for both dynamic and leakage power, we now plug in the appropriate voltages for all of the modes. The results are shown in Table 9.7.

Thus, we have completed the first part of the analysis. The second part is simply duplicating the analysis but with the new points for both the voltage guardband and the load-line points.

Part II: Our next job is to adjust both the guardband and the load line to see if the resultant power consumed is better or worse. The trade-off was to steepen the load line but to move the voltage guardband downward. If the system droop is still met by the two different conditions, then it is possible that the one that gives the lower power consumption would be the proper choice. However, it should be noted that this data is for only one particular workload. Normally, a designer runs through a number of workloads or benchmarks and then determines the average power consumed for both conditions.

The results are shown in Table 9.8 for the different guardband and load line.

Table 9.7 Part I computations.

Part I	Active power (W)	Leakage (W)	Total (W)	Total energy (mJ)
Mode I	1.767	0.640	2.41	0.481
Mode II	11.38	0.476	11.85	5.93
Mode III	6.422	0.562	6.98	27.9
Mode IV	0	0.656	0.656	1.97
Total	—	—	—	36.313

Table 9.8 Part II computations.

Part I	Active power (W)	Leakage (W)	Total (W)	Total energy (mJ)
Mode I	1.792	0.6519	2.44	0.489
Mode II	11.16	0.464	11.62	5.81
Mode III	6.422	0.562	6.98	27.9
Mode IV	0	0.671	0.671	2.01
Total	—	—	—	36.253

The results here show that even though the load-line slope was increased, the actual energy consumed for this workload was less. One may at first think that the difference is negligible. However, if we were to see this difference in every core over a 24 hours period and for hundreds of thousands of processors, such as in a data center, the energy saved could result in very significant savings financially. Thus, even a very small change of less than 0.5% can result in a huge impact overall.

Examining the data further, it is important to note that for some modes, though the power consumed was higher, the actual energy consumed overall was less. This was because the energy consumed for a few of the modes resulted in higher work done even though the power was larger for some of the shorter mode intervals. This is an important thing to note when analyzing power loss in a system with dynamic workloads. It is not always the highest power consuming mode that dominates the energy consumed in the system.

Many workloads with lower power consumption but with very long durations can consume higher energy in the system, and for battery-based platforms, this can be crucial. Thus, it is important to analyze the load behavior to ensure that the optimal PDN and system settings are applied. This is one reason why performing an analysis like this is important for both systems and power integrity engineers. These "what–if" analyses can result in lower energy overall and can show benefit to the final product.

In Section 9.6.2, we cover the losses due to the DC resistance of the PDN.

9.6.2 DC Losses in the PDN

Though the DC resistance of a PDN should be well known, its impact on power is often less of a concern to power integrity engineers. The reason is simply because normally the power loss in the interconnect has been small compared to the other losses. However, with changes to both load line and guardband, the DC path becomes more important. The reason is because the voltage across it increases, so does the current which will increase the power loss.

Very often designers try to match the load line to the resistance of the PDN in order to compensate. Though this does have the effect of matching the resistance to the load line, it does not necessarily result in the most optimal power loss in the system as we saw in Section 9.6.1. What dictates the current is the silicon load and if the voltage is increased along with the frequency (*p-state* of the core, for example), then significant amounts of energy may be consumed and the efficiency of the platform may be degraded. This is why it is important to understand not only the actual DC path but what the load and source are doing to impact it as well.

The DC path resistance is quite simple if one is examining just the resistance at room temperature. However, as we have seen, the resistance can increase by quite a bit due to board temperature because the resistance is essentially the copper path from the output of the VR to the load. Additionally, however, what is often neglected in such simple analyses is the fact that the resistance is truly a function of the current density in the planes of the copper. Moreover, both the positive going path (voltage) and ground paths are normally quite different since layout engineers normally secure a solid ground return for the system prior to routing the power planes.

Thus, at first glance, though the DC resistance of the PDN may seem simple to analyze, it can become very complicated in a hurry. In fact, the DC path from the output of the regulator normally is a conglomerate of multiple planes combined. This can be seen in the plan view of the diagram below, illustrating the different planes one must traverse in order to eventually reach the silicon. This path also comprises the package which is often constrained due to routing of the signals into and out of the device.

As shown in the diagram, there are multiple planes that get traversed. Moreover, the interconnects between the planes become important with regard to the resistance. Since the resistance is a function of the current density in the planes, the current can egress into and out of a group of vias funneling the current. If the group is too condensed, the actual resistance of the path will

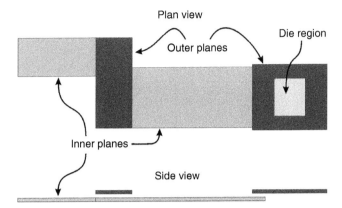

Figure 9.39 View of planes (without vias) for resistance estimate.

increase, which will result in higher loss. The solution to this is to place an array of vias such that the current path remains as large as possible. However, often this is not possible and, as shown in Figure 9.39, the current is forced into areas of the plane, resulting in unused copper and thus a higher resistive path. This can be true for the return path as well.

The other issue that often occurs in high-density designs is perforation of the planes themselves. This occurs when signals are forced to go through the planes above and below the power path, creating holes in the planes, which can result in a significant increase in resistance in that area. Perforation factors are normally assigned to such planes where the percentage of copper that is removed increases the overall resistance of the path for that segment. This is very evident within packages that have very high numbers of IO that must be routed from the top-side silicon down into the main motherboard in a design.

The next example illustrates what can happen in a simple DC analysis when an engineer examines the power path from a simplistic view and then from a more complex view. Using first principles, we can analyze both paths. Normally, engineers have field solvers at their disposal to extract the plane resistances. This can be important when trying to estimate accurately the DC loss in a particular PDN path. However, we show how one can get a reasonable result through good analytical techniques.

Example 9.5 A power integrity engineer is given a simple diagram of the power planes from a layout engineer and performs a quick resistance check on that path (see Figure 9.39).

Later the data is updated and after scrutinizing the PDN DC path, and the via locations, he comes up with a new diagram that resembles the one in Figure 9.40 along with the dimensions in Table 9.9.

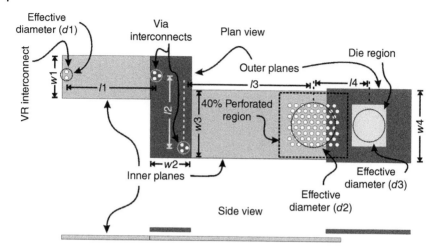

Figure 9.40 Detailed layout of power planes.

Table 9.9 Geometries for resistance calculations.

Metric	Value (mm)
L_1	15
L_2	11
L_3	27
L_4	8
D_1	3.5
D_2	9
D_3	6
W_1	10
W_2	8
W_3	12
W_4	15

Analyze both paths for the DC resistance and compare the results. How far off was the resistance for the more complex path than the simpler one?

Solution:
The first thing to do is to analyze Figure 9.39 in order to determine the resistance of the power path. This is just an estimate of course. We will assume pure copper at 20 °C and simply add up the values overall. We can use the formula in Chapter 3, which should give a reasonable approximation. Also, we have assumed that

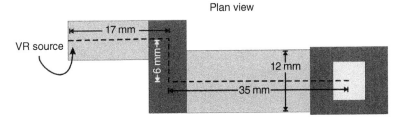

Figure 9.41 Mean free dimensions for planes.

the current density is constant, and thus from the voltage source on the right side, we would expect the DC resistance to be constant. Note that there are four planes that need to have the resistance computed. Figure 9.41 shows the approximate mean-free-path of the current.

We use Eq. (9.20) and Figure 9.41 to compute the simple resistance for the first plane and then determine the resistance for all of the planes.

$$R = \frac{l}{\sigma A} \tag{9.20}$$

Putting in the proper dimensions and computing, we get the first plane resistance based on Figure 9.41.

$$R_{Pl1} = \frac{17 \times 10^{-3}}{(5.7 \times 10^7)(0.0712 \times 10^{-3})(10 \times 10^{-3})} = 419\ \mu\Omega \tag{9.21}$$

Note that we converted the weight of the copper to mm (1 oz. copper is ~0.0014″). We now compute the rest of the plane resistances and assume the current is uniform. The results are in Table 9.10.

Now for the more complex resistance. We now can see that there is essentially a cluster of vias where the current is funneling in as shown in Figure 9.40. Thus, we need to compute the resistance based on the source and destination, assuming the source and resistance is a concentric circle (approximately). We can use the formula in Chapter 3 to compute the resistance from two points.

$$R_{pl1\_v} \cong \frac{1}{\sigma \pi t} \ln\left(\frac{s-a}{a}\right) \tag{9.22}$$

Table 9.10 Simple plane resistance calculation.

Plane resistance	Value (mΩ)
R_pl_1	0.419
R_pl_2	0.185
R_pl_3	0.719
Total	1.32 mΩ

The formula though assumes an infinite plane between the two points. In fact, the plane is finite, and there is a limited amount of copper between the source of the current in the plane and the vias on the other side. Additionally, we know that the current flows in a small region and that it is probably only uniform through a limited part of the plane. We compute the first resistance, assuming that the planes are at room temperature once again.

$$R_{pl1\_v} \cong \frac{1}{\pi(5.7 \times 10^7)(0.0712 \times 10^{-3})} \ln\left(\frac{15 - 3.5}{3.5}\right) \cong 0.093 \text{ m}\Omega \quad (9.23)$$

The result though is rather surprising. The reason is that we know that the resistance should be *higher* because the cross-sectional area is lower. However, if we examine the formula more carefully, the result is more expected than first thought. As we stated above, the formula assumes an infinitely large plane region. This is obviously a *wrong* assumption. So where do we go from here?

Since we know our formula is not correct for this problem, we need a better estimate for this. Since the planes are finite, we know that the boundary of the problem, our lowest or best-case resistance assuming the current is uniform throughout the planes, is given in Eq. (9.23). We may also assume that a reasonable worst case would be if the current was uniform from the via cluster boundary all the way to the next via boundary cluster. We could use this as an estimate. However, it would certainly be pessimistic. A more reasonable approach would be to assume the current is uniform at the middle of the plane only and then use a linear approximation to average the width. This is similar to the approximation shown in Figure 9.42.

The actual current envelope would be determined from a field solver. Thus, applying a simple linear approximation for the average width yields

$$w_{avg} \cong \frac{(3.5 + 10)}{2} = 6.75 \text{ mm} \quad (9.24)$$

Thus, the approximate resistance would then be

$$R_{pl1v} = \frac{15 \times 10^{-3}}{(5.7 \times 10^7)(0.0712 \times 10^{-3})(6.75 \times 10^{-3})} \cong 548 \text{ } \mu\Omega \quad (9.25)$$

We see quickly that this is much more in line with a current density that is higher than for a full plane width, which now shows that our resistance is

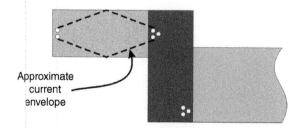

Approximate
current
envelope

Figure 9.42 Section of plane showing current envelope.

Table 9.11 Plane resistance calculation for complex geometries.

Plane resistance	Value (mΩ)
R_pl_1	0.548
R_pl_2	0.510
R_pl_3	1.07
R_pl_4	0.263
Total	2.39 mΩ

therefore higher. The other resistances we simply show the results but leave as exercises for the reader at the end of the chapter. In the second plane, one should notice that one via cluster is diagonal to the first. As a hint, the reader may assume that the horizontal separation here is about 4 mm. Placing all of the values now in a Table 9.11 yields the following computations:

Though one could argue that the approximation using the average width is certainly not exact, it is actually reasonably close to the actual envelope. Moreover, the results show that the resistance is significantly higher than the simple approximation where we ignored the current density due to the vias. Additionally, we did not make any assumptions about the perforation which would increase the resistance even further. For a given current going through this section of the PDN, the power loss would be higher by 80%. This example illustrates how important it is to use reasonable approximations for estimations of resistance and power loss.

9.6.3 AC Power Loss in PDN

The last item to consider in the power loss of the PDN is the alternating current (AC) losses. Normally, these would be the smallest components in the overall loss in the system. However, if there is a lot of workload activity that results in multiple droop events over a period of time, this loss can add up.

In most cases, the AC signal that is of most concern is the slowest one. This is because the power is lost in the equivalent series resistance (ESR) of the capacitors that are doing the bulk of the decoupling [6]. Because the other power is reactive, the energy is mostly transferred from one part of the network to the other. Unfortunately, some of this power is lost before it is delivered to the load.

If we examine a typical network near the load and assume the signal is sinusoidal for the most part, we can estimate the root mean square (RMS) power that is lost in this network. Very often, the ESR of the network (aggregate) increases as one gets closer to the load. This is typical because the HF capacitors

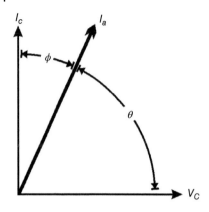

Figure 9.43 Angle between capacitor current and actual current relating to loss tangent.

are designed to have low equivalent series inductance (ESL) rather than ESR and some dampening will take place.

The magnitude of the impedance of a simple capacitor (neglecting the inductance when at low frequency) will be

$$|Z_C| = \left| \frac{1}{j\omega C} + R_{ESR} \right| \tag{9.26}$$

We can find the ESR in a capacitor (if it is not already specified) through the loss tangent. In an ideal capacitor, the current would lag the voltage by 90° as shown in Figure 9.43. However, due to the dielectric loss, there is a slight angle offset as shown in Figure 9.43, which results in the actual loss for the capacitor.

The loss tangent is usually specified in a data sheet and is related to the ESR through the following equation:

$$\tan \phi = \frac{R_{ESR}}{X_C} = DF \tag{9.27}$$

The loss tangent is often called the *dissipation factor* or *DF* for the capacitor. The reactance of the capacitor, X_c, is the impedance at a given frequency.

$$X_C = \frac{1}{j\omega C} \tag{9.28}$$

The ESR is actually only constant at a single frequency and is related to the dielectric material conductivity, permittivity, angular frequency, and the capacitance.

$$R_{ESR} = \frac{\sigma}{\epsilon \omega^2 C} \tag{9.29}$$

To find the power loss in the capacitor, we need to first find the total impedance for the capacitor. This may be found by first computing the ESR for the capacitor and the reactance. The impedance is fairly straightforward. The loss tangent though may be found from a data sheet or through measurement.

As an example, we can compute the reactance directly. Let us assume that we have a ripple voltage of 40 mvpp across a 1 µF ceramic capacitor at 1 MHz. Also, we assume the dissipation factor $DF = 0.025$ for this capacitor. The total impedance would then be

$$|Z_C| = \sqrt{\left(\frac{1}{\omega C}\right)^2 (1 + DF^2)} = 0.16 \, \Omega \tag{9.30}$$

The RMS current is then the RMS voltage divided by the impedance or

$$I = \frac{V}{Z_C} = \frac{0.04}{0.16\sqrt{2}} = 177 \, \text{mA} \tag{9.31}$$

The total power dissipated is then the RMS current times the ESR of the capacitor. In this case, it is

$$P_A = I_{RMS}^2 R_{ESR} = (0.177)^2 (0.004) = 126 \, \mu\text{W} \tag{9.32}$$

This does not sound like a lot of power loss, but one can imagine that with higher ripple and a different dielectric material, for example, the power loss could actually add up. Moreover, the ripple voltage is seen at different points in the PDN and with different values and frequencies across different capacitors. In one case, close to the load, the HF capacitors may see a 100 MHz ripple due to the load changing. Whereas near the voltage regulator, the voltage ripple could be at the frequency of switching (or a harmonic of it if the converter is multiphase) and result in an entirely different power loss. Either way, the PI engineer should be aware of the potential to affect the power loss of the system during operation. Moreover, as we have seen in Chapter 2, the voltage ripple of the converter is not just a function of the continuous mode operation of the converter.

Discontinuous mode operation can result in even higher ripple voltages, and for smaller platforms which typically operate over wide ranges of currents, the discontinuous mode losses may be higher in the capacitors overall then in continuous conduction mode (CCM).

One very straightforward way to compute the losses in a PDN is to look at the power losses in the frequency domain over an AC sweep, which we did in the previous chapter. The real question is what can be done if the losses are significant enough to change the performance of the system (efficiency) for engineers once the analysis is done? This is normally a very complex question as we have seen since changing portions of the PDN will affect not only the droop responses of the system but also the margins for the silicon teams at the load [7]. Thus, the trade-off may not be of any value especially if even a large percentage change in the PDN power loss results in a minor change in voltage margins could result in not only an increase in the overall efficiency of the system but could also result in lower performance (frequency) of the silicon.

One way to resolve this issue is to perform an analysis and then determine if changing portions of the capacitance for lower loss does not change the voltage margins across the frequency of operation. Sometimes, the quality of the capacitor may be improved without compromising the droop behavior such that the losses are reduced slightly. In this case, as long as other system aspects are not affected (area increase, reliability, etc.), the trade-offs may be worth it.

9.7 Summary

We have seen in the sections of this chapter that time-domain analysis is indeed one of the more critical analyses that a power integrity engineer can perform. The reason is that there is an indirect relationship between data integrity and power integrity that is crucial for the proper operation of the devices within a platform. When the droop begins to affect the load, delays between the clock and data in an active device can be affected, resulting in corruption of data and even system failure. Thus, bounding the droop in a system is critical to ensure proper operation. To help understand how droop affects the overall system, it is good to use standard nomenclature that defines the different droop durations. In this case, first, second, and third droop events were defined and discussed. In addition, droop event magnitudes are different depending upon what state the load is in. For dynamic high-performance loads, such as processors, the state of the device, for both frequency and voltage, will determine the limits of the droop event and the guardband in the system during that state. Normally, long droop events (third order droop) can be the least forgiving since they can cause the longest delays between clock and data. In order to understand this, power integrity engineers often run step responses in the time domain to determine their effects on the system. Higher frequency droop events are also important but are normally near the load and are of much shorter duration. However, multiple sequential load current changes can cause the droop the voltage to trend downward, especially if the excitation occurs at a specific time in a third-order droop event. Thus, examining the timing of such events is also important. Additionally, PI engineers should look for events where the system may be susceptible. Because the duration and frequency of the workload is somewhat indeterministic, it is good to examine certain current waveforms that can excite the PDN in a way where the droop may cause issues. This is particularly true for *p-states* where the margins may be smaller than expected.

There are analytical approaches to understanding the droop behavior as well which needs to be part of the toolset for the PI engineer. Boundary budget analysis is one of them, which helps engineers understand the limits of the droop in the system. Another critical aspect to the PDN which is often overlooked is the power loss. There are multiple mechanisms which can cause power loss in the system via the PDN. The one that is most critical is the effects the PDN has on the silicon voltage and load. Small changes in the voltage near the load can have large power changes. These losses manifest themselves as power lost due to the load itself but are caused by design choices in the system and the PDN. The typical losses that engineers examine are the DC losses of the interconnect. However, even these can be complex and are affected by the voltage source and the load state. Finally, AC power loss, which is often overlooked, can be critical mainly when a certain choice of decoupling is made. If there is significant ripple in a system, then depending upon the capacitor(s) that are chosen for the PDN, there could be extraneous losses in the design that are undesired. For analyzing the PDN, asymptotic analysis and Bode diagrams are quite useful especially with math programs, which enable engineers to break up the subsections and analyze their behavior together or independently.

After analyzing the PDN, prior to running time-domain analysis, engineers often need to try to optimize their designs. By examining the PDN as a function, one can not only determine the slopes and peak resonances but also use techniques to reengineer portions of their PDN to achieve a more optimal design.

In addition, it is important that engineers consider not only optimization but also where the PDN can change over time, temperature, and excitation. Monte Carlo analysis is very helpful in determining this key behavior. Power loss is also now an important aspect to PDN design where excitation at the right frequency and over a longer period of time may result in higher loss in the system than desired. It is best to perform a power loss analysis as a check in the system to ensure that certain frequencies of excitation cause larger than expected losses during operation.

Problems

9.1 Using Table 9.1 in Example 9.1 and parameterizing t_2 by varying it through the following values (10, 5, 2.5, 1 µs), determine the step response keeping the other variables the same. What changed? Why?

9.2 Using data from the previous problem, parameterize t_4 by varying the following values (124, 30, 15, 5, 2.5 µs). Determine the step response keeping the other variables the same. Compare with the baseline of 124 µs. Did the droop voltage change substantially? If so, where did it change?

9.3 What are the main issues with an increased droop at high frequency. List them and explain why?

9.4 Take the profile from Example 9.1, and using the same PDN/power source, look at the droop profile. If you were to alter this without increasing the amplitude of any of the currents in the profile, can you increase the droop? If so, how did you do this? Is the change feasible given the discussion on processor load generation capability?

9.5 Using the following values in Table 9.12, analytically determine the droop behavior and compare it with Figure 9.34.

9.6 Redo Example 9.4 using Tables 9.13 and 9.14.

9.7 Redo Example 9.5 using the values from Table 9.15.

Table 9.12 Values for analytical step response in Problem 9.5.

Definition	Parameter	Value
Inductor	L	$0.2\,\mu H$
Capacitor	C	$100\,\mu F$
Source voltage	V_S	$1\,V$
Load current	I_{ld}	$1\,A$

Table 9.13 Data for workload/current profile in Problem 9.6.

Parameter	Value	Mode
I_2	$0.05 \times I_{max}$	P1-initialization state
I_4	$0.9 \times I_{max}$	Mode 1 – high active state
I_3	$0.2 \times I_{max}$	Mode 2 – mid-active state
I_1	I_{min}	Leakage nonactive state
t_1	0.25 ms	P1-initialization state
t_2	0.45 ms	Mode 1 – high active state
t_3	4 ms	Mode 2 – mid-active state
t_4	3 ms	Leakage nonactive state

9.8 Determine the power loss and impedance for a capacitor switching at 5 MHz which has an ESR of 5 mΩ, capacitance of 2 µF, and inductance of 125 pH. Plot the impedance over frequency from 100 kHz to 300 MHz. Where is the resonance? What is the dissipation factor and loss tangent?

Table 9.14 Parameters for power loss problem in Problem 9.6.

Definition	Parameter	Value
Activity factor (init mode)	AF_init	0.05
Activity factor (high mode)	AF_h	0.6
Activity factor (low mode)	AF_l	0.35
Load-line slope	R_{LL}	0.044/0.051 Ω
Dynamic capacitance	C_{dyn}	10 nF
Frequency of operation	F	3 GHz
Guardband voltage	$V_{gb1/2}$	90/70 mV
V_{min1}/V_{min2}	$V_{min1/2}$	1.05/1.01 V
V_{max1}/V_{max2}	$V_{max1/2}$	1.2/1.21 V
Maximum current	I_{max}	20 A
Minimum current	I_{min}	1.0 A
Leakage power	P_{1_L}/P_{2_L}	0.4/0.6 W
Leakage voltage	V_{1_L}/V_{2_L}	0.6/1 V

Table 9.15 Geometries for resistance calculations.

Metric	Value (mm)
L_1	10
L_2	6
L_3	22
L_4	5.5
D_1	3
D_2	7
D_3	5
W_1	8
W_2	7
W_3	9
W_4	11

Bibliography

1 Krishna, A. (2008). *Understanding Setup and Hold violations in Digital System Design.* NCSU.
2 DiBene, J.T. (2014). *Fundamentals of Power Integrity for Computer Platforms and Systems.* Wiley.
3 Leng, J., Zu, Y., and Reddi, V. (2014). GPU voltage noise: characterization and hierarchical smoothing of spatial and temporal voltage noise interference in GPU architectures. In: *2015 IEEE 21st International Symposium on High Performance Computer Architecture* (7–11 February 2015). Austin, TX: University of Texas.
4 Pant, M. and Bowhill, B. (2012). *Era of Intelligent Power Delivery.* Intel.
5 Triquenaus, N. (2015). *Energy Characterization and Savings in Single and Multiprocessor Systems: Understanding How Much be Saved and How to Achieve it in Modern Systems.* HAL. https://hal.archives-ouvertes.fr/tel-01295567.
6 Fiore, R. (2007). *ESR Losses in Ceramic Capacitors.* American Technical Ceramics. http://www.atceramics.com/documents/notes/esrlosses_appnote.pdf (accessed September 2018).
7 Selli, G., Cocchini, M., Knighten, J. et al. (2007). Early time charge replenishment of the power delivery network in multi-layer PCB's. In: *2007 IEEE International Symposium on Electromagnetic Compatibility* (9–13 July 2007). Honolulu, HI: IEEE.
8 Schmitt, R., Huang, X., Yang, L., and Yuan, X. (2004). Modeling and hardware correlation of power distribution networks for multi-gigabit designs. In: *2004 Proceedings. 54th Electronic Components and Technology Conference* (4 June 2004). Las Vegas, NV: IEEE.
9 Mu, Z. (2008). Discussing impedance distribution with multiple stimulating sources in power distribution system design and simulation. In: *2008 IEEE-EPEP Electrical Performance of Electronic Packaging* (27–29 October 2008). San Jose, CA: IEEE.
10 Herrell, D. and Beker, B. (1998). Modeling of power distribution systems in PC's. In: *IEEE 7th Topical Meeting on Electrical Performance of Electronic Packaging* (26–28 October 1998). West Point, NY: IEEE.
11 Aygun, K., Hill, M., Ellert, K., and Radhakrishnan, K. (2004). Measurement to modeling correlation of the power delivery network impedance of a microprocessor. In: *Electrical Performance of Electronic Packaging* (25–27 October 2004). Portland, OR: IEEE.
12 Selli, G., Drewniak, J., Dubroff, R. et al. (2005). Complex power distribution network investigation using SPICE based extraction from first principle formulations. In: *IEEE 14th Topical Meeting on Electrical Performance of Electronic Packaging* (24–26 October 2005). Austin, TX: IEEE.

13 Archambeault, B., Cocchini, M., Selli, G. et al. (2008). Design methodology for PDN synthesis on multi-layer PCB's. In: *2008 International Symposium on Electromagnetic Compatibility* (8–12 September 2008). Hamburg: IEEE.

14 Mandhana, O.P. (2012). Decoupling optimization for IC-Package and PCB systems considering high performance microprocessor core and signal interface interactions. In: *2012 IEEE 62nd Electronic Components and Technology Conference* (29 May–1 June 2012). San Diego, CA: IEEE.

15 Paul, C.R. and Love, C. (2010). A brief SPICE tutorial. *IEEE EMC Society Newsletter*, Issue 225, Spring.

16 Cadence, Inc. (2009). *PSPICE Users Guide*, Cadence Design Systems. http://www.cadence.com (accessed September 2018).

17 Cadence, Inc. (2009). *PSPICE Reference Guide*, Cadence Design Systems. http://www.cadence.com (accessed September 2018).

18 DiBene, J.T. II, (2008). Integrated power delivery for high performance server based microprocessors. In: *International Workshop on Power Supply on Chip* (24–26 September 2008). Cork: PwrSoC.

19 Hockanson, D. and Dibene, T. (2007). Power delivery for high performance processor packages – Part I. In: *2007 IEEE International Symposium on Electromagnetic Compatibility* (9–13 July 2007), 1–6. Honolulu, HI: IEEE Inc.

20 Liu, X. and Liu, Y.-F. (2009). The extraction and measurement of on-die impedance for power delivery analysis. In: *2009 IEEE 18th Conference on Electrical Performance of Electronic Packaging and Systems* (19–21 October 2009). Portland, OR: IEEE.

21 Meijer, M., Pessolano, F., and Pineda de Gyvez, J. (2004). Technology exploration for adaptive power and frequency scaling in 90 nm CMOS. In: *Proceedings of the 2004 International Symposium on Low Power Electronics and Design* (11 August 2004). Newport Beach, CA: IEEE.

10

Silicon Power Integrity

10.1 Introduction

As discussed in the previous chapters, power integrity clearly encompasses many aspects and disciplines in the development of a computer product. From the voltage regulator through the platform distribution network and into the silicon, there is a plethora of technical problems to analyze. In this chapter, however, we discuss one of the more interesting and important challenges to power integrity that is starting to become ever more critical in the design of systems, that is, the power integrity on the silicon itself.

Silicon power integrity differs from platform power integrity in a number of ways. First, the frequency band of interest is normally much higher than that of the platform because the signals within the silicon are much faster, both in edge rate and periodicity. Also the generated load signals are less filtered. Second, the decoupling strategy and decoupling technology on a silicon device are very different from what is on a platform. The decoupling is normally due to silicon capacitance (often called transistor capacitance) or a type of metal capacitance (or sometimes MiM) or, for more complex systems, a variant of trench capacitance. Third, the distribution network is very different because of the smaller distances traveled and the interconnect technology. Within a silicon device, the traces are much thinner even for the power distribution, and the routing from layer to layer can be very complex and dense. These and other properties distinguish silicon power integrity from the platform.

It should be noted that interest in silicon power integrity is relatively recent. The reason, in part, is (in past designs) the lower complexity within the on-silicon routing and also the load transients. The change from monolithically powered devices (single rails) to multi-rail, and to higher localized currents (higher density currents), has brought on the need for a more detailed analysis of the efficacy of the power. Indeed, as more complex processing units and system on a chip (SoC) functions have been developed, the requirement for understanding the power integrity at the device level has increased. One of the biggest changes has been the trend from a single or small number of power

Power Integrity for Electrical and Computer Engineers, First Edition. J. Ted Dibene II and David Hockanson.
© 2020 John Wiley & Sons, Inc. Published 2020 by John Wiley & Sons, Inc.

rails to a trend toward the many. Very few advanced devices today require a single voltage rail to deliver power to it. By splitting the power into multiple domains on a single piece of silicon, the requirement to analyze the power delivery on die has grown significantly.

Another issue, as mentioned, is the increase in current density. As silicon processes continue to shrink, the current density inside the device becomes critical, and properly distributing the current (or charge) into regions effectively is a challenge in itself. Moreover, the large changes in dynamic current have resulted in potentially larger voltage droops, which can cause data faults as we have seen in the previous chapter. The current density problem has come about from the higher concentration of transistors in a region along with the advancement in silicon process technology.

The purpose of this chapter will be to discuss the issues that can occur on silicon in some detail. We will first start with how a high-performance device is generically constructed. This will entail understanding both the plan view (topology) of a high-performance device and the metal stackup. We will then move onto different decoupling technologies and strategies. Though there are a number of decoupling technologies available to the device designers today, each type of capacitor has both benefits and limitations in usage. We will also discuss routing between the package and the silicon and how that can affect power integrity. Often, the network from the off-silicon capacitance into the device can play an important role in how a functional unit will operate. Additionally, we will discuss how to analyze certain problems such as current density and local droop issues.

From an alternating current (AC) analysis perspective, the power distribution network (PDN) that is local is very different from that of the platform. Throughout the bandwidth of the platform PDN (normally below 100 MHz), the PDN is relatively flat and the perturbations that excite it are typically due to lower frequency current harmonics at the load. However, the portion of the PDN that is local can have significant droop behavior because of the fast di/dt events that can be excited along with high-frequency resonances in the 80–200 MHz regime. Thus, understanding this behavior is critical to silicon power integrity. We will therefore examine the local impedance near the silicon to examine these resonances.

The high-frequency resonance issue is exacerbated by the fact that many devices today have multiple blocks that can create large droops, and often the rails are split. To emphasize this a bit further, the increase in rails in a high-performance device is probably one of the key problems power integrity engineers face. As the silicon shrinks and the current densities increase, at some point, a limit must be attained to ensure proper design and physics are adhered to in the development process. As the power is brought into such complex devices, it will be important to consider the effects of power gating. Power gates offer a form of isolation from a given rail and can help in reducing power consumption overall for many platforms. However, once again, there can be issues with employing such circuits that should be discussed and how they affect the power integrity of a given domain. At the end of this chapter, we will discuss noise and how to reduce coupling between functional units in a device. Noise from other functional blocks on a silicon device as well as platform noise can affect multiple rails if the system aspects are not considered. It is also important to consider the development of the silicon itself along with the substrate and where currents can circulate from one section to another whether or not their grounds or power domains are tied together.

We start with a discussion of the device construction for a generic advanced silicon device in the next section.

10.2 Device Construction and Architecture Considerations

Before starting any analysis of a silicon power integrity problem, it is best to understand the basic construction of a device from the floor plan up through the stackup of the layers. Many high-performance devices today have upwards of a billion transistors in them and can have 10 or more metal layers for signal and power distribution. However, all systems are constrained by basic manufacturing and layout rules that must be adhered to. It is this combination of the design and manufacturing which governs the end device construction.

When power integrity engineers examine a particular piece of silicon, they normally start with the basic floor plan and the package capacitor placement. This entails looking at both the top and bottom sides of the package as well as the relative locations of the functional blocks. Figure 10.1 is a simple illustration of an SoC device with three simple domains. It should be noted that not all power domains in a high-performance silicon product require detailed examination of the power integrity. This is because some loads on a device are much less dynamic and have lower currents overall.[1] In the figure, we have outlined the areas of concern for possible power integrity evaluation (X marks the areas). Here, it is mainly the graphics and core processing regions that we wish to examine. This is typical for such high-performance devices. The main off-chip control and interface portion (general SoC) is normally comprised of many different types of analog and digital circuits, the aggregate of which normally does not create large dynamic events. Usually, the PI engineer does a check of this region for the PDN and then moves onto the more complex portions of the chip.

The plan (top) view shows where the main functional blocks are located and where one might place the capacitors. This is the side where the silicon resides.

---

1 Of course, this depends upon many factors, including the decoupling strategy and immunity to noise from other sources.

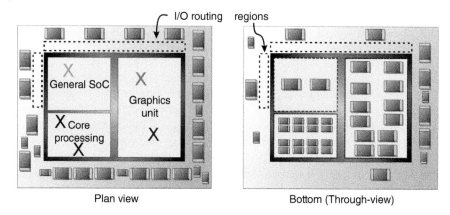

Figure 10.1 Simple example SoC plan and bottom views (X marks main areas of interest).

The core region has both large and small capacitors around the periphery. This is for both bulk and high-frequency decoupling (relative to the region of interest). The generic SoC region has fewer capacitors near the periphery due to the IO that needs to egress out from this area. The graphics unit has I/O, but it also requires some larger bulk capacitance.

The placement of the capacitors in each region is an important aspect to the layout. Note that there are a plethora of high-frequency capacitors placed on the bottom side of the package (through-view here rather than flipped for clarity). This is because this region can have relatively high current excursions. There will be some I/O going out to the package for this particular region but less so in this functional block relative to the other two. The general SoC region has much less capacitance underneath the package as well as on its periphery. This is to allow the signals to route out through the periphery of the package and onto the printed circuit board. These routing regions are marked in dashed box areas.

The stackup of the metal layers inside the silicon is part of the overall interconnect that ties the external capacitors to the load sources on die. Engineers who develop circuits or perform layout and design on a device are normally familiar with at least portions (if not most) of a design and understand the constraints to layout. In general, the upper metal layers in a device are dedicated to power distribution, while the lower metal layers are dedicated to routing signals. The reason can be seen in Figure 10.2.

The metal layers closest to the bumps on the device are typically used for power distribution and are normally placed in the middle of the functional block (in the plan view). This is because, as discussed previously, most of the IO signaling is on the periphery for escaping the silicon and package, allowing for easier distribution to other devices. Sometimes, there is ring or dedicated

Figure 10.2 Example cross-section and stackup of metal layers in silicon device.

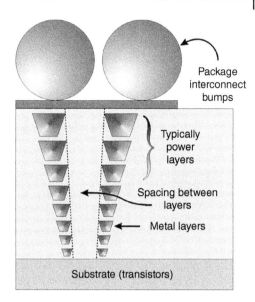

Package interconnect bumps

Typically power layers

Spacing between layers

Metal layers

Substrate (transistors)

Figure 10.3 Illustration of signal routing to I/O on device on periphery (top view).

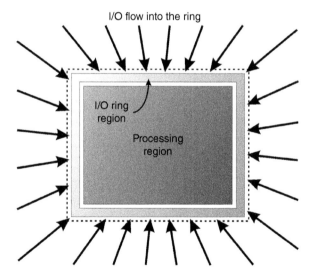

I/O flow into the ring

I/O ring region

Processing region

region on the periphery on one or more sides of a device, which allows for ease in signal escape as shown in Figure 10.3.

The reason why signals are routed to the periphery (mainly) rather to the middle of the silicon is because it is easier to route the signals from the outer layers of the package into the silicon from this point. If there are many bumps on a device, the signals must snake through the vias and the bumps along the way to get to their destination. The cross-section in Figure 10.4 illustrates this. In the

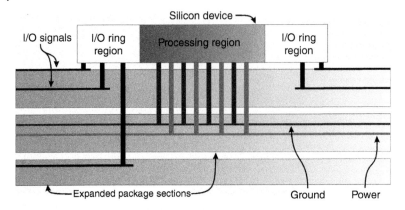

Figure 10.4 Cross-section of package and silicon regions (expanded).

package construction, the outer layers are normally finer, and the density of the signals can be higher. Thus, it is easier to route signals on these layers for this reason. The power and ground layers which are bulkier and carry the current are dedicated to the middle of the package where the metal is thicker. It is obvious now why the silicon layout is linked so closely with the package layout; both are designed to allow ease of routing for the finer IO signals while providing a low impedance path for the higher current power and ground regions.

Of course, the dedication of the layers in a package is highly dependent upon the number of I/O in a device, but in today's SoC's and processors, the number of I/O going into and out of a chip is normally in the thousands at least. Thus, congestion with signal and power routing is always an issue.

Conversely the power planes, or strips, as discussed, are normally routed in the innermost layers of the package as they egress into the silicon. The middle of the device usually has the bulk of the power and ground connections (Figure 10.4, middle). As we have seen, normally the package and the motherboard have inner layers which are dedicated to power distribution; thus, once the power is routed into the center of the silicon, the interconnect is then nearly vertically directed into the transistor load region. Because of this vertical path, it normally becomes the lowest impedance route (and lowest resistance) from the power source.

There are always trade-offs between power and signal integrity on a platform, however. Every layout and mask designer involved in either chip or platform layout and routing understands congestion between both power and signaling. This is because the layout of the device is not only limited by design, but also by manufacturing of the silicon itself.

The construction of a piece of silicon is complex and involves many steps. Over years of advancements in manufacturing and development, the computer-aided design (CAD) tools have become nearly as important as the

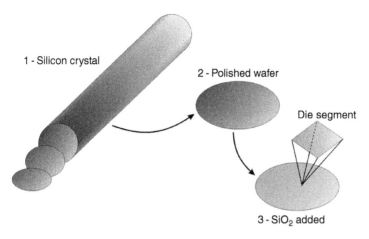

1 - Silicon crystal

2 - Polished wafer

Die segment

3 - SiO$_2$ added

Figure 10.5 Processing of wafers up to SiO$_2$.

construction of the silicon itself. Though these tools have enabled developers and layout engineers to create extremely advanced (and dense) silicon, there are still limitations in what a designer can do simply because of the manner in which the device is constructed.

Because of processing constraints, when constructing a wafer, the metal layers are built up from the substrate upward. To understand this, we take a moment to examine, at a very crude level, the process of constructing a wafer and the metal layers of the interconnect. A pictorial view of the construction is shown in Figures 10.5 and 10.6.

From the silicon crystal (ingot), wafers are cut and then prepared for processing. The wafer is initially polished in preparation for adding the layers. The first layer is SiO$_2$ (typically), which is added as a dielectric [1].

The next major step is photo-lithography where a light is projected onto the wafer through a mask which generates a pattern. The region that was not covered by the mask is then etched away to reveal the pattern. Additional layers such as polysilicon are added to the structure during deposition as well. Additional chemicals are then added to the substrate to create p- or n-type dopants into the silicon. From this point, additional metal layers are created through masks as part of the process repeats itself to build up the metals in each layer.

Once the masks are created to enable the creation of the transistors, the base is constructed through a photolithography process and then each layer is built up via another mask process until the entire wafer is constructed. Before a layer is added, the previous layer is polished (CMP, chemical metal polish) to ensure it is as flat as possible before depositing the next layer and interconnect. The process of adding metal layers to the wafer is called Back-end of Line (BEOL). Though the polishing helps to flatten the surface, it is still quite rough and the

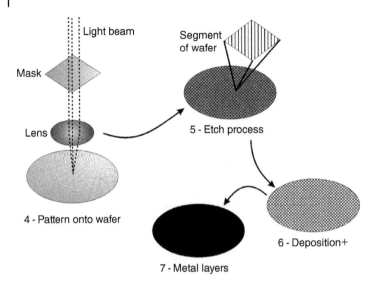

Figure 10.6 Next steps in silicon manufacturing.

tolerances between the wires become more difficult with each added layer. This is one of the key reasons why the metal layers become wider as one moves toward the bump interconnect of the chip. To ensure good yields, this spacing normally increases which reduces the density of the metal wires overall, but the height and width are normally larger which helps for power and clock trace routing [2]. The final steps are cutting the dies from the wafer and then packaging them. The final packaged parts are then mounted onto the system motherboards.

Between layers are vias, which join one layer to the other. The layers closest to the silicon are the densest and the finest and are thus most suited for signal routing. As the layers build up, they are normally constructed with thicker metal, and thus the distance between them grows larger to ensure separation due to tolerances. This is where the power layers are normally used to distribute into the lower silicon layers. Vias are still required to connect down into the transistors themselves, but normally, the thicker layers are used to distribute power laterally to minimize the resistance.

The key to a good power distribution on die is when the flow of current into the device results in minimal drop to the transistors. When a single rail supports one or more blocks on a piece of silicon, it is critical that the physical design team understands where the current densities may be the highest. This is because even when the planes are well distributed from the capacitors to the loads, it is easy to get *charge funneling* into an area, resulting in locally large voltage drops. This is even more critical as the transistor density increases, while the block that sinks the current decreases in size. We will examine the effects of current density into a transistor in more detail later on in the chapter.

Though process technology is becoming more condensed with every generation, there is a price to pay when delivering power to a functional block within a device. Unfortunately, this price is current density. For example, it is common for a core microprocessor to shrink with every generation of process technology, but at the same time keep, essentially, its current the same. As a result, though the power for that unit may have decreased, its peak and average current are likely to have stayed fairly constant over a number of generations. As the process technology continues to shrink and the currents tend to remain constant, the result can be very high current densities in areas of the processor, resulting in not only thermal and reliability issues but also power delivery problems.

As we saw in the previous chapter, delivering power from a wide plane into a particular region on the die does not guarantee that there will be no issues with voltage droop. This is because the smaller the sink of current, the higher the current density in that region. An example of this is a processing core on an advanced piece of silicon. Because of the architecture of most processing units, it is common for certain regions of the design to attain high current densities. Figure 10.7 shows a highly simplified example of a processing core architecture flow with standard execution and interfacing blocks.

Normally, there are many of these cores or units in a single high-performance device. The operation of an activity within such a processing unit is fairly straightforward. A very simple activity process is shown in Figure 10.8.[2]

When work is required, data and instructions are fetched from memory and then put into the scheduling unit. Once the instructions are taken from cache, they are then sent into the decode unit which determines the type of instructions and order for that particular workload. The instruction and data are sent into one of a number of execution units, depending upon the instruction type

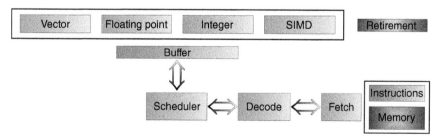

Figure 10.7 Simple diagram of processor flow.

---

2 Today's processors can have much more complex instruction flows including branching into other threads. The example in this section is simplified to illustrate what can happen with the ramp of current.

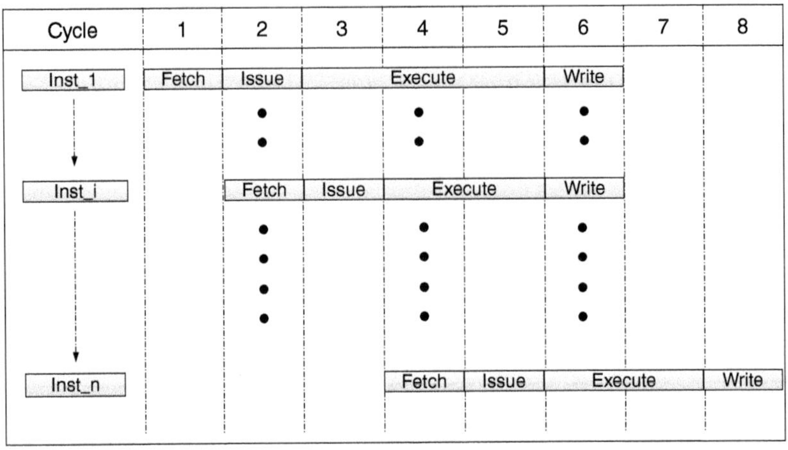

Cycle	1	2	3	4	5	6	7	8

Figure 10.8 Simple instruction process flow.

(floating point, integer, etc.). The output is then written back out to memory. Processors can perform multiple activities in parallel and can execute instructions in a non-serial fashion (out-of-order). Performance in a processing unit can be measured in a number of ways, but the most common is IPC or instructions per cycle. The higher the IPC the greater the performance.

However, the performance of the processing unit does not directly translate to where the current is or where the highest current densities will be. It is common to see very high current densities in the execution units when the unit is highly active. This can occur especially when the processor is running at its highest *p-state*. As was discussed in an earlier chapter, it can take a number of cycles before this ramp-up occurs. When an instruction is pushed into the pipeline, work is done on every cycle as the data and processing pushes the workload down the pipeline. However, the current is in part a function of how many transistors are active at any instant, and thus, it may take many cycles in order to activate enough transistors in an execution unit (or sometimes called a cluster) to cause the current to rise significantly. Once this occurs, if the charge has been somewhat depleted in that region, a voltage drop can occur. The topographical illustration in Figure 10.9 illustrates where the current density may be the highest in a processing unit at the peak of pipeline activity.

The points where the currents are largest may be seen in the three-dimensional version in Figure 10.10. As expected, the peaks are normally around the execution unit area, which means that the current densities will be highest there.

Figure 10.10 is somewhat generic, and the current densities are normalized for illustration purposes. There are other processing architectures where certain units may exhibit higher current densities than those shown (relatively

Figure 10.9 Example simple topography of core.

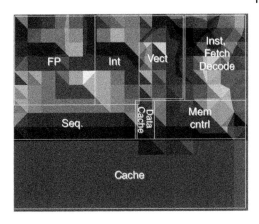

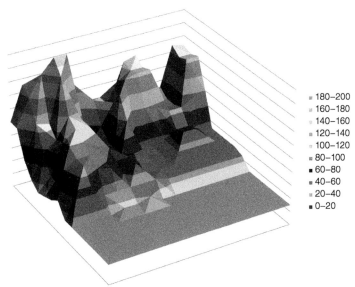

Figure 10.10 Relative current densities in a generic core processing unit.

speaking). The main point here is that the current can funnel into these regions, causing localized droops to occur. The key is to understand where these can happen and what decoupling and routing strategies are necessary to help mitigate them.

One of the first goals, besides proper modeling of these regions, is to ensure that the distribution impedance from external decoupling to internal decoupling in and about this area is adequate. This requires making sure that the on-silicon capacitance is *large* enough, and of *low* enough impedance, to make sure that when a large fast local current spike occurs, there is enough charge

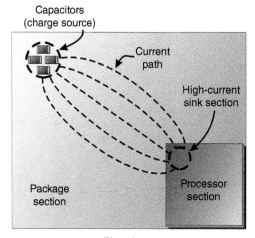

Capacitors
(charge source)

Current
path

High-current
sink section

Package
section

Processor
section

Plan view

Figure 10.11 Example current density on processor.

in and about the structure to minimize the droop event. In many cases, charge may be available from multiple sources on the die for the same rail. The question is, what will the lateral impedance be between one section and the other? To understand this requires investigating not only the local distribution network but also the decoupling into and out of the silicon.

Normally, the metal routing in the package is much bigger than the metal routing in the silicon. However, what governs a good distribution network is not only the resistance but also the inductance. As an example, we can see in Figure 10.11 where the current sink on the die is.

We can assume, as a worst case that this current sink will be essentially a point relative to the dimensions around it. To start with, we will assume that the local and adjacent capacitances are in locations relative to the source and sink. We will investigate two examples: one with a low resistance package solution and the other with a low inductive silicon solution. The cross-sectional constructions will therefore be different and will result in two different PDN's. The question is, which one will be sufficient to mitigate the high frequency (HF) droop? The next example illustrates this.

Example 10.1   An engineer is investigating two different designs based on Figure 10.11 and the cross-section in Figure 10.12. The generic characteristics of the PDN for both cases are shown in Figure 10.13. The values for the elements for the first case are shown in Table 10.1. Run a SPICE simulation given the current step of 5 A with a slew rate of 10 ns and find the maximum droop.

The second case has the same PDN but with different values, indicating that the package and silicon layer routing was different. Run this case in SPICE and

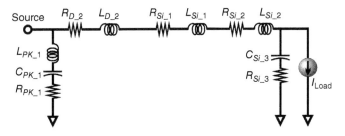

Figure 10.12 Cross-section of die and package.

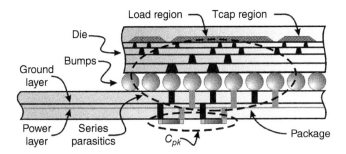

Figure 10.13 Schematic for local PDN distribution.

Table 10.1 Parameters for first analysis.

Parameter	Value
C_{pk1} (µF)	10
L_{pk1} (pH)	100
R_{pk1} (µΩ)	300
R_{d2} (µΩ)	500
L_{d2} (pH)	40
R_{si1} (µΩ)	700
L_{si1} (pH)	30
R_{si2} (mΩ)	2
L_{si2} (pH)	5
C_{si3} (nF)	80
R_{si3} (mΩ)	2

Table 10.2 Parameters for second analysis.

Parameter	Value
C_{pk1} (µF)	40
L_{pk1} (pH)	500
R_{pk1} (µΩ)	90
R_{d2} (µΩ)	200
L_{d2} (pH)	300
R_{si1} (µΩ)	100
L_{si1} (pH)	100
R_{si2} (µΩ)	300
L_{si2} (pH)	50
C_{si3} (nF)	80
R_{si3} (mΩ)	2

determine the local droop (Table 10.2). Which one is greater? What were the differences in the characteristics of the droops for the two cases?

Solution:
We set up the SPICE model and run the simulation for the first high-frequency step. Since this is a lumped element simulation, we can group the series elements together to simplify the analysis. Thus, we sum up the series elements for the first analysis and put them into one R–L model. The two values will be

$$R_{S1} = 3.2\,\text{m}\Omega, \quad L_{S1} = 155\,\text{pH} \tag{10.1}$$

The package capacitance is 10 µF. The other simulation will have the following properties:

$$R_{S2} = 0.6\,\text{m}\Omega, \quad L_{S1} = 950\,\text{pH} \tag{10.2}$$

The package capacitance is 40 µF. All of the other parameters are the same. We now run the first SPICE model and determine the results.

Figure 10.14 is our SPICE circuit. Because we are using an ideal source, there may be some anomalous affects due to the discontinuities at the ends of the pulse but for now we are only after relative differences in the droop.

The first simulation shows the voltage with the current step. Note there is significant ringing. This is clearly due to the inductance of the package. This is important to note since it is only 155 pH but is still causing this ringing (Figure 10.15).

The maximum droop appears to be in the 63 mV range though which would normally be acceptable for many operation modes. When we now change the

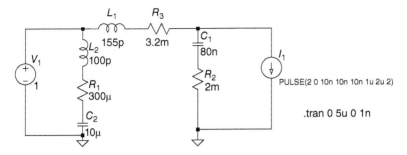

Figure 10.14 Simple circuit diagram of SPICE model.

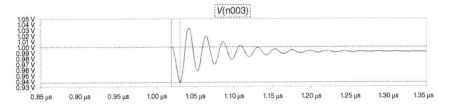

Figure 10.15 Voltage droop for first case.

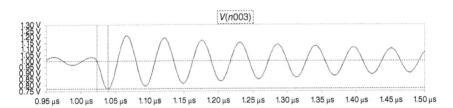

Figure 10.16 Voltage droop for second case.

parameters for the second case, we can see the effects of the larger inductance (Figure 10.16).

The second case is very revealing. It is clear that even though the series resistance went down, the inductance dominated the droop. This shows the droop to be ~250 mV. In fact, it is nearly 5× higher. This begs the question, what is the reason for this large increase in voltage droop and why is the inductance such an important quantity here? To answer this, we can simply examine the reactance of the series effects.

If the rise time of the signal dominates the harmonic content, we can approximate the magnitude of the impedance. The rise time of a signal is related to the frequency of the signal via the following equation:

$$f_{BW} \cong \frac{1}{2t_r} \tag{10.3}$$

Plugging in the rise time now, we approximate the bandwidth as

$$f_{BW} \cong 50\ \text{MHz} \tag{10.4}$$

The reactance for the second case is then

$$X_{Lpk2} \cong \omega L_{pk} = 2\pi(50\ \text{MHz})(955\ \text{pH}) = 300\ m\Omega \tag{10.5}$$

Compare this now with the series resistance of 3.2 mΩ, which is 100$\times$ smaller. This explains now why the voltage drop across this region is so large for the second case. We can see why it is important to mitigate the parasitic inductances in a package. It should also be noted that the small network before this inductance will have some effect on reducing the edge of the signal, but clearly, the capacitance would need to be increased significantly in order to have a reasonable effect.

The previous example illustrated three important items. First, the frequency content of the current spike is important when investigating localized droops. Second, it is critical to examine both the inductive portion of the interconnect and the resistive part. And third, placement and type of capacitance are very important when examining on-die high current loads. In the example, we did not discuss the type of capacitance that would be available to the engineer. Choosing the correct capacitance and amount and how it is interconnected between the load and external package capacitors can be a complex endeavor. In the next section, we discuss different types of capacitance that are available to engineers who develop silicon devices while we investigate their effects on the final design.

10.3 On-die Decoupling

Decoupling on silicon is the last defense in ensuring that the efficacy of the power distribution to the load is maintained. To mitigate the HF transient droops that come from the dynamic loads on silicon, such as core processors, it is the job of the local capacitance on silicon. Capacitance that is very near or on the silicon device is normally very different from the capacitance of a typical multilayer ceramic chip (MLCC) device that may be on the package or motherboard. The biggest difference is not only performance but also the density. Capacitance within the silicon is normally many times smaller in density (capacitance) than an MLCC capacitor that is located on the package or motherboard. The plot below shows different silicon capacitor densities relative to that of a common MLCC rated at a similar voltage level. Note that all of the capacitive technologies are much smaller in density than the common MLCC. One of the main reasons is that there is more vertical space to construct the MLCC ceramic device, whereas there is limited volume in

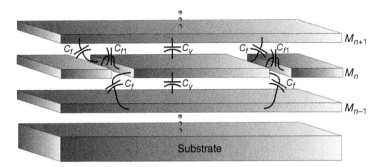

Figure 10.17 Simple metal capacitance in a device.

the silicon die for the others. Another difference is that the construction of an MLCC is very different from that of a silicon device (in most cases). Details on how MLCC devices perform were described in Chapter 3.

The lowest density capacitance is typically the simple metal type, which is basically using the existing silicon dielectric and adjacent metal routing layers to create a capacitor. This density is very small and is a function of not only the area under between the two metal layers but also the fringing fields to other metals that surround it. This is shown in Figure 10.17. The metals above and below the main capacitor as well as to the sides increase the overall capacitance. The center portion of metal layer M_n is normally the cathode or the positive node of the metal cap. The other metal layers are then the anodes. The cathode is tied to the main power layer for the device and thus also connects down into the power source for the transistor loads. Likewise, the other metal layers are tied to the local ground or return paths for the transistors.

The total capacitance comprises the sum of the vertical metal capacitance along with the fringing capacitance.

$$C_M \cong 2C_V + \sum_{i=1}^{n} C_{fi} \tag{10.6}$$

The metal capacitance vertically can be easily approximated given the area of the central metal region as a simple metal plate capacitor.

$$C_V \cong \frac{\varepsilon_r \varepsilon_0 A}{d} \tag{10.7}$$

where ε_r is the permittivity of the dielectric between the metal layers, typically silicon dioxide which has a dielectric constant of ~3.97 [2]. The area A is the physical area of the center metal layer and d is the dielectric thickness between the metal layers. The thickness varies depending upon the process technology. Depending upon the layer, it can be as low as 1 nm in the lowest metal layers all the way to 8–10 nm in the higher ones (mainly for advanced processes such as 28 nm or smaller).

The fringing capacitance (per unit length) may be approximated using a simple formula assuming the spacing is large relative to the thickness of the metal.

$$C_{f\_l} \cong \varepsilon_r \varepsilon_0 \ln \left(1 + \frac{t}{d} \right) \tag{10.8}$$

where t is the center metal thickness. For closely spaced wires, the capacitance may be approximated using the simple parallel plate equation (turned on its side).

Though this type of capacitor may be used for power decoupling, it is normally only used for small circuits that require very small but high-quality capacitance, and its main function is filtering rather than power decoupling. Metal capacitance can be very important for a number of circuits that use small filtering for analog decoupling. Circuits such as high-speed clocks and oscillators frequently use metal decoupling capacitors because of their availability and high quality. The other reason for using such capacitance in circuits is that the capacitor itself can be referenced to a different point rather than tied to ground. As we will see in a moment, some capacitor types must be tied to the main ground structure in order to get the most capacitance out of the area it occupies. As one can tell from the figure, it is also possible to isolate the capacitance by moving layers away from other possibly metal layers, which might have more noise on them. With regard to power integrity, some of these critical circuits have isolation from other power layers, that is, moving certain devices (both passive and active) away from them. For the PI engineer, this can involve looking into the susceptibility of a sensitive power plane due to noise from another. Understanding the design overall in a system is one reason why it is important for PI engineers to have a full system view of both the platform and layout of the devices on it.

Another type of capacitive technology that is used is the MiM or metal-in-metal capacitor. This is more of a dedicated parallel plate capacitor that often uses a slightly higher dielectric substrate than silicon but is also imbedded in the upper metal layers. It has been used for power decoupling for IO, but because its density is also relatively smaller than other capacitors and is also limited, large current transients can deplete the capacitor quite quickly.

One of the more common devices interspersed throughout logic regions, including high-performance processing units, is the transistor or T-cap. Essentially, this is a group of transistors, which have been biased to become strictly capacitors as shown in Figure 10.19. Transistor capacitance is very good for locally decoupling regions between circuits. As indicated in Figure 10.18, its density is normally much higher than other on-silicon technologies. Unfortunately, transistor capacitance is limited by process technology. The voltage rating on the transistor will be the same as that of any similar transistor in the device, and thus it can suffer from breakdown and avalanche effects just like any other CMOS device.

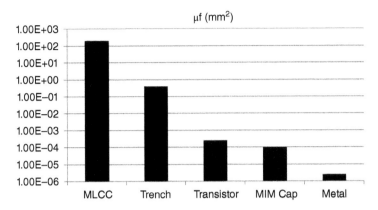

Figure 10.18 Approximate relative capacitor densities (note log scale).

Figure 10.19 Circuit diagram for transistor capacitor.

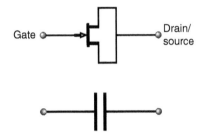

Transistor capacitance has good density, but it also can suffer from lateral routing parasitics unless it is right next to the active load area. It also has issues with voltage spikes though this is normally constrained by the power supply decoupling. Because transistor capacitance is used heavily due to its availability and ease of use in logic designs, it is worthwhile describing its properties in some detail here.

The circuit representation for a simple transistor capacitor is shown in Figure 10.19. The gate is normally biased by some voltage, while the drain and source (often interchangeable here) are tied together.

A simple representation of the metal-oxide semiconductor (MOS) capacitor is shown in Figure 10.20. The electrical representation is shown 90° from the cross-section to illustrate the dimensional direction for the flow of electrons and holes. The gate is tied to a highly doped P+ polysilicon layer, which is conductive. The insulator is the oxide layer which is typically SiO_2 (silicon dioxide). The semiconductor is the N-doped region.

There is a voltage applied across the gate and the body V_{GB}. As one would expect, the capacitance is dependent upon the voltage applied across it at the gate. The behavior of the transistor capacitor is typically more complex (at a macro level) than most capacitances [3]. The reason is that its operation

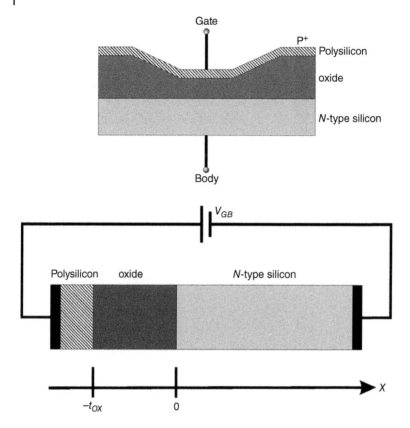

Figure 10.20 Simple cross-section of MOS capacitor.

depends upon the transport of electrons and holes from one part of the transistor to the other. There are three main regions in which the MOS capacitor may operate in: inversion, depletion, and accumulation. A simple graph of the capacitance as a function of gate bias voltage is shown in Figure 10.21.

The charge on the capacitor varies as a function of frequency which is why there are mainly two different values on the left side of the chart for the capacitance: high-frequency C_{HF} and static C_s. As the frequency increases, the charge density decreases which is why for power integrity it is important to understand the operational mode of the capacitor to ensure that proper amount of capacitance is assessed. In most cases, the MOS capacitor is used in the *strong inversion* or accumulation mode due to the operation of the local logic for the device.

As shown in the figure, there are two voltages which separate the three regions. There is the threshold voltage V_T that separates the inversion from the depletion region. The MOSFET can be operated in this region where

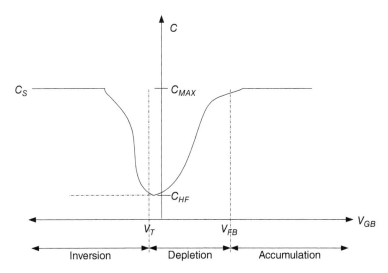

Figure 10.21 Regions of operation for MOS capacitor.

$V_{GB} < V_T$. In this region, the surface inverts its conduction type from n-type to p-type. In this region, the electrons are pushed away from the SiO_2 (silicon dioxide) interface, creating a positive space charge. Essentially, this layer now has no mobile carriers. In fact, there are still electrons here, but they decrease exponentially from the surface as they move toward the bulk region.

The middle region is between the threshold voltage V_T and the *flatband* voltage V_{FB}. The flatband voltage occurs when there is no net charge in the gate of the MOS capacitor. When the threshold voltage is below V_{FB}, a negative charge exists in the interface between the poly-gate and the oxide. Thus, a positive charge exists on the other interface of the semiconductor or the N portion of the substrate. The surface is thus depleted of its mobile carriers, which results in a positive space charge in the region.

The last region is on the right where $V_{GB} > V_{FB}$. This is where a positive charge is on the metal gate, and therefore a negative charge exists on the semiconductor. It is called *accumulation* because the electrons accumulate at the surface. The charge density on the device is a function of the gate voltage as shown in Figure 10.22.

The capacitance of the MOS cap is a function of the thickness of the oxide and the dielectric constant.

$$C_{ox} \cong \frac{\varepsilon_{ox}}{t_{ox}} \tag{10.9}$$

It is important to ensure that when estimating the capacitance in a transistor device, a proper operational mode is used. Equation (10.9) only applies for strong inversion or accumulation mode operation. Normally, this is not an

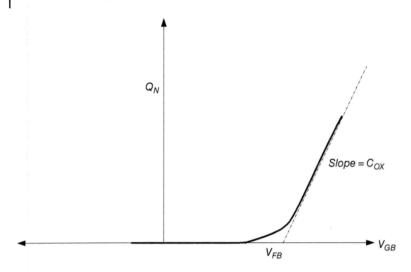

Figure 10.22 Capacitance in accumulation region.

issue, but as the voltage begins to sag in a certain region along with the gate bias voltage – especially for very advanced process nodes – shifting from accumulation to depletion is possible.

Another capacitor technology that is found external and sometimes internal to the device is silicon *trench* capacitance. As shown in the Figure 10.18, this capacitance has much higher capacitance per unit area than the others. The fundamental reason is that micro-trenches are used in the silicon to make cylinders where a cathode and an anode may be formed to create tiny little cylindrical capacitors. When ganged together, the capacitance density can be very large. Often this capacitor is built with a different technology than the main device process and then mounted close to or on the device.

As with all silicon-based capacitors, there are inherent limits to the technology that must be considered before designing them into a final product. As previously mentioned, the metal cap has limitations simply because of its small value. Thus, in general, it is not very good for large current transients. On the other hand, MiM capacitance is slightly larger but requires extra processing steps. Also, because of the lateral metal layers egressing into the MiM window area as discussed previously, the resistance can be relatively large. Thus, unless the window for the capacitors is small, resistance could be an issue for large current steps as well.

Though the trench capacitance has a much higher density, the trench resistance can be very high, resulting in a large IR drop when the current spikes hit nearby. This also slows the charge from moving from the terminals on the trench "cones" into the silicon load itself. Moreover, the additional packaging of the trench capacitor can cause vertical stacking and cost issues in the final product.

Normally, it is not one type of capacitor that is designed to solve such high di/dt issues but a combination of multiple devices. Even in such cases where space is limited, designers still make use of MLCC's which are placed as close as possible on the package or PCB to the load to mitigate the droop events. Though the lateral parasitics may cause a droop between these devices and load, the combination of on-silicon and off-silicon capacitance is intended to help mitigate these localized droop events. The key is in understanding the routing constraints within the layers of a silicon device and package to help mitigate these local voltage drops. In the next section, we discuss the important parasitics from routing on a device substrate.

10.4 Device Metal Routing Revisited

In the design of a silicon device, such as an SoC or a processor, the metal routing for the power is an important consideration. Physical design engineers work closely with mask designers to help route the power to the proper places on the die and to place the capacitance that is necessary for decoupling the loads. Depending upon the process technology and design constraints, certain metal layers are dedicated for power routing as discussed previously. As we have seen, due to manufacturing tolerances, the metal layers get thicker and wider as they get closer to the external contacts of the silicon.

There are very specific design rules that are used by the CAD systems to help prevent failures when the device is out in the field. These design rules include minimal metal densities as well as keep out rules for the manufacturing and reliability of the device in the long term. However, it is very difficult to predict how the device will behave with respect to power integrity based on the type of logic underneath the metal. The tools do not have insight into the external capacitance or the architecture of the silicon below. Thus, it is up to the power integrity engineer to help give guidance in such areas.

Figure 10.23 shows a functional unit, similar to the ones discussed earlier, which relates to a processor subsection. The architects are usually familiar with the activity of such a region and can somewhat predict the current draw into this area. The question will be, what will actually happen to the voltage droop in this section once the current pulse hits? If the lateral parasitics are large, even though the local capacitance is of high quality, charge may not be able to traverse from the next adjacent capacitors simply because of the impedances between the external capacitors and the local ones.

Thus, the goal here is to analyze the layers and distribution from one region to the next. In this case, the charge may come from many areas on the die. In many ways, the capacitance acts like a lateral charge or circular capacitor

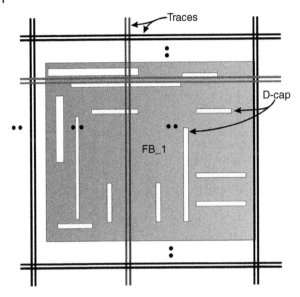

Figure 10.23 Functional block with decoupling capacitance interspersed.

where the charge comes in from all sides. If the charge is moving circularly (inward toward the load), the rate of the change of that charge is found from the following well-known simple relation:

$$I(t) = \frac{dQ}{dt} = \frac{dV}{dt}C \tag{10.10}$$

where both the voltage here is a function of time. The capacitor in this case may be represented as a radial structure or

$$C \cong \frac{\varepsilon_0 \varepsilon_r \pi (r_o - r_i)^2}{d} \tag{10.11}$$

where r_o is the outer radius and r_i is the inner radius of the load. The radial capacitor acts differently from a standard parallel plate capacitor in that the charge flow from the outer capacitors in the device flow radially toward the load [6]. This is more of an apparent action mainly because the load, as we have seen, appears as more of a point source, and thus if the capacitors are charged around the structure and are essentially at equipotential points radially, the currents will flow inward as shown in Figure 10.24.

If we assume a continuous plane for the power and ground connections, we may estimate the time constant from a given ring into the center of the load given the resistance of the metal. This is shown in the next example,

Example 10.2   An engineer is estimating the capacitance of a section of silicon and assumes a density based on the representation in Figure 10.24. Given the dimensions in Table 10.3 along with the metal resistance, compute the time constant from outer ring to the edge of the load.

Figure 10.24 Capacitance on-die appearing as "rings" of capacitance due to load.

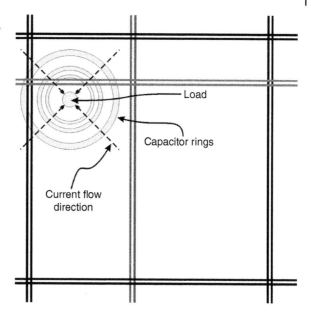

Table 10.3 Parameters for Example 10.2.

Parameter	Value
R_1 (μm)	500
R_2 (μm)	480
R_{load} (μm)	75
T_m (μm)	3
Temp (°C)	20

Solution:
The first goal is to compute the capacitance of the outer ring. The area of the ring is calculated as

$$A \cong \pi\,(r_o - r_i)^2 = \pi(500\,\mu\text{m} - 480\,\mu\text{m})^2 \cong 1260\,\mu\text{m}^2 \tag{10.12}$$

The capacitance for the T-cap is then

$$C \cong \frac{0.25\,\text{fF}}{\mu\text{m}^2} \times 1260\,\mu\text{m}^2 \cong 314\,\text{fF} \tag{10.13}$$

The metal resistance may be computed from the following relation at 20 °C:

$$R = \frac{l}{\sigma A} \cong \frac{(r_o - r_i)}{\sigma(2\pi r_a)t} \cong 1.86\ m\Omega \tag{10.14}$$

Thus, the time constant is now simply

$$\tau \cong RC = 0.585 \text{ fs} \tag{10.15}$$

One might compare this with placing a large capacitance adjacent to the load and then computing the time constant from that capacitor to the load. Because the resistance will be higher (as will the capacitance), the time constant can be very large. Thus, it is clear that there is some advantage to distributing the capacitance within the logic region, even though the size of the capacitance may be an issue for large loads.

Let us revisit the circular capacitor for a moment and determine why this is a reasonable representation for the capacitance on-die. For most very dense silicon loads, the current density in a particular region is small relative to the dimensions of the entire silicon functional block. This is particularly true in a processing unit. Thus, in essence, the load looks almost like a point source relative to the rest of the structure. Because of this, and because of the construction of the capacitance in the unit, the charge on the capacitors appears as units that are distributed radially from this point source. This is very different from the static representation where the capacitance is interspersed among the logic and the metal layers cross over the silicon orthogonally with respect to the power and ground layers near the top of the routing stack.

However, because of the fast change of current (di/dt) that occurs when a load occurs in the region, the flow of current moves, for the most part, radially toward the center as if the charge were distributed that way as well. This also gives the appearance that the charge on the capacitors appears as rings of capacitors where metal layers above connect each capacitor ring from one section to the other.

Because the change in load is so fast (typically in nanoseconds), there is often a delay between the charge that restores these capacitors and the source from the external capacitors that reside external to the silicon. If this is the case, then the capacitors actually lose their charge as the current increases, and the flow of current can result in the shrinking of capacitance.

However, the most important aspect to this has to do with how the capacitors are actually constructed and what other circuitry is connected to these capacitors. As we saw in the previous section, transistor capacitance does not behave like simple parallel plate capacitance. The reason is the capacitors themselves are made from transistors. We saw from Figure 10.21 that the capacitance changes as a function of frequency and voltage, which determines the region of operation that the capacitor is in. This is an important distinction between what engineers are used to when representing the capacitance in a system.

The other important aspect to the capacitor has to do with how the power is interconnected to the load. In Section 10.7, we will discuss on-die power gating, which is very common in many high-performance SoC's and processors today. On-die power gates connect the load to the transistor capacitance. Thus, the current must flow through these gates when the devices are on and the load is active. The power gates can be resistive, and because they are normally interspersed, like the capacitance, among the logic, as the current funnels into the load, two effects can be observed. First, the charge is depleted first from the capacitance on the side where the logic is. This has the effect of having to recharge this capacitance through the capacitance on the opposite side of the gate, which means it must go through this higher resistance path. Second, there is a temporal and dimensional effect to this charge motion, meaning the current that flows into the load comes from a localized region, and thus the gate and path resistance can be higher on the initial surge, causing both a voltage change in the capacitor and the time-dependent shift in how the charge gets replenished. This effect then changes how the capacitance on-die is normally represented.

Going back now to our equation for the radial capacitor, we find that it is now both a function of voltage and time, regardless of the spatial dependence. Thus, the charge on the capacitor can be represented by the function below.

$$Q_R = C(V(t))V(t) \tag{10.16}$$

Taking the derivative of both sides gives us the current

$$\frac{\partial Q_R}{\partial t} = \frac{\partial}{\partial t}[C(V(t))V(t)] = V(t)\frac{\partial C(V(t))}{\partial V}\frac{\partial V(t)}{\partial t} + C(V(t))\frac{\partial V(t)}{\partial t}$$

$$= \frac{\partial V(t)}{\partial t}\left[\frac{V(t)\partial C(V(t))}{\partial V} + C(V(t))\right] \tag{10.17}$$

where we have omitted the function for t in the last part of the equation on the right. Equation (10.17) says that the current flowing from the capacitance here is a function of the change in capacitance and voltage in time.

When simulating the effects on die now, we need to represent the capacitance, not by a static constant value but by a function that changes depending upon the voltage and the time shift locally. The next example illustrates this.

Example 10.3   A power integrity engineer is asked to simulate the same structure as in Example 10.1 but now determines that the capacitor on the silicon is no longer constant but a function of voltage. Given the function below for the transistor capacitance, re-simulate the network shown in Figure 10.12,

$$C(t) = \left\{\sin\left(2\pi\left[\frac{t}{10\,\mu s}\right]\right) + 1\right\} \times 80\,\text{nF} \tag{10.18}$$

Solution:
The engineer is essentially changing one value in the circuit in Figure 10.12 (C_{si_3}). Essentially, the capacitor on die is changing as a function of frequency; that is, the capacitor has a value of 80 nF at its maximum but goes to zero at

certain frequencies. In this case, we know that for integer values of π, the capacitance goes to zero.

We use the charge function call in SPICE to represent the capacitor.

$$q = x \times \left(\sin \left(2\pi \left[\frac{time}{10 \ \mu s} \right] \right) + 1 \right) \times 80 \times 10^{-9} \qquad (10.19)$$

where x is the voltage across the capacitor. We leave everything the same in the circuit. The capacitance is plotted in Figure 10.25.

Basically, the capacitance goes to zero when

$$time = (4n - 1) \times 2.5 \ \mu s \qquad (10.20)$$

The plot for various points of the voltage across the capacitor can be seen by running the SPICE. First, note at 5 μs the high-frequency ringing. This is because the capacitance has essentially gone to zero (Figure 10.26).

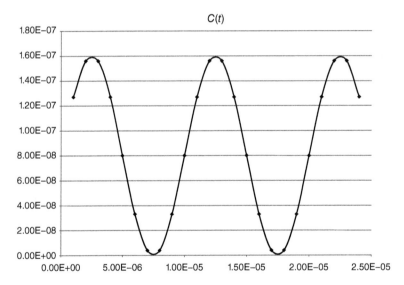

Figure 10.25 Capacitance as function of time.

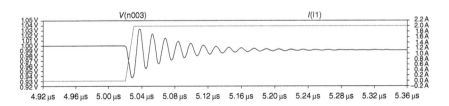

Figure 10.26 High-frequency ringing due to capacitance going to zero.

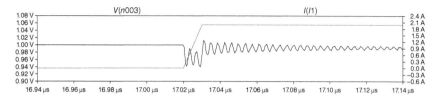

Figure 10.27 Droop due to capacitance (maximum value) in circuit.

However, we can see that there is a different effect when the capacitor is in the circuit. This can be seen in the Figure 10.27. Note the difference in the ringing, especially after the first two droop events. Also, note the fast slew rate due to the current (Figure 10.27).

The question will always be, how would the droop and ringing affect the logic circuits that are being powered locally. As has been discussed previously, data integrity is a complex problem and is normally beyond the scope of the power integrity engineer to analyze. The most important thing is to be able to determine whether or not the voltage droop at a given frequency is within the bounds of the system requirements. If this portion of the design is satisfied, it is up to the other development engineers to ensure the efficacy of the functioning logic to operate correctly.

The circular capacitor is only a representation of how the system operates and the fact that the capacitance itself is not ideal. One can imagine that if the charge depletion was fast enough, then the capacitance near the load would actually shrink physically as the charge was pulled away from the other regions. The reason why this is the case for this particular capacitance is because the radial capacitance changes as the charge begins to funnel in toward the center of the load. Because the reduction in charge density can be fast relative to recovery from the external capacitances and because of the capacitance here being mainly transistor based capacitance, it can be time dependent.

To model this capacitance correctly requires modeling the structure as a MOS capacitor as discussed earlier. However, a reasonable approximation to the operation of this local capacitance may be made if we approximate the local region with a time- and voltage-dependent circular capacitor. We saw earlier that a circular plate capacitor may be easily estimated (statically) from the following relation:

$$C_R = \frac{\epsilon_{r0}\pi r^2}{d} \tag{10.21}$$

where r is the radius of the capacitor and d is the separation. However, in this case the radius is a function of time. Thus, a more accurate relation would be

$$C_R(t) = \frac{\epsilon_{r0}\pi r^2(t)}{d} = K r^2(t, V) \tag{10.22}$$

where K is simply a constant. The radius itself can essentially change as a function of both time and voltage, and we can represent the solution as a decaying exponential, or

$$r^2(t) \cong V_0 e^{-mt} \tag{10.23}$$

Thus, our capacitor is essentially a decaying function in time. The constant m may be determined through examining the reduction in voltage on the capacitor based on the frequency of the load change. Once again, this is only an approximation but may be used to understand the shrinkage in the capacitance near the load region and thus the amount of capacitance that is left over after a fast di/dt event. Once the time constant is determined, the capacitor may be represented as a function in SPICE.

The voltage drop would be dependent upon the impedance from the outer portion of the circular region to the inner portion. If the plane was solid, then we can easily predict the drop. Unfortunately, this is not the case since the metal layers, as we have seen, crisscross from one layer to the next, and the percentage of metal that fills this region is dependent upon the process and the particular layer of the device. However, since we normally have a reasonable idea about the utilization factor from the design team from layer to layer, we can estimate the resistance assuming it was solid and knowing the percentage of metal that is filled in that layer. By using these assumptions, we can estimate how our circular capacitor would operate. This is examined in some problems at the end of the chapter.

In the next section, we reexamine the effects of the AC response of the PDN but in this case focusing on the local portion of the PDN.

10.5 The Localized Impedance Network

Because of the locale of the functional blocks on silicon, it is instructive to analyze the impedance locally to determine how it behaves relative to the rest of the network. We chose here a single core in anticipation of the discussion of the next section on multi-rail devices.

Though this was discussed in previous chapters, it is worthwhile mentioning again. The reader may wonder why lumped element models are used here rather than more complex distributed or extracted models from sophisticated CAD tools. There are a few answers to this question. First, for engineers who work in groups where these tools are readily available and are part of the process for that company, it is usually good practice to continue to use these tools to extract parasitics and simulate their problems using an integrated toolset. This is particularly true when delivering a final product or report out that requires a certain accuracy. Moreover, there are a number of very good texts that focus on how to use these tools, and the reader can usually refer to the online product manuals for guidance should there be any questions regarding their context.

However, plugging in values and pushing buttons will not grow the fundamentals in an engineer's background. This is especially true when it comes to power integrity where the foundation is rather broad and requires good scientific analytical skills and knowledge. One of the key reasons is that without understanding the physics behind a certain problem, it is difficult to cultivate a strong base. By solving certain problems and understanding the fundamentals behind those problems, an engineer's ability to solve more complex problems broadens. Moreover, the time required to solve those problems also shortens, making them more efficient and valuable to an organization.

By choosing to use lumped element models where the accuracy is sufficient for certain analyses, engineers can gain faster insights into the problem by examining the lumped circuits behind the problem. This is certainly more difficult with large distributed models where making multiple changes to see subtle differences can make it difficult to identify the cause of a certain issue. Thus, as we examine these local phenomena, we use simplified models for the load and the network to be able to delve into these problems more closely. Also, the process is then focused on the behavior of the circuit rather than trying to decipher complex subsystems which is not the point of the exercise.

Getting back now to the circuits in question, there are important differences between the platform PDN and that near the die. Because of the type of decoupling, that is on-silicon and local to the package, there is normally a large difference in the impedance between the platform PDN and the local one. This is particularly true for blocks where the rails are split and the routing metals are small relative to the platform. The goal here is to perform a simple AC analysis on a subsection to be able to quantify this behavior. Certainly, with simple lumped element networks, running the full simulation for the platform from the power supply is beneficial since all of the resonant effects are in play. However, by running a segment of this PDN, in particular the local die region, often yields insight into its behavior.

In our case, we want to run the network all the way to 1 GHz to determine how the impedance changes. We start with a network shown in Figure 10.28. We note that the local capacitance is very small for the FUB from our previous discussions. Also, we note that the inductance on die is also very small. However, the series and shunt resistance elements on die can be relatively large. This will certainly have an effect on how the impedance will look overall.

Referring to the diagram, the load itself can be lumped for power integrity analysis; it is sometimes worthwhile (especially with power gates) to analyze different loads within the same simulation as we shall later see. The main capacitance at the load is due to the local decoupling. Here we will represent it as an "ideal" capacitance for simplicity to start. The series resistance is due to

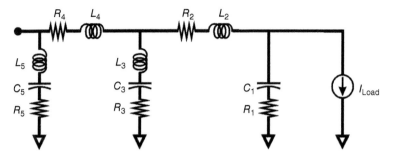

Figure 10.28 Simple circuit diagram for local PDN.

the Rds_on of the transistor capacitance, which we can also model as a single value. The next parasitics have to do with the series interconnection to the next adjacent capacitance. This capacitance is often slightly larger than the local capacitance in that it can come from different portions of the die. There are then two additional series elements, which are important before we get to the package capacitance. First, there is a distribution interconnect that is on-silicon that routes from the previous capacitance out to the interconnection of the die. The second portion is due to the package routing itself. They are considered distinct only because they are often very different in construction. For example, the silicon route may be relatively high in resistance but low in inductance (see Example 10.1). The package route may be just the opposite, of relatively high inductance and low resistance. Understanding both is important for the PI engineer in order to properly model the effects.

The main question is where to decouple the platform PDN from the local one? This is usually done by taking a subset of the package decoupling and splitting it. The reason is because there will always be some reservoir of charge available to feed the silicon level loads and capacitance. However, there is normally reasonable impedance between these capacitors and that of the platform. Thus, we chose to split the PDN here.

There is an important aspect to this split, however: the charge density distribution. First, most engineers start with trying to model the complete structure. With the CAD tools available today, this is certainly possible. However, what normally happens in such a modeling effort is that the engineer sees behavior that cannot easily be explained. Thus, by starting with a subset of the PDN and the given capacitance, it is usually easier to perform "what–if" analyses to find out what structures and regions are causing the issues.

A simple way to do this is to develop a model with most of the key interconnects in place and then keep the bulk of the capacitance very small (out of the circuit) for the first initial runs, except for a subset of the local PDN capacitance. In our case with the circuit in Figure 10.12, this is what we will do. We will add additional networks, once we determine the initial behavior of the circuit. As

Table 10.4 Parameters for Example 10.4.

Parameter	Value
C_1 (nF)	10
R_1 (mΩ)	2
R_2 (mΩ)	5
L_2 (pH)	5
L_3 (pH)	100
R_3 (mΩ)	5
C_3 (μH)	1
L_4 (pH)	300
R_4 (mΩ)	2
C_5 (μF)	1
R_5 (mΩ)	2
L_5 (pH)	500

stated earlier, this may also involve adding additional loads with different step functions and phase relationships relative to the main current step.

The following example illustrates what the AC impedance looks like for a local PDN.

Example 10.4   Determine the local impedance of a local PDN given the values in Table 10.4 and the lumped element model in Figure 10.28.

Solution:
It is straightforward to set up a simple AC sweep with a 1 A current load as the source. We will generate the plot in log scale up to 1 GHz and start at 1 MHz. The SPICE circuit is shown in Figure 10.29.

The frequency sweep is given in the SPICE run below. Note that the AC impedance has a very large resonance close to 165 MHz. This resonance is a function of the LC circuit between the on-die capacitance and the series-shunt inductance. If the series impedance is large enough, it becomes difficult for the capacitance to have an impact on the resonance at this frequency. The biggest effect on this resonance (peak and frequency) is the on-die capacitance. This can easily be seen if we just move the capacitance by factors of 2 up and down (Figure 10.30).

By changing the capacitance by a factor of 2 greater, we see the result below (Figure 10.31).

The impedance profile shifted downward by about 50 MHz, and the peak reduced by about 5 dB. Thus, one can see that the on-die capacitance can have

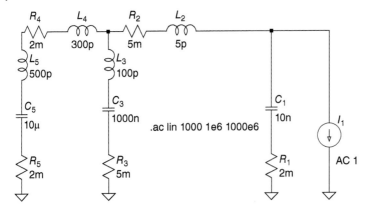

Figure 10.29 Equivalent Spice circuit for analysis.

an enormous effect on the resonance point. The other more important thing to note is that this will also have a large effect on the localized droop due to the stimuli of the load. If there is a large di/dt event due to a functional block locally, the slew rates frequency content as it relates to this resonance will certainly influence the result.

It is not simple to increase the local capacitance on silicon mainly because there is limited space and the cost for doing this can be prohibitive on large devices. However, there are additional things that can be done to shift resonances around and reduce their effect locally. One of them is to try and place external capacitance as close as possible to the load (vertically integrated) to try and add some quality capacitance in parallel. Normally, the inductance of these devices tends to dominate, but other silicon-level capacitance can have lower inductance which can help to reduce these resonances.

After the analysis of the local structure is completed, it should be straightforward to combine it with the platform-level PDN and then re-simulate as a complete power distribution path. However, as shown above, it is often instructive to separate them to begin with. The next section discusses multiple rails where different loads can have varying effects on the PDN while performing a local analysis can yield insights into the larger PDN issues.

10.6 Multi-rail vs. Single Rail Power Discussion

When it comes to bringing power into a device, it is important to consider not only the flow of current from the external capacitors (package, motherboard, etc.) to the on-silicon capacitance but also how the rails are built. The best scenario to moving charge from the large capacitors to the silicon capacitance and

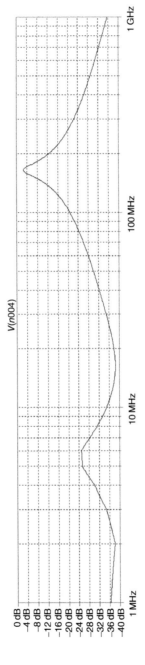

Figure 10.30 AC sweep of impedance.

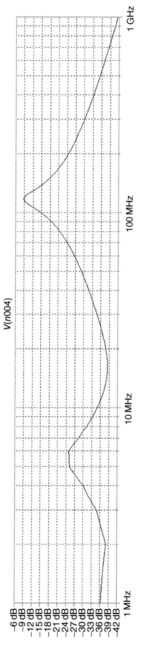

Figure 10.31 AC sweep increasing C_1 by 2×.

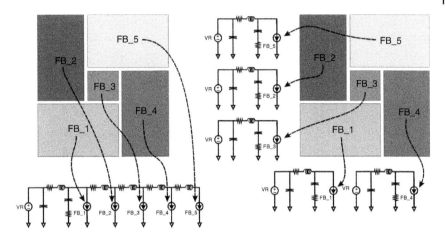

Figure 10.32 Single vs. multi-rail representation.

load is a complete plane of very low impedance between the two points, as well as a large region for the current sink. Even better, where the charge ingresses into the load on multiple sides we have charge flow resembling that of a circular capacitor. Consequently, a very low impedance path and (in most cases) a low voltage drop between the capacitance and the load is the result. This is the case with a single rail load where one has the luxury of being able to (under certain packaging constraints) bring charge in from more than one side of the device.

However, as we have seen, most devices today require more than one rail to support the power. The reason architects choose to do this may be many, as we have discussed. However, one of the main reasons is usually performance improvements and power savings. Unfortunately, multi-rail devices are constrained from where the power can come in. If there is no gating on die (which we will discuss in the next section), the flow of charge is normally limited to a narrow region where the next adjacent charge source is. This is where the analysis can become complex. Figure 10.32 illustrates the basic difference between a single- and multi-rail power distribution.

The single-rail representation clearly shows that if the power source is singulated, the flow of current into each region will be unequal; that is, there will be an extra impedance that must be added from the nearest load point to the next one in the best of cases. This is because the main DC power source is normally located on only one side of the silicon rather than having components on all of the sides. Also, it explains why most platform engineers place capacitors circumscribing the silicon rather than in one location to help bring the flow of charge into the loads from the lowest impedance path possible.

The layout engineer will do their best to try and minimize the impedance, but once the planes get into the functional block region in the silicon, lateral drops will occur due to metal routing that is normally much higher in resistance (less metal) laterally than what would normally be on a package. On the left-hand side of the figure, it is shown the worst-case scenario where there is an additional impedance for each load. Moreover, the power source has a bank of external capacitors that must feed all the loads. Because of this, the capacitors are normally sized to meet the maximum demand. To do this, a combination of HF and bulk capacitance is used. This can take up a significant region on the platform.

On the right-hand side of the figure, each block is powered by its own voltage regulator. Thus, the impedance from each source is distinct and known. If the placement of the power sources is such that one is able to locate the converters as close to the loads as possible, then normally the result is a lower impedance path for each to the load. From a strictly electrical perspective (ignoring cost and area for a moment), this can result in a better power delivery solution overall.

Normally, in the case of a multi-rail power distribution, the ground planes are shared, but the power planes are not. This creates a problem for routing, and thus potentially large voltage drops may be evident. It is especially difficult if the external capacitance cannot be placed near regions of the die where *moating* takes place. Moating is where a functional block is isolated physically in the middle of a device, and the flow of charge from one part of the platform into the load region can be difficult. Normally, mask designers ensure that their design rules for routing traces into the device are always met. This is true with the package and the platform people as well. However, reliability is just one facet of a satisfactory design. It is normally up to the power integrity engineer to check whether or not the efficacy of the power distribution is good. This requires building a sufficiently detailed model of the distribution through a given region, however. When an engineer is assigned to analyze this problem, it is usually good practice to sit down with floor planners (physical design team) of the chip to make sure they know where each rail is routed to. Then, once this is understood, it is good practice to analyze the sections that could be of greatest risk. This requires going through not only the routing in the X and Y dimensions but also the Z.

As we will see in the last section in this chapter, noise plays an important role in the construction of a device with multiple rails being brought onto the silicon. The main issue is normally coupling through a low impedance path (often the ground or substrate), which can cause sensitive circuits to malfunction. Normally, there are two main types of rails which are brought into a device such as an SoC: high performance and low performance/low noise. Normally the high-performance circuits are digital in nature, and though they too are sensitive to noise as we have seen, it is not usually the main power source which is an issue; rather it is the high-frequency, high-current switching that causes the noise locally. Concerning the other type of rail, the low-performance/low-noise types usually have significant decoupling around them because they are usually sensitive to noise pickup. These can be sensitive and mixed analog-digital or purely analog in nature. Often they involve amplifiers, which have gain, and though virtually all of them have some type of attenuation or filtering, noise can couple into them from various sources.

From a power integrity perspective, this is where the engineer needs to understand how the high-performance PDN filtering has been developed all the way through to the device and into the substrate. Most silicon has some connection through the substrate with the transistors. However, the ground systems are connected, and usually there is an impedance between different functional blocks on the die; currents can circulate from one section to the other through the substrate and the ground.

In the next section, we examine the effects due to on-die gating which is a common circuit technique employed in many advanced silicon devices today.

10.7 On-die Gating

Though bringing in multiple power rails into a device is challenging as we have seen, there are options architecturally which can help reduce the complexity. One of them is to use what is called *on-die power gating*. Figure 10.33 shows a simple example of what an on-die power gate looks like from a circuit

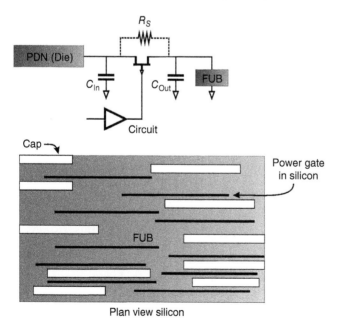

Figure 10.33 Simple illustration of power gate transistor instantiated in FUB.

perspective. A large transistor (PMOS normally) is placed in series with the power distribution and is used to gate-off the load from the main rail. The advantage here is that one does not need to bring in multiple rails to support all of the loads. Moreover, when not in use, the other functional blocks may be turned off by their own power gate as well.

The power gate (PGT) may be instantiated within the FUB between the logic along with the local capacitance. The capacitance resides on both the input and output side of the PGT. When the PGT is turned on, it is usually controlled through a driver, which prevents a large in-rush of current coming into the local capacitance and logic. The purpose for the slew rate control is to prevent a voltage drop from occurring on the input side of the PGT, which could affect the other FUB's that are tied to the same point. The reason for instantiating the PGT throughout the block rather than simply placing it on the side is because there can be significant delays in the charge going laterally through the interconnect, and voltage drops across the gate can occur which could be seen at different points in the logic. Placing multiple smaller gates within the block in a distributed fashion helps to mitigate these local drops to the logic.

However, the disadvantages can be many. First, the large power gates take up significant room on the silicon. Because the real estate is very expensive, even a few percent of added silicon can be very costly to the overall product. Second, there is always an intrinsic voltage drop across the power gate when active. This loss adds to the overall guardband in the system and contributes to the power loss when devices are active. Third, when off, the power gate leaks current. The leakage is normally not an issue to systems which are server based or when activity is high. However, for mobile devices, the leakage can add up and reduce battery life since the amount of time the functional block is active is very low relative to being inactive. This can result in unwanted downtime for the platform.

The other problem with the power gate has is related to charging and discharging of the local capacitance within the device itself. When the device (transistors in a functional block) is turned on, because it is so large, the in-rush current may be high which causes a significant amount of charge to be transferred from the main power decoupling (local) to the device. This is shown pictorially in the circuit below. When this occurs, a local voltage drop may occur in other functional blocks, which are active. Thus, the rate of in-rush current is controlled in the power gate to mitigate this. However, this slows up the turn-on time of the power gate which adds to the latency in turning on the functional block (Figure 10.34).

For the power integrity engineer, it is important to be able to appropriately model the effects due to power gating. From the previous sections, we know there is a localized effect due to the functional block funneling current into a region when the block switches. This localized fast di/dt event can funnel into the region, and the capacitance can behave like a circular capacitor. In

Figure 10.34 Power gate
with local inrush.

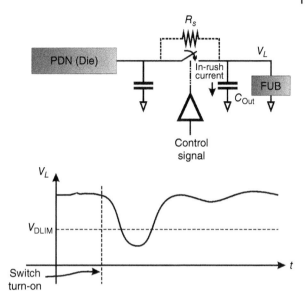

addition, if the capacitance it transistor based, we know that there can be frequency-dependent effects that can reduce the charge on the capacitor and limit its effectiveness. Now that we have added a series device into the mix that is also active, we have a third issue to address: the modeling of the power gate itself. Fortunately, when the device is active, the voltage at the gate is normally controlled through a separate power source where (typically) a charge-pump or other power source keeps the gate voltage steady. Thus, variations in the behavior of the power gate are normally kept to a minimum. This simplifies modeling the power gate where the device is operated in either the saturation or linear region. Thus, we can model the device as a simple series resistor when it is on. However, what happens when we add in the power gate to our circular capacitor model when the load is very localized?

Figure 10.35 illustrates what happens. Since we essentially have "rings" of capacitance now relative to the load, we see that the power gates local to the load are less and less. This means the resistance actually *increases*, as we near the load. This is because there are both less power gates near the load region *and* less metal. From a static perspective, this appears as a series of R–C networks where the time constant remains essentially the same as one moves physically from the load back to the periphery of the circular capacitor. However, what does change is the fact that both the capacitance that is available to the load is *less* and the resistance connecting to the capacitors is *larger* than desired.

This is one of the key disadvantages to using power gates in such systems. As process technologies become denser and as system designers and architects utilize more logic in these regions, there will continue to be a conflict between

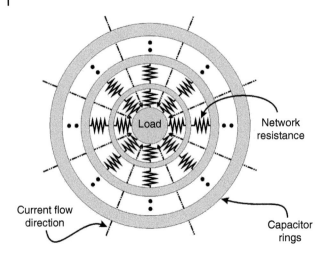

Figure 10.35 Effect of current flow into load region with power gates.

the usage of power gates in these logic regions and the ability to keep speed paths for the circuitry within certain design margins. One of the problems is that current density for transistors do not appreciably change with process, even though logic density does. This means that though engineers will see a benefit with moving from one process to another, the size of the power gates will remain essentially unchanged. This indicates that relative to the logic density for a fixed area, the size of the power gates will grow. In the end, this should prove that circuit and system designers will need to get a better understanding about the way they solve power delivery local to the die.

We revisit the charge and current density issues in Section 10.8.

10.8   Discussion of System-Level Issues with Charge and Current Density

Much of this chapter has been about how charge and current density localized in the die region can affect local droops for the power integrity engineer. The bulk of the problem is due to the load itself having a higher density in current relative to the electrical structures around them; that is, the rate of change of charge and current is increasing at a rate that is not necessarily being kept up by the physics of the components around them (storage elements).

One of the potential issues as we discussed in the previous section was the use of power gates and the increase in localized resistance relative to this load behavior. However, there are additional issues that advanced silicon processes are enabling. As we saw earlier, the transistor-based capacitor is a very common device that is used to decouple locally the logic around it. However, this capacitance is far from ideal. It is both frequency and voltage dependent. This means that the local charge density will decrease with higher and higher localized di/dt events. To counteract this, designers can attempt to increase this capacitance. Since it scales with process, to a first order, one might believe that this should solve the problem. Nevertheless, there is another issue that is negating this increase in density: *system margins.*

As the logic advances forward, voltages are decreasing relative to the *p-states* from one generation to the next. This is by necessity since to keep the overall power constrained, it is critical that voltage scaling continues forward. However, we now know that transistor capacitance is a function of voltage as well. Moreover, as one moves toward different regions of operation (from accumulation down to weak inversion for example), the capacitance decreases significantly. Thus, as the voltages decrease, we should expect to see a decrease in the charge density on the local capacitance.

The one small benefit to this decreasing voltage is that for the lower *p-states*, the di/dt events also decrease. Thus, because both frequency and current go down, this helps with the basic problem and margins in the system. However, it is clear there is an effect from shrinking the capacitance locally. Moreover, the power gates themselves also are affected by the decreasing voltage. As the PGT gate voltage decreases, along with V_{ds}, the on resistance will eventually start to increase again. Moreover, as this resistance increases, so does the drop across it when the logic is active. Essentially, the effect is very much like the load line behavior in a VR, except that the voltage drop across the PGT eventually becomes inversely related to the current draw instead of proportional to it.

These effects, in aggregate, can certainly cause issues for designers and architects moving forward. One of the techniques that may be used is to attempt to operate the power gates in a manner where the gate voltage is highly biased; that is, for a PMOS device, the gate voltage can be driven below ground to try and increase the conduction through it. Of course, generating a negative voltage on-die is very challenging. The other more realistic method is to, of course, increase the number of power gates in the region. In effect, the goal would be to skew the number of power gates for certain known high di/dt sections of the die. This is illustrated in Figure 10.36.

The advantage, of course, is that the localized drop in this region due to the power gates overall will be less. The disadvantage is that timing paths may be more difficult to meet as well. The result may be a compromise in performance and power delivery for this local part of the functional block.

Another, more advanced solution is to bring in separate power supplies to each block instead of using power gates. This removes the drop due to the power gate and frees up more room on the die. Unfortunately, creating multiple power sources external to the main SoC die or even near or on is a very big challenge in itself, and there are many system issues associated with it. However, the benefits

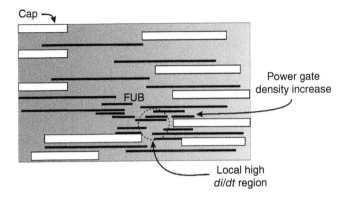

Figure 10.36 Illustration of region with added power gates.

are very clear with regard to power delivery and even power savings since each power supply can be individually controlled.

An additional solution here is to improve the capacitance in this region. Designers are already using other capacitance to help solve this problem as we discussed in an earlier section. Often this capacitance is off the main die and has much higher density. Stacking or vertically integrating the capacitance helps tremendously in reducing the voltage drop between it and the local blocks.

As process technology advances, eventually, the transistor density itself becomes the limiting factor due to both thermal and tunneling effects. However, before this occurs, power delivery may become the main limiter. A few of the main reasons may be due to a combination of packaging, transistor density, and impedance between capacitance in the system.

One way to look at this is to examine the increase in current density as a function of transistor area. We also have a reasonable idea of the activity of a functional block inside of a device. If we simply shrink the devices down (say an execution unit inside of a processing core), then we can start to see the issues that can occur moving forward. First, we would expect the current density to rise geometrically since eventually voltage scaling will be limited by the noise in the system (see Section 10.9). Second, since routing density locally will also go down, we can expect the resistance to increase from this FUB to the local capacitance that is intended to decouple it. And third, if local power gates are used, we would expect an additional drop from the aggregate side of the power gate to the load side, which also increases the impedance. The result would eventually be that when a large current step occurred, the local voltage drop could be enormous.

The analysis of this problem for PI engineers will also be challenging. One thing that will be essential toward understanding these effects is to be able to properly model both the capacitance and the power gate activity. This means that power integrity engineers must begin to get inside of the silicon itself and learn to understand not only the load behavior but also the silicon circuits that link the load to the capacitance. As one moves back toward the package and the peripheral capacitors, it will be essential that a very detailed view of the distribution path be developed in order to complete the entire picture for the simulation.

From a system perspective, the movement of charge on the die can be very important in small regions of the die when examining local power distribution effects. When a local portion of the die has a very high current density, it normally results in what engineers call a *hot spot*. This is because the area heats up literally from the sudden density of charge flow into a particular region and the surrounding medium is unable to cool it quickly enough to prevent that region from increasing in temperature. With the addition of thermals, for the power integrity engineer analyzing the problem on the silicon, this presents an additional problem. The charge density into a local area can result in a very large voltage drop, and though the time duration of this drop may be small, the local effect may be large enough to cause data errors.

Thermal time constants are normally very large relative to electrical time constants. However, metal resistances can heat up along with the silicon, which slows up the response of local transistor capacitance which, we have seen, is frequency and voltage dependent. If we combine all of these factors together, it is easy to see that localized voltage drops can occur within the structure due to many factors.

One of those important factors has to do with charge depletion from multiple load effects. Up until now, we have examined only single loads in a system. However, when power gates are involved, the charge source is on the other side of the gate (opposite the load), and when a fast di/dt event occurs in one, it can affect the other. The main reason has to do with localized charge being depleted in the region where the two loads are occurring. Figure 10.37 shows how a load event in one area can affect the voltage in another.

First, we see that when a current step occurs in one functional block where the charge is depleted not only from the local load area but also from the capacitors that are sourcing charge to all of the other loads. This capacitance now needs to be replenished to support the other loads. If none of the devices are active, then this should not be a problem. However, what if they *are* active and a load event occurs on an adjacent block? The result is that the second load can have a larger relative drop in it as compared to the first one. Mainly because the local charge has been depleted. If the impedance from the external package devices is too large relative to the time constant to recharge the capacitors, then it is possible that errors can occur. The next example illustrates the effect.

Example 10.5   An engineer is examining the effects of multiple loads occurring in two adjacent blocks. Run the SPICE file below for the values given. Using

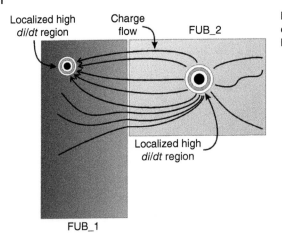

Localized high di/dt region

Charge flow

FUB_2

Localized high di/dt region

FUB_1

Figure 10.37 Localized di/dt events coincident in different FUB's.

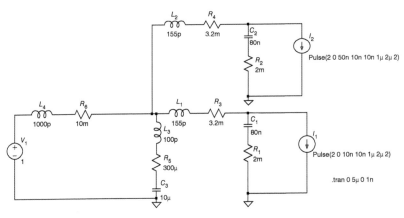

Figure 10.38 SPICE diagram for simulation.

Figure 10.38 and values in Table 10.5, run the simulation where a di/dt event occurs 40 ns after the first one. Rerun the simulation for values where the time between them is 30, 20, 10, and 0 ns.

Solution:
We can set up the simulation to run all of them at the same time. However, this will make examining the results difficult. Thus, we run the four simulations separately and examine the results.

It is clear that the first simulation shows that there are some resonant effects going on between the two droop events near each of the loads; that is, the charge depletion from one region is affecting the other. In this case, the droop in the top circuit is much worse than the lower one. This is with the separation of 40 ns between events (Figure 10.39).

Table 10.5 Parameters for Example 10.5.

Parameter	Value
C_1, C_2 (nF)	80
R_1, R_2 (mΩ)	2
R_3, R_4 (mΩ)	3.2
L_1, L_2 (pH)	155
L_3 (pH)	100
R_5 ($\mu\Omega$)	300
L_4 (nH)	1
R_6 (mΩ)	10

The next simulation is with 30 ns separation. Note that even though there is only 10 ns difference from the first simulation, a slight difference is found between the two. However, the maximum droop appears the same at our victim block (900 mV) (Figure 10.40).

In the next droop event, we see a markedly different result, however (Figure 10.41). Instead of an increase in the droop, we see a large decrease. This is because the two circuits are interacting with each other, and the resonance between them shows an interference in waveforms.

It should be clear to the reader now why one performs sweeps when running simulations of these types. It is easy to see how an engineer could underestimate the problem by not running multiple time-dependent simulations on the same circuit (Figure 10.42).

The last run shows why this is important. Note that there is now some alignment between the two droop events showing the victim circuit is now below 900 mV worst case. It usually takes some time to zero in on the worst-case resonant, and sometimes it is not in perfect alignment (Figure 10.43).

We show the last one to illustrate this. Note that the lowest point does not appear to be any worse than the previous simulation.

In the case of multiple sources being activated with a single rail, local phenomena are normally more of an issue than the global PDN effects. However, in the case of noise, often the opposite is true. Section 10.9 discusses general noise issues close to and on the device.

10.9 Noise

In all computer systems, noise is an inherent issue that must be dealt with in order to prevent platform errors. For the silicon, this is even more critical

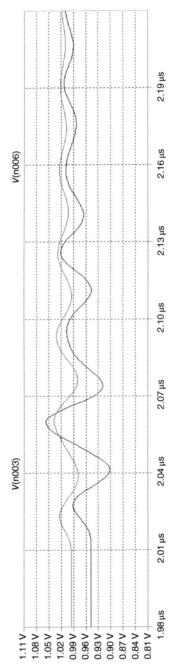

Figure 10.39 Droop events with 40 ns separation.

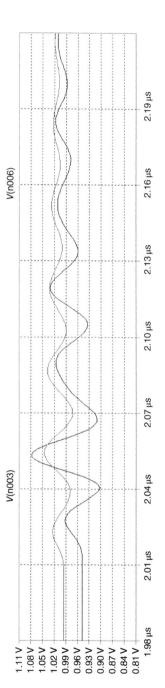

Figure 10.40 Droop events with 30 ns separation.

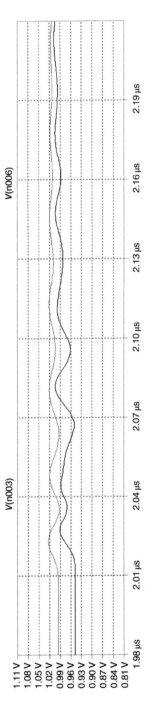

Figure 10.41 Droop events with 20 ns separation.

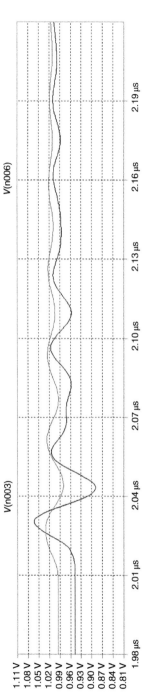

Figure 10.42 Droop events with 10 ns separation.

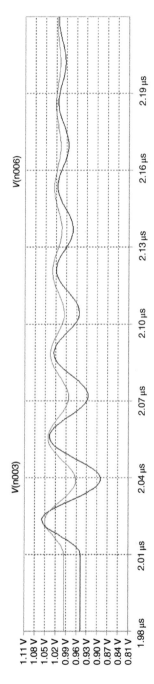

Figure 10.43 Droop events with no separation.

because as we have seen, noise that couples into the logic can result in data faults which is obviously unacceptable. From a power integrity perspective, the PDN is not only intended to deliver high-quality power to the silicon but also designed to mitigate noise that is generated from the system that can affect the device.

Good layout practices as well as proper use of decoupling helps to reduce the noise in most systems, but often coupling occurs that can cause functional issues. There are a number of other sources of noise, which can cause issues in a system. These are shown below:

- General power bus noise
- Coupled power bus noise
- Local coupled noise
- Conducted emission noise

Most sources of noise have their origin in some region of analog or digital circuitry. Thus, from a design perspective, it is certainly best to start to limit it there. However, often system designers are unaware of where this noise comes from and normally find out very late in the development process. This is where the power integrity engineer usually must help to solve the issue.

The first noise source is generated from the power supply itself. There are two sources, or frequencies, which are important here. There is first the noise due to switching of the converter itself. This is normally induced by the switching frequency of the converter. However, because of the change in duty cycle, even with fixed frequency designs, harmonics can generate more noise in areas then others. Normally, the ripple from a converter does not disturb the main logic that it is intended to deliver power to, if designed correctly. However, the noise can reflect to the ground, which is often common to all of the other rails in a system. Figure 10.44 illustrates this phenomenon. First, the noise is generated at the power source. It is then seen on both the ground and power planes as a differential noise.

The loop includes multiple inductive paths as can be seen in the figure. It is normally the inductive coupling that induces the noise in the circuit. Since the noise starts out as differential in nature, it is reflected in the ground path. However, depending upon the layout and the coupling, the noise may become common mode as well. The noise can easily conduct to the ground system to a victim circuit labeled silicon load 2. This could be a more sensitive logic or analog circuit than the source path. Because of this, it is possible that the noise could be amplified through the circuit and thus cause it to misfunction. A good example of a circuit type is a phase lock loop. The noise here could couple into oscillator portion of the circuit and cause the phase-locked loop (PLL) to mis-lock or not lock at all. Thus, it is important for the PI engineer to consider how to mitigate this type of coupling.

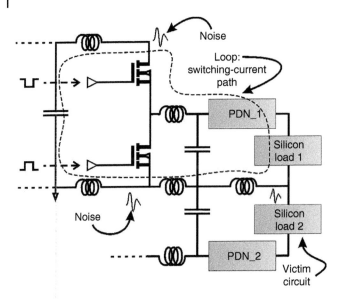

Figure 10.44 Noise generation due to power source.

Depending upon the amplitude and frequency of this noise, the capacitors, which are normally chosen to attenuate certain frequencies near the load and package, are used here to steer the currents and reduce this noise. In addition, many engineers place both low-frequency and high-frequency capacitors near the power source to create local loops to shunt the noise back to its source as shown in Figure 10.45.

Normally, these capacitors are tuned to certain frequencies that are generated near the power source. Often the power supply designer adds these since they have the most knowledge about the type of noise that may be generated in the design. However, sometimes, this noise manifests itself long after the design has been completed and only during the test phase at the platform is the noise seen. Very often, as we have discussed, it is exacerbated by a certain choice of layout on the platform and the package. If that is the case, the power integrity engineer is usually involved since at this point, there will be interaction with the PDN that cannot be avoided. In most cases, the addition of capacitors to a design results in lower noise that was there previously. Unfortunately, due to constraints in packaging and area on the platform, it is very difficult to add capacitors and normally, especially for mobile platforms, designers are forced to replace capacitors with different ones to help mitigate the noise. This is where system noise can manifest itself in ways that were not foreseen. Capacitors, as we have examined in previous chapters, are fairly complex devices and given the proper excitation can generate noise on platforms from very HF all the way

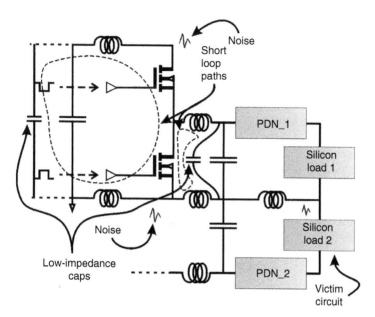

Figure 10.45 Added low impedance caps to create short loop paths near converter.

down in the audio regime. Thus, when replacing a capacitor in the system, it is best to examine the impact to the PDN before the final change is submitted.

Coupled power bus noise differs from general power bus noise in that as we have seen that the general power bus normally is driven by the power supply low or mid-frequency affects and the noise couples in through the ground plane inductively. Coupled power bus noise may couple electromagnetically through the inductive and capacitive elements and couple directly into other power planes. A good example of this is high-frequency noise from the local power FET's in a switching power supply coupling magnetically into the common ground plane, which happens to be close to sensitive logic or analog circuitry. This is different from general bus noise, which normally affects the circuitry that it is powering. Once the noise is coupled into the ground system, it can propagate through low-impedance paths into the silicon devices. Normally, through careful routing, high-frequency noise from the power supplies can be mitigated. It is normally a function of placement of the planes and shunting of the HF noise of the power devices that reduce this coupling effect.

Local coupled noise is fundamentally a load phenomenon; that is, the load itself is the source of the noise and the victim is a local portion of circuitry that is susceptible. Our example of the PLL voltage-controlled oscillator (VCO) circuit is a common one. The load from a logic block switches, and the HF noise is picked up in the ground plane. From a PDN system simulation perspective, it is difficult to see how this noise affects the overall power distribution.

SPICE is a purely differential mode simulation platform that does not model common-mode affects electromagnetically. However, it is possible to assign different impedance models in a SPICE analysis to show where the currents can flow. The PI engineer can actually give some insight into the coupling (though they may not provide the solution here) to the system designers. Normally the victim circuits are either not known or not fully understood by the PI engineer, and thus it is usually guidance that is given to the chip design team with regard to such coupling effects.

By far the most complex of noise coupling is due to conducted emission noise. Though in many cases, noise sources do not normally cause data integrity issues, they can in fact reduce voltage margins in the system and prevent a product from passing certain emission standards. Though power integrity engineers are normally not responsible for certification, they certainly have insights into where power-related noise may be causing problems.

A very common example of this is a simple capacitor. If a given capacitor has an impedance and a signal is injected at a frequency that is exciting it, then the capacitor may cause inductive circuits around it to radiate noise, or possibly conduct noise back into the system. Normally, this means that the particular device that has been chosen may not be doing the job adequately. If the product is tested in an anechoic chamber, then the electromagnetic compatibility (EMC) engineer can lend much insight into the cause. However, normally, this type of testing occurs far into the development process. Prior to the prototype phase, the PI engineer has an opportunity to model certain power source behavior. The modeling often does not involve electromagnetic radiation, but can in fact show induced currents into impedance paths that are within the proposed layout and design.

10.10 Summary

In this chapter, we discussed some of the key aspects of silicon power integrity. What separates this area from that of the platform are a number of important factors. First, the PDN is normally defined from the package capacitance all the way to the load. Second, the PDN itself comprises a much different type of capacitance (overall) which includes one or more different types of silicon capacitance. Third, the frequency of excitation is much higher than what is seen back into the platform due to the filtering that takes place as one moves toward the power source away from the load. And finally, the goal of the PDN design is much different than that of the platform overall, being mainly focused toward reducing high-frequency harmonics.

The construction of the silicon is also quite important when analyzing the problem. When examining the construction of the device, it is important to include not only the topology of the layers but also the topography of the functional blocks. Each functional block may be constructed differently and has very different current densities than those of others. Thus, understanding which portions of the chip create larger, more dense current surges is the first step. The second part of the construction is the metal layers which reside above the silicon. The layers are constructed from very thin metal strips rather than large planes, and though the inductance is normally quite low on silicon, the resistance is higher due to the way these layers are built which is why it is advantageous to move the package capacitance as close as possible to the actual load vertically to reduce this voltage drop when a current step occurs. The design of the metal layers is very much dictated by not only the design of the circuits and logic underneath but also by the manufacturing process. As one traverses vertically from the transistors upward, the metal gets wider and thicker normally, which means that most of the power distribution is done in the upper metal layers rather than in the ones nearest to the transistors. This is mainly a consequence of how manufacturers must develop these chips from the base wafer all the way to final packaged parts in order to maintain reliability and yields.

Within the construction of the silicon, designers must place capacitance to locally decouple the functional blocks. There are different types of on-silicon decoupling that are available to the designer and may be used for the local PDN. The four main types are transistor capacitance or T-cap, MIM, simple metal capacitance, and trench. The transistor capacitance is the most used due to its availability, cost, and ease of use. However, transistor capacitance does not operate as normal parallel plate capacitance does. Because it is based on making a capacitor from a transistor, it is governed by the physics of the transistor as well. As we have seen, this capacitance is frequency dependent, so when analyzing power integrity problems, engineers need to keep in mind what the actual mode of operation of the capacitance is, just like a transistor. The capacitance is also voltage dependent as discussed in the chapter, which means that as the voltage changes (lowers), the capacitance can be reduced. The other capacitors that are available are simple metal capacitance which basically operates like parallel plate capacitance (with fringing fields as well), MiM capacitance which is constructed also within the metal layers but may use a higher dielectric constant and a thinner dielectric for higher density, and finally, trench-based capacitance. Trench capacitance is normally constructed on a cheaper silicon device, and though it is very high in capacitance density, because of its construction, its resistance may also be high.

Because of the way functional units behave, current densities from the loads can often make the capacitance behave like a circular construction; that is, the currents can flow inward toward the load rather than laterally from one point to the other. If the capacitance is transistor based and the load changes are very fast, this can actually reduce the apparent capacitance as the load is switching. This time dependency of the capacitance must be modeled by the power integrity engineer to ensure they are properly determining the actual effects on the silicon.

Because with today's high-performance devices there are normally more than one rail brought into the die, it is important to consider the routing effects from the platform and package all the way to the silicon. In particular, the topology of such devices includes large numbers of I/O, which can interfere with the power distribution getting to the silicon loads. Multiple rails benefit the system designers and architects as they are able to isolate functional blocks and through power management reduce the power consumption of the design overall. However, it also raises the complexity of the design over a single rail, and for the power integrity engineer, often can create a complex PDN and analysis to determine where the high current locations are and possibly the higher droop regions.

However, once on-die, the power may be routed to islands or groups of functional blocks that are separated by on-die power gates. These are essentially large transistors distributed around the FUB with the goal of allowing them to be somewhat isolated electrically for power management. Power gates work in concert with the multi-rail system by isolating loads from each other and allowing them to be powered down. However, power gates also create issues for power distribution because they create current funneling regions in the silicon and also limit the ability of charge to get from one side of the gate to the other due to their resistance. Understanding the power gate's construction both electrically and physically is necessary for PI engineers to allow them to develop the proper tools and analysis to understand how the PDN is affected.

Though PGT's have clear benefits on silicon for power management, there can be issues with them with respect to power integrity. One clear potential issue is charge funneling. Even with distributed power gates, the current can funnel through these resistive gates in such a way as to concentrate through a local group of gates which can limit the ability to recharge the capacitance on the other side of the gates. This can result in voltage droops that are local, which can cause issues with data integrity. Modeling the power gates correctly is critical to ensuring the efficacy of the power delivery.

Though the power integrity engineer can help mitigate issues locally by working with the silicon designers, there is always the potential for noise coupling onto the die. Noise can couple into the device from multiple sources including other power rails; localization from switching on a main rail where power gates have been placed; ground systems where the impedances are such that a low impedance path from a source has passed through to a receptor rail; and through other devices in the system. Even capacitors can cause unwanted noise if they are not properly chosen for the function. Thus, it is critical that the power

Table 10.6 Parameters for Problem 10.1.

Parameter	Value
C_{pk1} (μF)	10
L_{pk1} (pH)	200
R_{pk1} ($\mu\Omega$)	50
R_{d2} ($\mu\Omega$)	200
L_{d2} (pH)	200
R_{si1} ($\mu\Omega$)	70
L_{si1} (pH)	100
R_{si2} ($\mu\Omega$)	200
L_{si2} (pH)	30
C_{si3} (nF)	80
R_{si3} (mΩ)	2

integrity engineer be involved with much of the development upfront in both the system design and main SoC or device to ensure a robust power delivery.

Problems

10.1 Run Example 10.1 with the following values in Table 10.6. Examine the changes between the results from the example. What were the differences in the maximum droop values? The length of the ringing? Can you explain these differences?

10.2 Determine the capacitance for a plane metal capacitor using Eqs. (10.7) and (10.8) for the following values ($t = 1\,\mu$m, $d = 200$ nm, $w = 20\,\mu$m, $l = 400\,\mu$m) for a simple metal capacitor. How big (area wise) would this need to be to achieve 100 nF?

10.3 Determine the charge from Eq. (10.18) by integrating it with respect to time.

10.4 Determine the radial capacitance for a load given the following values (Table 10.7).

10.5 Redo Example 10.5 using the values in Table 10.8.

Table 10.7 Parameters for Problem 10.4.

Parameter	Value
R_1 (μm)	250
R_2 (μm)	120
R_{load} (μm)	75
T_m (μm)	3
Temp (°C)	20

Table 10.8 Parameters for Problem 10.5.

Parameter	Value
C_1,C_2 (nF)	80
R_1,R_2 (mΩ)	2.7
R_3,R_4 (mΩ)	5
L_1,L_2 (pH)	200
L_3 (pH)	170 pH
R_5 (μΩ)	400
L_4 (nH)	1.3
R_6 (mΩ)	13

Bibliography

1 Baker, R.J. (2005). *CMOS Circuit Design, Layout, and Simulation*. Wiley.
2 Microelectronics processing technology, Harvard Lecture, 2001.
3 Sawyer, S. (2016). *MOSFET Device Physics and Operation*. Rensselaer Polytechnic Institute. http://homepages.rpi.edu/~sawyes/Models_review.pdf.
4 Fonstad, C. (2009). *MOSFET's in the Sub-threshold Region*. Massachusetts Institute of Technology. https://ocw.mit.edu/courses/electrical-engineering-and-computer-science/6-012-microelectronic-devices-and-circuits-fall-2009/lecture-notes/MIT6_012F09_lec12_sub.pdf (accessed October 2018).
5 Roozeboom, F., Dekkers, W., Jinesh, K., et al. (2008). Ultrahigh density trench capacitors in silicon. IEEE PowerSoc.
6 Norgren, M. and Jonsson, B. (2009). The capacitance of the circular parallel plate capacitor obtained by solving the Love integral equation using an analytic expansion of the kernel. *Progress in ElectroMagnetics Research*.
7 Mukhopadhyay, S., Mahmoodi-Meimand, H., Neau, C., and Roy, K. (2003). *Leakage in Nanometer Scale CMOS Circuits*. IEEE.

8 Elgharbawy, W.M. and Bayoumi, M. (2005). Leakage sources and possible solutions in nanometer CMOS technologies. *IEEE Circuits and Systems Magazine*: 6–17.

9 Ferré, A. (1999). On estimating leakage power consumption for digital CMOS circuits. PhD dissertation. Tech. Rep., Universitat Politècnica de Catalunya.

10 Roy, K., Mukhopadhyay, S., and Mahmoodi-Meimand, H. (2003). *Leakage Current Mechanisms and Leakage Reduction Techniques in Deep-Submicrometer CMOS Circuits*. IEEE.

11 Moore, S. (2016). Breaking the multicore bottleneck. *IEEE Spectrum* 53: 16–17.

Appendices

A.1 Introduction to SPICE

This appendix is a brief introduction to Simulation Program with Integrated Circuit Emphasis (SPICE) as it pertains to some of the lumped element circuit analyses that a power integrity engineer may encounter in their work. A few reference studies are given in the Bibliography section including one from Paul [1]. Though in Paul's paper the emphasis was on EMC-related problems and he had assumed PSPICE as the main tool, there are still a lot of similarities to the analyses, and the reader is encouraged to review his work and other references in addition to this particular tutorial.

SPICE is a circuit simulator which has been around since the 1960s and has become a staple in the industry for complex circuit-related analyses. Though there are many commercial-based versions of the tool, there are still many license-free programs available. In this text, we used LTSPICE since many students cannot afford commercial tools. However, any generic SPICE program may be assumed. The LTSPICE online Help manual has details about the syntax for every device with examples and information which can help the reader in navigating through their analyses [2]. It is highly recommended that the student look this reference over in detail to help with their learning.

Today, because of the ease of the graphical user interface, SPICE is now even more intuitive and easier to use than ever. Virtually all of the versions of SPICE have a GUI of some sort, allowing for a "click-and-connect" method for constructing a circuit. Indeed, this is certainly the preferred way to model and run simulations, and the examples in the text here were based on that method. However, for those who are new to SPICE, gaining an insight into how SPICE works and, in particular, the syntax for particular models, it is often more instructive to illustrate the analytical methods through the development

Power Integrity for Electrical and Computer Engineers, First Edition. J. Ted Dibene II and David Hockanson.
© 2020 John Wiley & Sons, Inc. Published 2020 by John Wiley & Sons, Inc.

of what is called a "SPICE deck." This is a text file that is essentially a small program that the SPICE engine interprets and is able to compute the output data for users to analyze. Thus, we will be introducing SPICE using this methodology first. We will later give some examples using the GUI for LTSPICE to further expound on the process.

To start using LTSPICE, the reader may download the file from the Linear Technology website. There are Windows versions for both the Mac and the PC. Once the program is loaded, a user can create an ASCII text file to create their SPICE deck and save it as a .cir file. To run the file, they can open up the LTSPICE program and choose File>Open and choose the .cir file they have just created. LTSPICE will recognize the file and run it directly.

To start this very brief introduction, we will discuss the SPICE deck, the sources and loads, and passive elements, along with a brief discussion on transistor format (MOSFET), time steps and other parameters. Throughout the introduction, we will also give some simple examples for the reader.

A.1.1   The SPICE Deck

When SPICE was first invented, and before we had advanced computers that could do sophisticated graphics and operate at multi-GHz speeds with multiple processors, the entry system for the SPICE program was the SPICE deck itself. Indeed, even the output was a text-like file that emulated a graph of the data. Today, we have graphical user interfaces that allow us to easily build circuits and view the data graphically. However, the program is still essentially based on the inputting of a SPICE circuit thus arranged based on node voltages and currents from Kirchhoff's equations. Thus, it is very instructive to go through the format and illustrate by example how SPICE operates.

A typical SPICE deck contains the following format:

```
                     ★ ★ ★
      *  Title and notes on SPICE deck
      <  Circuit and Model Definitions  >
      <  Execution Definitions  >
      <  Output Definitions  >
      <  End statement  >
```

The executable code is defined within the brackets, e.g. < **executable** >. Other than the **Title** and the **End** statement, the order may be varied as the developer sees fit. Though it is not necessary, but like any piece of code, it is good to put the following information into the **Title** section:

- type of simulation and purpose
- author and date
- project (if applicable)

The reasons for doing this may be obvious to many, but it is worthwhile mentioning again. First, some SPICE decks can end up being very complex and large, and thus finding the actual *simulation* callouts and definitions can often take time to find in the deck. The *purpose* of the simulation helps one to understand the reason for running this. The reason why *author* and *date* are used is self-explanatory. Documenting the *project* also helps to categorize why the model was built in the first place. Proper documentation also allows one to reuse the SPICE deck in the future.

The circuit and model definition are the models for the circuit and the actual node circuits themselves.

A simple example SPICE deck of an inverter DC sweep is shown below.

A SPICE Deck Example

```
        * Inverter SPICE Deck DC Sweep - testing for
power consumption in
        * sub-threshold operation for transistor
node Z.
        * J. Ted DiBene II - 10/17
        * Project: Power modeling for Power
Management of Processor X
        *
        * Model Calls for standard CMOS transistors
        M1 2 1 3 0 PMOS L=1u W=30u
        M2 2 1 0 0 NMOS L=1u W=10u
        * node voltages and circuits
        vgs 1 0 1.8
        vdd 3 0 1.8
        cload 2 0 0.1f
        * DC sweep
        .dc vgs 0 1.8 0.1
        *model calls BSIM 14
        .model PMOS PMOS level=14
        .lib C:\Users\ted\Documents\LTspiceXVII\lib
\cmp\standard.mos
        .model NMOS NMOS level=14
        .lib C:\Users\ted\Documents\LTspiceXVII\lib
\cmp\standard.mos
        .END
```

Note that some additional notation is added within sections to more clearly call out what is going on with the executable statements and circuit. This file has a circuit section (node voltages and circuits), an executable simulation (DC

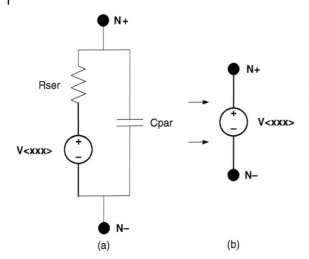

Figure A.1 Typical voltage source symbol with parasitic elements for notation (a) and normal circuit instantiation as shown in circuit schematic (b).

sweep), and model calls which one may interpret to be part of the circuit definitions. Note that there are no output statements defined. Often a .PRINT or .PROBE statement would be added. Also note that the order of the circuit, executable statements, and output statements are random as long as they are defined between the Title and End statements. We will go into more of the details about these statements in the following sections.

A.1.2 Sources and Loads

The sources and loads for SPICE have very clear definitions. In the previous example, static voltages were used (v_{gs} and v_{dd}) for supplies for the transistors. In the case of v_{gs}, the voltage was swept from 0 to 1.8 V in 0.1 V increments using the .DC statement. A static DC voltage symbol is shown in Figure A.1 as it would in the LTSPICE manual.

The sources for a power integrity simulation usually involve, for simple analyses, a static voltage source, or simply a ground as described in Chapter 9. Obviously, if it is critical to understand the effect that a VR source has on the power distribution network (PDN), building a model such as that described in Chapter 3 is required. The syntax for a voltage source (static) is shown below.

Voltage Source Syntax

```
Vxxx n+ n- <voltage> [AC=<amplitude>] [Rser=<value>]
  [Cpar=<value>]
```

In most cases, it is not required to specify the series resistance or the parasitic capacitance. The default values for those are zero if not specified. Also, if the source is a static voltage only, then only the value for the voltage and the node connections need be specified. An example is shown below:

Voltage Source Example

```
V1 10 11 5
```

This basically specifies the voltage source V1 across nodes 10 (positive side) and 11 (ground) with a value of 5 V.

For power integrity problems, the loads may be slightly more complex. In the textbook, we used only current sources for loads. Furthermore, we only specified two types: piecewise linear and AC. The piecewise linear current source was used for performing step responses or transient analyses, while the AC source was for frequency analysis. The syntax for the current source, virtually identical to the voltage source, is shown below:

Current Source Syntax

```
Ixxx n+ n- <current> [AC=<amplitude>] [Rser=<value>]
[Cpar=<value>]
```

In the case of transient analyses in power integrity, either a periodic pulse or piecewise linear waveform is used. The syntax for a piecewise linear waveform, along with an example, is shown below:

Piecewise Linear Current Source Syntax

```
Ixxx n+ n- PWL( t1 i1 t2 i2 t3 i3....)
```

PWL Current Source Example

```
I1 14 11 PWL( 0n 0 10n 0 20n 1 120n 1 130n 0 1m 0 )
```

The example specifies a current source I1 between nodes 14 and 11 that has a piecewise linear current source that starts at time 0 and has a zero current until 10 ns where it rises to 1 A at 20 ns is constant for 100 ns until at 120 ns it falls to 0 A again at 130 ns. It remains at 0 A until 1 ms. The notation for the time or even the current is not necessary but helps to clarify which is which.

An AC source is even simpler to specify as the below exampleillustrates:

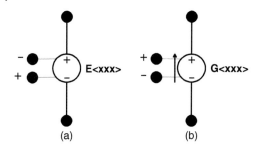

Figure A.2 (a) Voltage-controlled voltage source and (b) voltage-controlled current source.

AC Current Source Example

```
I3 7 0 AC 1
```

This defines a current source I3 between nodes 7 and 0 (zero is ground), which has an AC amplitude of 1 A. Note that there is no DC offset here, nor is the frequency indicated. The frequency is specified, for an AC sweep, in the .AC callout which defines the type of analysis run. As described in Chapter 8 for frequency-domain analysis, it is common to define an AC source of 1 A and simply look at the voltage since the impedance is now simply the voltage divided by 1 A.

There are a number of controlled sources that may be applied to either an AC or transient analysis (mainly for transient cases). We will touch on these only briefly as it is recommended again to the reader to refer to the Help menu in LTSPICE for more details. These are often used in the case where there is concern over effects that might occur from a simple PWL or step-response call function. The effects are usually related to "Gibbs" phenomena. Normally, these effects in SPICE are minimal for most analyses.

The reason is that trying to emulate the load behavior perfectly is virtually an impossible task and is normally unnecessary. However, there are times when one wishes to control the load with either a voltage or a current indirectly. Two common sources are thus the voltage-controlled voltage source and the voltage-controlled current source. The models for both are shown in Figure A.2.

There are also current-controlled sources available. The syntax for the voltage-controlled voltage source is given here.

Voltage-controlled Voltage Source Syntax

```
Exxx n+ n- nc+ nc- <gain>
```

where nc (x) are the nodes for the controlled source. A function or lookup table may also be used for the transfer function.

Figure A.3 Passive component symbols: (a) resistor, (b) inductor, (c) capacitor.

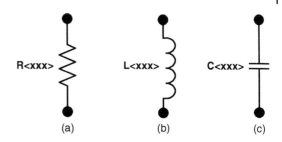

R<xxx>   L<xxx>   C<xxx>

(a)   (b)   (c)

A.1.3 Passive Elements

The most often used components for PI modeling are the passive elements. In most cases, these are kept static as well unless an engineer is performing a Monte Carlo analysis or is varying a device such as a capacitor based on excitation. The passive device symbols are shown in Figure A.3.

The syntax, for example, of a resistor is the following:

Resistor Syntax

```
Rxxx nX nY ,<value> tc1, tc2,…] [temp=<value>]
```

The nodes can be defined between any two points in the SPICE deck. If the resistor is temperature dependent, then at specific times in the SPICE sweep (tc<n>, etc.) the resistor value can change according to a function specified in Ref. [2]. An example of a typical resistor definition is given here.

Typical Resistor Example

```
R11 4 8 1K
```

This simply defines resistor 11 between nodes 4 and 8 with a static value of 1 kΩ.

Inductors are more complex than resistors and may comprise multiple circuit elements in them if desired. The syntax for an inductor is given here.

Inductor Syntax

```
Lxxx n+ n- <value> [ic=<value>] [Rser=<value>]
  [Rpar=<value>] [Cpar=<value>] [m=<value>]
  [temp=<value>]
```

The first <value> is the inductance itself; the second is the initial condition, similar to that used in a Laplace transform. In this case, it is the initial current through the inductor if any is defined. In a transient simulation, sometimes the

UIC (use initial conditions) is flagged, which will then assign an initial current through the inductor.

If there are series or parallel parasitic resistances (Rser and Rpar), they may be defined here, as well as a parallel capacitance (Cpar). The **m** value is the mutual inductance elements (quantity), and finally, the temperature may be defined, or if not specified, the default is 20 °C.

For more complex inductive element, there are two forms available for non-linear inductors: a behavioral model and Chan's model [2]. The behavioral model is simply a hyperbolic tangent function that emulates the shape of the flux through the inductor at a given time. However, the Chan model is more complex and uses parameters around the hysteretic loop for an inductor where the coercivity (Hc) and magnetic flux density (B) may be defined to emulate the magnetic properties of the inductor. For most power integrity simulations, even the behavioral model may not be needed. However, if an engineer is building a closed loop model and the droop behavior may be affected, it is sometimes necessary to use a more advanced model as defined in Chapter 2. For more information, see the Help section in LTSPICE [2].

A typical inductor call is the following:

Typical Inductor Example

```
L7 3 5 15u
```

This simply defines a 15 µH inductor (number 7) between nodes 3 and 5. A slightly more complex instantiation is defined below.

Inductor Example with Flux Definition

```
L7 3 5 Flux=240u*tanh(3*x)
```

This is an inductor which varies as a function of the current "x" with a hyperbolic tangent function for the flux density B.

The capacitor syntax is similar to the inductor syntax:

Capacitor Syntax

```
Cxxx n1 n2 <value> [ic=<value>] [Rser=<value>]
 [Lser=<value>] [Rpar=<value>] [Cpar=<value>]
 [m=<value>] [RLshunt=<value>] [temp=<value>]
```

The circuit is described in the LTSPICE Help manual. In this case, the capacitance value that is defined first is main capacitance in the circuit between Rser and Lser and RLshunt. The series resistance is normally, for PI simulations, the effective series resistance (ESR) of the capacitor. The Lser is the ESL. For more complex models, there is a parallel resistance that goes across the entire device. Normally, this is quite large though it may have an effect depending upon the circuit simulation. The parallel capacitance also goes across the entire circuit (n1 and n2) in parallel with Rpar but is normally quite small. The shunt resistance is also typically large. The m value is for the number of instantiations of this device in its entirety in parallel.

A.1.4 Transistor Formats

Though transistors are less commonly used than passive elements, there are instances where they are applicable. An example is a T-cap where one wishes to model (roughly) the variation in the capacitance on-die of a transistor capacitor.

Most SPICE modeling programs have a library of MOSFET models for power transistors that are more than adequate for advanced power integrity modeling. Here, we go over the simple syntax and discuss what an engineer may use for an example model.

In the beginning of this introduction, we showed a SPICE deck example which had a MOSFET call in it. The syntax for a simple MOSFET is shown here.

MOSFET Syntax

```
Mxxx Nd Ng Ns Nb <model> [m=<value>] [L=<len>]
  [W=<width>] [AD=<area>] +[AS=<area>] [PD=<perim>]
  [PS=<perim>] [NRD=<value>] [NRS=<area>] [off]
+[IC=<Vds, Vgs, Vbs>] [temp=]
```

The first item is simply the MOSFET number. The next items are simply the connections for the drain, gate, source, and bulk (substrate) of the MOSFET respectively. The next call is the model which is likely the most important portion of the circuit. There are generic models such as NMOS and PMOS which are intrinsic to the SPICE library. Then there are those which are custom or specific models that the user has available to them in other libraries.

The next item is the number of devices or multipliers in parallel (m). The channel length and width of the device are in meters, while the area of the drain and source diffusions (AD and AS) are in square meters. PD and PS are the perimeters of the drain and source junctions, while NRD and NRS are the number of "squares" of the drain and source diffusion regions. The OFF call is for an initial condition relative to a DC analysis while the initial conditions for the bias voltages are for use with UIC and the .TRAN analysis call. The .TEMP call is optional and overrides the call in the .OPTION call.

As discussed in the introduction, a simple MOSFET call was used. Another example of a MOSFET call is shown below.

MOSFET Call Example

```
M13 2 1 3 0 PMOS L=0.7u W=23u
*model call BSIM 49
.model PMOS level=49
```

The .model call must always accompany a MOSFET call. The simple model call simply states that for a PMOS model MOSFET number 13, the drain is connected to node 2, the gate to node 1, and the source to node 3. The bulk is connected to ground. The length and width are 0.7 and 23μ respectively. The models can be very complex and the reader is referred to the references at the end of the Appendix for more information.

A.1.5   Analysis Calls, Frequency/Time Steps, and Initial Conditions

Most engineers who run SPICE are familiar with simulation definitions for different analyses. These are essentially the executable statements that define what type of simulation the user wants. The two most commonly used for PI simulations are the frequency sweep and the transient analysis (the third one is a DC analysis which uses the DC statement which was illustrated earlier). The frequency analysis call starts with a .AC statement, whereas the transient begins with a .TRAN statement.

The .AC statement is very valuable for frequency sweeps, especially for PDNs where the engineer wishes to determine impedance and resonances around.

.AC Call Syntax

```
.AC <oct,dec,lin> <Nsteps> <Startfreq> <endfreq>
```

The first item after the .AC statement is whether or not the frequency steps will be in terms of an octal, decade, or linear sweep. In this text, we have used nearly exclusively the decade step call simply because it is easier to see in terms an impedance sweep. The number of steps comes next, which is the number of steps that occur per decade for example. The last two statements are the start and end frequencies. Typically, we have used a starting frequency of 1 kHz and then ended in the 1 GHz maximum for an impedance sweep. An example call is shown below:

.AC Call Example

```
.AC DEC 10 1K 1G
```

There is an additional call with the .AC list that allows one to step through a list of explicit frequencies. For more information, the reader should see the LTSPICE manual.

The .TRAN statement is used for transient analyses such as the time-domain simulations used here for power integrity. The call syntax is as below:

.TRAN Call Syntax

```
.TRAN <Tstep> <Tstop> [Tstart [dTmax]] [modifiers]
```

After the .TRAN statement, the user defines the time step. For regions close to the die, usually this time step is set to 1 ns or shorter. The stop time is defined which can be anywhere from 100 ns to 10 ms or longer depending upon the power integrity simulation that most PI engineers are used to. A modifier could be, for example, UIC (use initial conditions) for, say, an inductor or capacitor in the call function. A typical .TRAN call is shown below:

.TRAN Call Example

```
.TRAN 1ns 10us 0 1ns UIC
```

This basically states to start a transient analysis with a 1 ns step and run to 10 µs while also using initial conditions (defined in various circuits) starting at 0 and incrementing the time step every 1 ns as well.

A.1.6 Some SPICE Examples

We now give a few simple SPICE examples for the reader. As a general rule, when building a SPICE deck, it is a good idea to draw out the circuit first, label the circuit elements and nodes and then define the execution and output statements for what is desired.

The first SPICE deck is an AC sweep of a portion of a PDN. It is the same circuit as shown in Chapter 10. This is shown below with the plot.

Example SPICE Deck for AC Sweep

```
* Example SPICE deck for AC sweep
```

```
    * Author: J. Ted DiBene II - Project:
 Impedance_1, LF section - 6/2018

    * Circuit Definition
    * First shunt impedance (Z1) - connected to cur-
rent source

    C1 1 2 10n
    R1 2 0 2m

    * First series impedance (Z2) - connected to cur-
rent source

    L2 1 3 5p
    R2 3 4 5m

    * Second Shunt impedance Z3

    L3 4 5 100p
    C3 5 6 1m
    R3 6 0 5m

    * Second series impedance (Z4)

    L4 4 7 300p
    R4 7 8 2m

    * Third shunt impedance (Z5)

    L5 8 9 500p
    C5 9 10 10u
    R5 10 0 2m

    * AC current source load

    Iac 1 0 AC 1 0

    * AC sweep

    .AC DEC 50 100K 1G
    .PROBE V(1)
    .END
```

There are a few simple things to note here about the SPICE deck. First, note that each impedance element has its separate definition between comments. Though this is not required, it helps immensely for reuse and debugging down the road. Second, the .PROBE statement has a voltage defined for its output, which makes it easier when running and rerunning the deck. It should be noted that LTSPICE has its own text editor though one could use virtually any deck. To run this or any deck in Windows, the user simply can find the icon for LTSPICE, select the *.cir file and the LTSPICE will come up with two windows: the first is the output shown in Figure A.4, and the second the deck as shown in this example.

The next example is a transient analysis. This is the same circuit as in the previous example except that we have added a series inductor and DC source to represent the converter, and we have altered the analysis for a transient run rather than an AC sweep.

Example SPICE Deck for Transient Analysis

```
* Example SPICE deck for Transient Analysis

* Author: J. Ted DiBene II - Project: Step
response - 6/2018

* Circuit Definition
* First shunt impedance (Z1) - connected to cur-
rent source

C1 1 2 10n
R1 2 0 2m

* First series impedance (Z2) - connected to cur-
rent source

L2 1 3 5p
R2 3 4 5m

* Second Shunt impedance Z3

L3 4 5 100p
C3 5 6 1m
R3 6 0 5m

* Second series impedance (Z4)
```

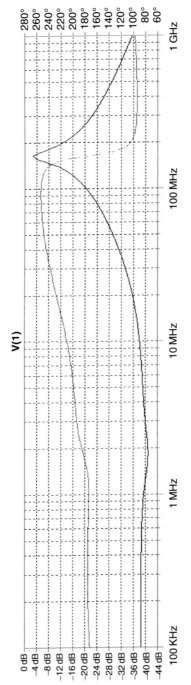

Figure A.4 AC sweep for example circuit.

```
L4  4  7  300p
R4  7  8  2m

* Third shunt impedance (Z5)

L5  8  9  500p
C5  9  10  10u
R5  10  0  2m

* Adding series inductor and DC voltage source

L6  8  11  10u
Vs  11  0  1V

* PWL current source and transient run with
.PROBE output V(1) and V(8)

Itran 1 0 PWL(0 0 1u 0 1.01u 1 20.01u 1 20.02u 0)

* Transient analysis

.TRAN 1n 22u
.PROBE V(1), V(8)
.END
```

There are a few things to note about the previous SPICE deck. First, it is clear that the modifications to the first AC analysis deck were very easy additions, allowing us to change the analyses using essentially the same circuit PDN. This can be an advantage if one does not have a GUI to work with. Second, the transient analysis statement here was quite simple and went just slightly beyond the time constraints of the current stimulus. Thus, it is easy to see that it is not required to put together a complex statement in order to get relevant information. The PWL current step was also fairly straightforward and gives a simple step response that lasts 20 μs with a 10 ns risetime. The output is shown in Figure A.5.

B.1 Quasi-Static Fields

Bleeding-edge signal integrity involves a very broad spectrum of frequencies given the digital nature of the protocols. Power Integrity is often considered more of an analog discipline, even though the converters are switching. The

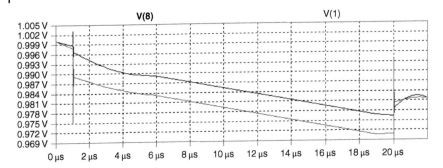

Figure A.5 Transient analysis example.

frequency dependencies come from the transients that happen in the demand for current, which understandably require higher frequency content in order to keep the voltage sufficiently stable. That said, engineers have learned through experience that certain regions in the circuit path only have a limited bandwidth that can be used to supply the necessary transient current. Consequently, those regions can be adequately modeled and investigated using lower frequency (quasi-static) equations. Such equations are much easier to use and can have relatively easy closed-form solutions and simple equivalent circuits.

What follows is not a mathematically rigorous proof of determining static or quasi-static equations, but rather a discussion that is based on intuition and engineering judgment. Of the four generally considered Maxwell's equations, there are two that are ripe for our perusal: Faraday's law and Ampere's law.

Faraday's law in the time domain can be given in differential form by

$$\nabla \times \vec{E} = -\frac{\partial \vec{B}}{\partial t} \tag{B.1}$$

where $\vec{E}$ is the electric-field intensity vector and $\vec{B}$ is the magnetic-flux density vector. The lower the frequency, the smaller the differential with respect to time. As the frequency gets closer to static, the right-hand side of the equation gets negligibly small, giving us the static equation.

$$\nabla \times \vec{E} = 0 \tag{B.2}$$

If we are comfortable with the idea of the curl, this would intuitively indicate that the electric field circulating around an arbitrary point is zero. The concept is more obvious looking at the integral form of the time-domain expression for Faraday's law given by

$$\oint_C \vec{E} \cdot \vec{dl} = -\iint_S \frac{\partial \vec{B}}{\partial t} \cdot \vec{ds} \tag{B.3}$$

where C is a closed path in space and S is the area inside the closed path. Again, the differential with respect to time yields smaller values as the frequencies get lower. In the limit Eq. (B.3) becomes

$$\oint_C \vec{E} \cdot \vec{dl} = 0 \tag{B.4}$$

Equation (B.4) is typically more intuitive to people. Essentially, the equation indicates that a closed-loop line-integral around a fixed point results in a zero. Essentially, there are no sources along or inside the path that can create an external electric field.

Notice that the equations given only changed subtly in the frequency domain. The derivative with respect to time essentially gets replaced with a $j\omega$ practically speaking. As discussed, the radial frequency ω gets diminishingly small, yielding the same results given in Eqs. (B.2) and (B.4).

Not unexpectedly, Ampere's law can be treated in the same fashion, with very similar results. Continuing this exercise, Ampere's law in the time domain can be expressed in differential form by

$$\nabla \times \vec{H} = \vec{J} + \frac{\partial \vec{D}}{\partial t} \tag{B.5}$$

where $\vec{H}$ is the magnetic-field intensity vector, $\vec{J}$ is the current-density vector (amperes per unit area), and $\vec{D}$ is the electric-flux density vector. The current-density vector represents the conduction current, while the time differential of the electric-flux density vector is displacement current. Again, assuming lower frequency evaluations, the derivative with respect to time becomes negligibly small, leaving

$$\nabla \times \vec{H} = \vec{J} \tag{B.6}$$

The same limit response can be applied to the integral form of Ampere's law:

$$\oint_C \vec{H} \cdot \vec{dl} = \iint_S \left(\vec{J} + \frac{\partial \vec{D}}{\partial t} \right) \cdot \vec{ds} \tag{B.7}$$

As with Eq. (B.3), the differentiation becomes a minimal contributor as the transients become longer in duration (and are therefore comprised of lower frequency content). Equation (B.7) is then reduced to the "static" equation of

$$\oint_C \vec{H} \cdot \vec{dl} = \iint_S \vec{J} \cdot \vec{ds} = I_{enc} \tag{B.8}$$

Notice that in Eq. (B.8) the right-hand side of the equation reduces simply to the total current that penetrates the area enclosed by the closed path C.

Astute observers will notice that a non-zero time differential is evident in Eqs. (B.6) and (B.8). By definition, current is the change in charge with respect to time. However, we look at this as a *constant* change in charge with respect to time, and therefore, the *current* is actually static, or close to it. The ultimate result of this development is that we have equations that can be used to determine the electric and magnetic behavior of a system with slow transients (low frequencies).

We can then model the capacitances and inductances along the path of power delivery into the package with reasonable accuracy. Assuming that the PCB can pass transients up to 100 MHz, the materials and distances inherent to typical PCBs can be modeled with the equations we just developed. Capacitors can be mounted (modeled as series C, L, and R) and the distance to the next capacitor can be included as a lumped-element inductor. Interplane capacitances can also be placed appropriately in the model for the higher frequency responses. The inductances can be calculated using the developed equations, which are far easier than using full Maxwell's equations with time retardation included.

C.1 Spherical Coordinate System

The spherical coordinate vector A may be defined as below.

$$A = A_r \hat{u}_r + A_\theta \hat{u}_\theta + A_\phi \hat{u}_y \tag{C.1}$$

where the unit vectors correspond to the unit vectors in Figure C.1.

To convert from rectangular to spherical, or conversely, may be accomplished through the following formulae:

$$x = r \sin \theta \cos \phi \tag{C.2}$$

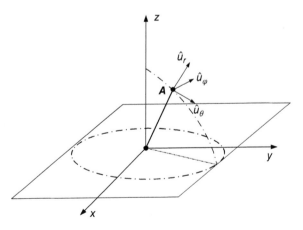

Figure C.1 Spherical coordinate system.

$$y = r \sin \theta \sin \phi \tag{C.3}$$

$$z = r \cos \theta \tag{C.4}$$

$$\theta = \tan^{-1}\left(\frac{\sqrt{x^2 + y^2}}{z}\right) \tag{C.5}$$

$$\phi = \tan^{-1}\left(\frac{y}{x}\right) \tag{C.6}$$

A thorough overview of the three different coordinate systems may be found in Paul [1].

D.1 Vector Identities and Formula

$$A + B = B + A \tag{D.1}$$

$$A \cdot B = B \cdot A \tag{D.2}$$

$$A \times B = -B \times A \tag{D.3}$$

$$(A + B) \cdot C = A \cdot C + B \cdot C \tag{D.4}$$

$$(A + B) \times C = A \times C + B \times C \tag{D.5}$$

$$A \cdot B \times C = B \cdot C \times A \tag{D.6}$$

$$A \times (B \times C) = (A \cdot C)B - (A \cdot B)C \tag{D.7}$$

$$\nabla \cdot (\nabla \times A) = 0 \tag{D.8}$$

$$\nabla \times \nabla \psi = 0 \tag{D.9}$$

$$\nabla \cdot (A + B) = \nabla \cdot A + \nabla \cdot B \tag{D.10}$$

$$\nabla \times (A + B) = \nabla \times A + \nabla \times B \tag{D.11}$$

$$\nabla \cdot (\psi A) = A \cdot \nabla \psi + \psi \nabla \cdot A \qquad (D.12)$$

$$\nabla \times (\psi A) = \nabla \psi \times A + \psi \nabla \times A \qquad (D.13)$$

$$\nabla \cdot (A \times B) = B \cdot \nabla \times A + A \cdot \nabla \times B \qquad (D.14)$$

$$\nabla \times \nabla \times A = \nabla (\nabla \cdot A) - \nabla^2 A \qquad (D.15)$$

E.1 Summary of Common Relationships Among Coordinate Systems

E.1.1 Variable Translations

Table E.1 Variable translations.

Variable	Rectangular (x,y,z)	Cylindrical (ρ,φ,z)	Spherical (r,θ,φ)
Rectangular (x, y, z)			
x	x	$\rho \cos \phi$	$r \sin \theta \cos \phi$
y	y	$\rho \sin \phi$	$r \sin \theta \sin \phi$
z	z	z	$r \cos \theta$
Cylindrical (ρ, φ, z)			
ρ	$\sqrt{x^2 + y^2}$	ρ	$r \sin \theta$
ϕ	$\tan^{-1} \dfrac{y}{x}$	ϕ	ϕ
z	Z	z	$r \cos \theta$
Spherical (r, θ, φ)			
r	$\sqrt{x^2 + y^2 + z^2}$	$\dfrac{\rho}{\sin \theta}$	r
θ	$\tan^{-1} \dfrac{\sqrt{x^2 + y^2}}{z}$	$\tan^{-1} \dfrac{\rho}{z}$	θ
ϕ	$\tan^{-1} \dfrac{y}{x}$	ϕ	ϕ

E.1.2 Coordinate Translations

Table E.2 Coordinate translations.

Variable	Rectangular $(\hat{x}, \hat{y}, \hat{z})$	Cylindrical $(\hat{\rho}, \hat{\phi}, \hat{z})$	Spherical $(\hat{r}, \hat{\theta}, \hat{\phi})$
Rectangular $(\hat{x}, \hat{y}, \hat{z})$			
$\hat{x}$	$\hat{x}$	$\hat{\rho}\cos\phi - \hat{\phi}\sin\phi$	$\hat{r}\sin\theta\cos\phi + \hat{\theta}\cos\theta\cos\phi - \hat{\phi}\sin\phi$
$\hat{y}$	$\hat{y}$	$\hat{\rho}\sin\phi + \hat{\phi}\cos\phi$	$\hat{r}\sin\theta\sin\phi + \hat{\theta}\cos\theta\sin\phi + \hat{\phi}\cos\phi$
$\hat{z}$	$\hat{z}$	$\hat{z}$	$\hat{r}\cos\theta - \hat{\theta}\sin\theta$
Cylindrical $(\hat{\rho}, \hat{\phi}, \hat{z})$			
$\hat{\rho}$	$\dfrac{\hat{x}x + \hat{y}y}{\sqrt{x^2 + y^2}}$	$\hat{\rho}$	$\hat{r}\sin\theta + \hat{\theta}\cos\theta$
$\hat{\phi}$	$\dfrac{-\hat{x}y + \hat{y}x}{\sqrt{x^2 + y^2}}$	$\hat{\phi}$	$\hat{\phi}$
$\hat{z}$	$\hat{z}$	$\hat{z}$	$\hat{r}\cos\theta - \hat{\theta}\sin\theta$
Spherical $(\hat{r}, \hat{\theta}, \hat{\phi})$			
$\hat{r}$	$\dfrac{\hat{x}x + \hat{y}y + \hat{z}z}{\sqrt{x^2 + y^2 + z^2}}$	$\dfrac{\hat{\rho}\rho + \hat{z}z}{\sqrt{\rho^2 + z^2}}$	$\hat{r}$
$\hat{\theta}$	$\dfrac{\hat{x}z\dfrac{x}{\sqrt{x^2+y^2}} + \hat{y}z\dfrac{y}{\sqrt{x^2+y^2}} - \hat{z}\sqrt{x^2+y^2}}{\sqrt{x^2 + y^2 + z^2}}$	$\dfrac{\hat{\rho}z - \hat{z}\rho}{\sqrt{\rho^2 + z^2}}$	$\hat{\theta}$
$\hat{\phi}$	$\dfrac{-\hat{x}y + \hat{y}x}{\sqrt{x^2 + y^2}}$	$\hat{\phi}$	$\hat{\phi}$

E.1.3 Curl Equation Expansions

Table E.3 Curl equation expansions.

System	Expansion
Rectangular $\nabla \times \vec{A}(x, y, z)$	$\hat{x}\left(\dfrac{\partial A_z}{\partial y} - \dfrac{\partial A_y}{\partial z}\right) + \hat{y}\left(\dfrac{\partial A_x}{\partial z} - \dfrac{\partial A_z}{\partial x}\right) + \hat{z}\left(\dfrac{\partial A_y}{\partial x} - \dfrac{\partial A_x}{\partial y}\right)$
Cylindrical $\nabla \times \vec{A}(\rho, \phi, z)$	$\hat{\rho}\left(\dfrac{1}{\rho}\dfrac{\partial A_z}{\partial \phi} - \dfrac{\partial A_\phi}{\partial z}\right) + \hat{\phi}\left(\dfrac{\partial A_\rho}{\partial z} - \dfrac{\partial A_z}{\partial \rho}\right) + \hat{Z}\dfrac{1}{\rho}\left(\dfrac{\partial}{\partial \rho}(\rho A_\phi) - \dfrac{\partial A_\rho}{\partial \phi}\right)$
Spherical $\nabla \times \vec{A}(r, \theta, \phi)$	$\hat{r}\dfrac{1}{r\sin\theta}\left[\dfrac{\partial}{\partial \theta}(A_\phi \sin\theta) - \dfrac{\partial A_\theta}{\partial \phi}\right] + \hat{\theta}\dfrac{1}{r\sin\theta}\left[\dfrac{\partial A_r}{\partial \phi} - \sin\theta\dfrac{\partial}{\partial r}(rA_\phi)\right]$ $+\hat{\phi}\dfrac{1}{r}\left[\dfrac{\partial}{\partial r}(rA_\theta) - \dfrac{\partial A_r}{\partial \theta}\right]$

E.1.4 Divergence Equation Expansions

Table E.4 Divergence equation expansions.

System	Expansion
Rectangular $\nabla \cdot \vec{A}(x, y, z)$	$\dfrac{\partial A_x}{\partial x} + \dfrac{\partial A_y}{\partial y} + \dfrac{\partial A_z}{\partial z}$
Cylindrical $\nabla \cdot \vec{A}(\rho, \phi, z)$	$\dfrac{1}{\rho}\dfrac{\partial}{\partial \rho}(\rho A_\rho) + \dfrac{1}{\rho}\dfrac{\partial A_\phi}{\partial \phi} + \dfrac{\partial A_z}{\partial z}$
Spherical $\nabla \cdot \vec{A}(r, \theta, \phi)$	$\dfrac{1}{r^2}\dfrac{\partial}{\partial r}(r^2 A_r) + \dfrac{1}{r\sin\theta}\dfrac{\partial}{\partial \theta}(\sin\theta A_\theta) + \dfrac{1}{r\sin\theta}\dfrac{\partial A_\phi}{\partial \phi}$

E.1.5 Del-Operator Expansions

Table E.5 Del-Operator expansions.

System	Expansion
Rectangular $\nabla(x, y, z)$	$\hat{x}\dfrac{\partial}{\partial x} + \hat{y}\dfrac{\partial}{\partial y} + \hat{z}\dfrac{\partial}{\partial z}$
Cylindrical $\nabla(\rho, \phi, z)$	$\hat{\rho}\dfrac{\partial}{\partial \rho} + \hat{\phi}\dfrac{1}{\rho}\dfrac{\partial}{\partial \phi} + \hat{z}\dfrac{\partial}{\partial z}$
Spherical $\nabla(r, \theta, \phi)$	$\hat{r}\dfrac{\partial}{\partial r} + \hat{\theta}\dfrac{1}{r}\dfrac{\partial}{\partial \theta} + \hat{\phi}\dfrac{1}{r\sin\theta}\dfrac{\partial}{\partial \phi}$

E.1.6 Laplacian Expansions

Table E.6 Laplacian expansions.

System	Expansion
Rectangular $\nabla^2(x, y, z)$	$\dfrac{\partial^2}{\partial x^2} + \dfrac{\partial^2}{\partial y^2} + \dfrac{\partial^2}{\partial z^2}$
Cylindrical $\nabla^2(\rho, \phi, z)$	$\dfrac{1}{\rho}\dfrac{\partial}{\partial \rho}\left(\rho\dfrac{\partial}{\partial \rho}\right) + \dfrac{1}{\rho^2}\dfrac{\partial^2}{\partial \phi^2} + \dfrac{\partial A_z}{\partial z}$
Spherical $\nabla^2(r, \theta, \phi)$	$\dfrac{1}{r^2}\dfrac{\partial}{\partial r}\left(r^2\dfrac{\partial}{\partial r}\right) + \dfrac{1}{r^2 \sin\theta}\dfrac{\partial}{\partial \theta}\left(\sin\theta\dfrac{\partial}{\partial \theta}\right) + \dfrac{1}{r^2\sin^2\theta}\dfrac{\partial^2}{\partial \phi^2}$

F.1 Some Notation Definitions

There are a number of voltage and current notations that are used throughout this text and others. Most of them have their origins from early on from the IEEE in the 1960s [3]. Though engineers use them liberally when putting together their circuits, many do not know their meaning. In this text, we have tried to use the standard conventions, and in this appendix, we define standard letter conventions to maintain clarity.

In general, from Ref. [3] lowercase letters for current voltage and power (i, v, p) are designations for instantaneous values. Uppercase letters are designated for maximum, average, and root-mean-square (rms) values. Uppercase subscripts represent DC values and instantaneous total values.

For example, in circuits where we represent the DC output of DC–DC converter, we use the letter designation V_O, where the capital letter "V" represents the voltage and the uppercase "O" represents the output node of that circuit. One of the more common representations for a DC voltage is the notation VCC and VDD. VCC would be defined as the voltage across the common collector for a bipolar circuit. The second letter in the subscript indicates that it is a supply voltage. The first subscript represents the node at which the voltage is measured. The same goes for VDD where the convention in this case is used for CMOS-based circuits which are most common today. The term VSS is normally designated as the ground terminal for many CMOS circuits. In the case where there are different grounds, the source terminal is then designated. Often the measurement of the voltage from VDD to VSS is used to determine the correct voltage across the two terminals. Again, the second subscript designates it as a supply terminal.

The convention extends to currents as well where the current into IDD would be the supply current into the drain terminal of a CMOS circuit, as an example.

Though there are numerous other conventions used for semiconductor circuits, these appear to be the most commonly used ones by engineers involved in power integrity analysis.

G.1 Common Theorems

A number of theorems prove very useful in calculating the necessary pieces for verifying power integrity. The proofs for these can be found quickly online or in other texts that the reader most likely has on a shelf. For completeness, several of these are included below, should they prove necessary for your calculations.

Stokes' theorem: $\oint_C \vec{f}(\vec{r}) \cdot \vec{dr} = \int\int_S \nabla \times \vec{f}(\vec{r}) \cdot \vec{ds}$

Binomial theorem: $(a+b)^n = \sum_{k=0}^{n} \dfrac{n!}{(n-k)!k!} a^{n-k} b^k$

Divergence theorem: $\int\int\int_V (\nabla \cdot \vec{f}(\vec{r})) dV = \oint_S (\vec{f}(\vec{r}) \cdot \hat{n}) ds$

Taylor series: $f(a) = \sum_{n=0}^{\infty} \dfrac{f^{(n)}(a)}{n!} (x-a)^n$

Bibliography

1 Paul, C.R. (2010). A brief spice (Pspice) tutorial. *EMC Society Newsletter*.
2 LTSpice. *Help Topics* (Part of LTSpice program).
3 (1964). IEEE standard letter symbols for semiconductor devices. *IEEE Transactions on Electron Devices* 11 (8): 392–397.

Index

Power Integrity for Electrical and Computer Engineers, First Edition. J. Ted Dibene II and David Hockanson.
© 2020 John Wiley & Sons, Inc. Published 2020 by John Wiley & Sons, Inc.